W9-BBP-049

ISNM

INTERNATIONAL SERIES OF NUMERICAL MATHEMATICS
INTERNATIONALE SCHRIFTENREIHE ZUR NUMERISCHEN MATHEMATIK
SÉRIE INTERNATIONALE D'ANALYSE NUMÉRIQUE

VOL. 41

General Inequalities 1

Proceedings of the First International Conference
on General Inequalities
held in the Mathematical Research Institute at Oberwolfach, Black Forest
May 10–14, 1976

Edited by
E. F. Beckenbach

Allgemeine Ungleichungen 1

Abhandlung zur ersten internationalen Tagung
über Allgemeine Ungleichungen
im Mathematischen Forschungsinstitut Oberwolfach, Schwarzwald
vom 10. bis 14. Mai 1976

Herausgegeben von
E. F. Beckenbach

1978
Birkhäuser Verlag Basel
und Stuttgart

CIP-Kurztitelaufnahme der Deutschen Bibliothek

General inequalities = Allgemeine Ungleichungen.
- Basel, Stuttgart: Birkhäuser.
NE: PT
1. Proceedings of the First International
Conference on General Inequalities: held
in the Math. Research Inst. at Oberwolfach,
Black Forest, May 10–14, 1976/ed. by E.
F. Beckenbach. - 1. Aufl. - 1978.
(International series of numerical mathematics; Vol. 41)
ISBN 3-7643-0972-5

NE: Beckenbach, Edwin F. [Hrsg.]; International
Conference on General Inequalities ‹01, 1976,
Oberwolfach›; Mathematisches Forschungsinstitut
‹Oberwolfach›

ISBN 3-7643-0972-5

Vorwort

Die erste internationale Tagung "Allgemeine Ungleichungen" fand vom 9. bis 15. Mai im Mathematischen Forschungsinstitut Oberwolfach statt. Da Herr Bellman leider aus gesundheitlichen Gründen nicht teilnehmen konnte, hat sich glücklicherweise Herr Beckenbach (Los Angeles) bereiterklärt, zusammen mit den Herren Aczél (Waterloo, Onterio) und Aumann (München) die Tagung zu leiten. Als Tagungssekratär wirkte Herr Ger (Katowice). Die 27 Teilnehmer kamen aus Europa, Amerika und Australien. Erfreulicherweise hat sich eine größere Zahl von Teilnehmern aus Polen eingefunden; leider konnten keine Wissenschaftler aus Jugoslawien, Rumänien, der Tschechoslawakai und Ungarn kommen. Die Tagung wurde von Herrn Aczél eröffnet und von Herrn Aumann geschlossen.

Zu der durch den Titel gegebenen allgemeinen Thematik haben die Teilnehmer in sehr verschiedene Weise beigetragen. Trotzdem bildeten sich einige Schwerpunkte heraus: Funktionalungleichungen, inbesondere iterativen Typs, konvexe und verallgemeinert konvexe Funktionen, Differentialungleichungen, Ungleichungen der Funktionentheorie, Funktionalanalysis und Geometrie. Ferner wurden Anwendungen von Ungleichungen auf Differentialgleichungen, Physik, Warscheinlichkeits-, Informations- und Wirtschaftstheorie gebracht. Die als besondere Tagungspunkte augesetzten "Bemerkungen und Probleme" haben sich als äußerst fruchtbar erwiesen, ja sogar in einem Fall zur vollständigen Lösung eines aufgeworfenen Problems geführt.

Die Tagung hat nicht nur gezeigt, daß das Netz von Ungleichungen, das die gesamte Analysis durchzieht, verdient in eigenständiger Weise und im Sinne allgemeiner Methoden behandelt zu werden, sondern auch bestätigt, wie viel schneller und wirksamer der Ideenaustausch ist, der durch die Vorträge und durch den persönlichen Kontakt der Wissenschaftler untereinander zustande kommt, als die unpersönliche, nur literarische Information.

Die Tagungsleitung und die Teilnehmer danken dem Mathematischen Forschungsinstitut für die Iniziative, Tagungen dieser Art in das Programm aufzunehmen. Es ist der einhellige Wunsch der Teilnehmer, daß die zweite Internationale Tagung über "Allgemeine Ungleichungen" in 1978, womöglich 6. bis 12. August, in Oberwolfach stattfinden soll.

Mit großem Bedauern wurde der Wunsch von Herrn Aczél zur Kenntnis genommen, an der Leitung der weiteren Tagungen über "Allgemeine Ungleichungen" nicht mehr mitzuwirken. Ihm wurde allgemeiner Dank für seine wesentliche Mithilfe am Gelingen dieser Tagung ausgesprochen. Die von der Leitung vorgeschlagene Zuwahl von Herrn Kuczma (Katowice) wurde lebhaft begrüßt.

Alle Beteiligten haben die hervorragenden Arbeitsbedingungen des Instituts und die freundliche Betreuung zu schätzen gelernt.

Genehmigt:

Janos Aczél

G. Aumann

E. F. Beckenbach

TEILNEHMER

J. ACZÉL, University of Waterloo, Ontario, Kanada

G. AUMANN, Technische Universität, München, BR Deutschland

H. BAUER, Universität Erlangen-Nürnberg, BR Deutschland

E.F. Beckenbach, University of California, Los Angeles, USA

W. BENZ, Universität Hamburg, BR Deutschland

T. BISZTRICZKY, University of Calgary, Alberta, Kanada

D. BRYDAK, Pädagogische Hochschule Kraków, Polen

L.L. CAMPBELL, Queen's University, Kingston, Ontario, Kanada

B. CHOCZEWSKI, Berg- und Hüttenakademie, Kraków, Polen

J. CURTISS, University of Miami, Florida, USA

W. EICHHORN, Universität Karlsruhe (TH), BR Deutschland

K. ENDL, Universität Giessen, BR Deutschland

R. GER, Schlesische Universität, Katowice, Polen

T. HOWROYD, University of New Brunswick, Fredericton, Kanada

F. HUCKEMANN, Technische Universität, Berlin

H.-H. KAIRIES, Technische Universität Clausthal-Zellerfeld, BR Deutschland

M. KUCZMA, Schlesische Universität, Katowice, Polen

O. MACEDOŃSKA, Technische Universität, Gliwice, Polen

A. OSTROWSKI, Universität Basel, Schweiz

J. RÄTZ, Universität Bern, Schweiz

R. REDHEFFER, University of California, Los Angeles, USA

D.K. ROSS, La Trobe University, Bundoora, Australien

D.C. RUSSELL, York University, Downsview, Ontario, Kanada

S. SCHAIBLE, Universität Köln, BR Deutschland

B. SCHWEIZER, University of Massachusetts, Amherst, USA

D.R. SNOW, Brigham Young University, Provo, Utah, USA

W. WALTER, Universität Karlsruhe (TH), BR Deutschland

WISSENSCHAFTLICHES PROGRAMM DER TAGUNG

Montag, 10. Mai

Frühsitzung von 9:30 bis 10:45 Uhr — Vorsitz: J. ACZÉL

M. KUCZMA: Nonnegative continuous solutions of a linear functional inequality

D. BRYDAK: On functional inequalities in a single variable

________ von 11:00 bis 12:00 Uhr — Vorsitz: R. GER

B. CHOCZEWSKI: A functional inequality in a single variable

Bemerkungen und Probleme

Nachmittagsitzung von 16:00 bis 17:30 Uhr — Vorsitz: W. EICHHORN

S. SCHAIBLE: Second-order characterizations of pseudo-convex functions

T. HOWROYD: Functional inequalities

Bemerkungen und Probleme

* * * * * *

Dienstag, 11. Mai

Frühsitzung von 9:30 bis 10:30 Uhr — Vorsitz: L.L. CAMPBELL

W. EICHHORN: Inequalities and functional equations in the theory of the price index

D.K. ROSS: On Turán-type inequalities

________ von 10:45 bis 11:45 Uhr — Vorsitz: R. REDHEFFER

K. ENDL: On a general theorem of Favard

Bemerkungen und Probleme

Nachmittagsitzung von 16:00 bis 17:30 Uhr — Vorsitz: K. ENDL

E.F. BECKENBACH: What Hadamard Overlooked

F. HUCKEMANN: Inequalities stemming from quadratic differentials

Bemerkungen und Probleme

* * * * * *

Mittwoch, 12. Mai

Frühsitzung von 9:30 bis 10:30 Uhr — Vorsitz: J. RÄTZ

J. ACZÉL: Picardus ab omni naevo vindicatus (On the axiomatics of vector addition)

H.-H. KAIRIES: Convexity in the theory of the gamma function

__________ von 10:45 bis 11:45 Uhr Vorsitz: A. OSTROWSKI

R. REDHEFFER and W. WALTER: Inequalities of parabolic type

Bemerkungen und Probleme

* * * * * *

Donnerstag, 13. Mai

Frühsitzung von 9:30 bis 10:30 Uhr Vorsitz: E.F. BECKENBACH

G. AUMANN: Induction with inequalities involving nonquasiarithmetic means

D.R. SNOW: Quadratic functionals and Rayleigh's principle by equivalent problems

__________ von 10:45 bis 11:45 Uhr Vorsitz: B. SCHWEIZER

J. RÄTZ: Some remarks on quadratic functionals satisfying a subsidiary inequality

Bemerkungen und Probleme

* * * * * *

Nachmittagsitzung von 16:00 bis 17:30 Uhr Vorsitz: M. KUCZMA

L.L. CAMPBELL: Bound on the measure of a set in a product space

W. BENZ: On characterizing Lorentz transformations

Bemerkungen und Probleme

Freitag, 14. Mai

Frühsitzung von 9:15 bis 10:15 Uhr Vorsitz: T. HOWROYD

D.C. RUSSELL: Solutions of the Hausdorff moment problem

R. GER: On almost subadditive functions

__________ von 10:30 bis 11:30 Uhr Vorsitz: G. AUMANN

B. SCHWEIZER: An inequality for distribution functions and Wald-betweenness

Bemerkungen und Probleme

PREFACE

These Proceedings contain, in full or abstract form, each of the twenty-one papers presented at the First International Conference on General Inequalities, held at the Mathematical Research Institute, Oberwolfach, Black Forest, in May, 1976. Also included are papers by others who were invited but were unable to attend.

Two noted authorities in the field sent regrets that they were unable to contribute. Professor J.E. Littlewood wrote:

> I stopped Mathematics at 86 and am
> now 91 and really can't contribute.

Professor George Pólya expressed a like sentiment somewhat more whimsically:

> I would love to contribute a paper
> for the Oberwolfach Proceedings you
> are planning, but unfortunately I have
> no new material. I am past 88, and
> "even the prettiest girl in Paris
> cannot give more than what she has."

We wish all the best for each of these illustrious and venerable colleagues.

The papers in this volume have been grouped by the editor into five roughly coherent sections. In addition, there is a section on Remarks and Problems.

Sketches of scenes around the Institute, which appear on the title pages of the sections of the book, were graciously provided by Irmgard Süss, who also, along with Susan Aczél, Liddy Aumann, Alice Beckenbach, Margaret Ostrowski, Diane Snow, and Professor and Mrs. Otto Haupt, added greatly to the interest and content of stimulating conversations at meals and other social events during the Conference.

The editor is deeply grateful to Elaine Barth of the U.C.L.A. Mathematics Department for expert editorial consultation and technical advice, and to members of her typing pool, Julie Honig, Connie Jurgens, and especially Debra Remetch, for their careful and excellent preparation of the typescript; and he sincerely thanks Mr. C. Einsele of Birkhäuser Verlag, Basel, for kind expressions of interest and encouragement.

E.F. Beckenbach, Editor
University of California, Los Angeles

CONTENTS

FUNCTIONAL INEQUALITIES

DIFFERENTIAL AND INTEGRAL INEQUALITIES

GEOMETRIC AND TOPOLOGICAL INEQUALITIES

Mean Values and Classical Inequalities

Spuren der wandelnden Denker in Schnee und Gras
unsrer Auen

Sind uns ein freundliches Sinnbild und festigen
unser Vertrauen,

Dass, wenn Natur und Geist ihre bindenden
Kräfte entfalten,

Menschliche Einheit sich baut gegen trennende
Schicksalsgewalten.

A NOTE ON INEQUALITIES

Richard Bellman
Department of Mathematics
University of Southern California
Los Angeles, California 90007
U.S.A.

ABSTRACT. In this note, it is shown that backward induction gives an equality for the arithmetic-geometric mean inequality.

1. INTRODUCTION

One idea in the theory of inequalities is that every inequality is a consequence of an equality. Often, it is not easy to find the relevant equality. See [1], [2]. We shall now show that backward induction gives such an equality for the arithmetic-geometric mean inequality.

2. BACKWARD INDUCTION

We begin with the simple result

$$a_1^2 + a_2^2 = 2a_1a_2 + (a_1 - a_2)^2 , \tag{1}$$

$$\begin{aligned} a_1^2 + a_2^2 + a_3^2 + a_4^2 &= 2(a_1a_2 + a_3a_4) \\ &+ (a_1 - a_2)^2 + (a_3 - a_4)^2 . \end{aligned} \tag{2}$$

Now in (2) we replace a_1 by a_1^2, and so on, obtaining

$$\begin{aligned} a_1^4 + a_2^4 + a_3^4 + a_4^4 &= 2(a_1^2a_2^2 + a_3^2a_4^2) \\ &+ (a_1^2 - a_2^2)^2 + (a_3^2 - a_4^2)^2 . \end{aligned} \tag{3}$$

Applying (1) on the right-hand side of (3), we get

$$a_1^4 + a_2^4 + a_3^4 + a_4^4 = 4(a_1a_2a_3a_4) + 2(a_1a_2 - a_3a_4)^2 + (a_1^2 - a_2^2)^2 + (a_3^2 - a_4^2)^2. \tag{4}$$

We now use equality (4) for a_5, a_6, a_7, a_8 and add, obtaining a similar result. Replacing a_1 by a_1^2, and so on, yields the desired equality for eight variable.

Continuing in this fashion, we obtain the result for any power of 2. In the usual way, we now specialize to obtain the equality for any integer.

This is related to the Artin result, first conjectured by Hilbert.

Observe that we have required only that all the quantities appearing are real. When we specialize, we have to keep that in mind.

3. DISCUSSION

The result for three variables may also be obtained from the well-known factorization

$$x^3 + y^3 + z^3 - 3xyz = (x + y + z)(x^2 + y^2 + z^2 - xy - xz - yz) \quad . \tag{5}$$

Equation (5) shows two things. In the first place, the desired representation may not be unique. In the second place, the arithmetic-geometric mean inequality holds in a wider region than is ordinarily assumed.

REFERENCES

1. E. F. Beckenbach and R. Bellman, Inequalities, Springer Verlag, Berlin, 1961; 2nd Edition, 1965; 3rd Edition, 1970.
2. G. H. Hardy, J. E. Littlewood, and G. Pólya, Inequalities, Cambridge Univ. Press, London and New York, 1934; 2nd Edition, 1952.

THE MANY LIMITS OF MIXED MEANS, I

John Todd
Department of Mathematics 253-37
California Institute of Technology
Pasadena, California 91125
U.S.A.

ABSTRACT. The question of the convergence and the limits of several sequences similar to the classical arithmetic-geometric mean sequences of Gauss, but where arbitrary choices of the determination of the square roots involved are made, is examined.

1. INTRODUCTION

It is fairly well known how Gauss expressed the common limit of the sequences $\{x_n\}$, $\{y_n\}$ defined by

$$x_{n+1} = \tfrac{1}{2}(x_n + y_n) , \quad y_{n+1} = (x_n y_n)^{\frac{1}{2}} , \tag{1}$$

with $x_0 \geq y_0 \geq 0$ given, in terms of a complete elliptic integral. It is much less familiar, despite the works of v. Dávid [17], Geppert [6], [7], [8], and the editors of Gauss' Werke [5], how Gauss determined the limits of (1), allowing for arbitrary choices of the determination of the square roots. We shall discuss the corresponding results in some other cases:

$$x_{n+1} = \tfrac{1}{2}(x_n + y_n) , \quad y_{n+1} = (x_{n+1} y_n)^{\frac{1}{2}} , \tag{2}$$

$$x_{n+1} = [x_n(x_n + y_n)/2]^{\frac{1}{2}} , \quad y_{n+1} = [y_n(x_n + y_n)/2]^{\frac{1}{2}} , \tag{3}$$

$$x_{n+1} = \tfrac{1}{2}(x_n + y_n) , \quad y_{n+1} = (x_n x_{n+1})^{\frac{1}{2}} , \tag{4}$$

which for convenience we call the Borchardt case, the logarithmic case, and the lemniscate case, respectively. The last two were introduced by B. C. Carleson [2] in a uniform account of many cases by means of invariant integrals.

The source of the results is that in (1) and (2) the reciprocal of the common limit, in (3) the reciprocal of the square root of the common limit, and in (4) the reciprocal of the square of the common limit, can be expressed as an algebraic integral. In (1) the integral is a complete elliptic integral with modulus k^2 depending on x_0, y_0; in (4) it is incomplete but with $k^2 = \frac{1}{2}$; while in (2) and (3) the integrals are elementary.

By homogeneity, in all cases we may restrict attention to the case

$x_0 = 1$, $y_0 = z$. The common limit can be shown to be an analytic function of z, and so, too, is the integral. The many values of the common limit corresponding to different determinations of the signs of the square roots coincide with the many values of the integral corresponding to different paths of integration, which are obtained by adding multiples of the periods. We expect later to discuss this approach more fully, but in the present paper we give a more arithmetic approach. It is possible to parametrize the algorithms; it was probably for this purpose that in (1) Gauss introduced the ϑ-functions; trigonometric or hyperbolic functions suffice in (2); powers suffice in (3); in (4) we use ϑ-functions (in the lemniscate case, $k^2 = \frac{1}{2}$) or lemniscate functions themselves. In the Gauss case, the step from the n-th to the (n+1)-th case involves a doubling of the ϑ-parameter; in (2) it is accomplished by a halving, and in (3) by taking a square root, of the argument; while in (4) we require a division of the argument by $(1 + i)$. The change of sign in the Gaussian case can be accomplished by the transformation

$$\tau \to \frac{\tau}{1 + 2i\tau};$$

in the other cases, a change of sign can be obtained by a translation in the argument by a half period.

We conclude this Introduction with a summary of the single computation of Gauss [5, X_1, 218-219]:

x_n	y_n
3.0000	1.0000
2.0000	1.7321
1.8660	1.8612
1.8638	1.8636
⋮	

$M(3,1) = 1.8636167$

x_n	y_n
3.0000	2.8284
2.9142	2.9130
2.9136	2.9136
⋮	

$M(3,2\sqrt{2}) = 2.9135822$

x_n	y_n
3.0000	1.0000
2.0000	-1.7320
.1339	-1.8612i
.0670 + .9306i	.3531 + .3531i
⋮	

$M_1(3,1) = .2469962 + .6318686i$

$M_1^{-1}(3,1) = .5365910 - 1.3728776i$

$= M^{-1}(3,1) - 4iM^{-1}(3,2\sqrt{2})$

The general result, in the usual elliptic-function notation (cf. Whittaker and Watson [18]), is

$$[M(1,k)]^{-1} = (2/\pi)\int_0^1 [(1-x^2)(1-(1-k^2)x^2)]^{-\frac{1}{2}}\,dx = (2/\pi)\ K'(k^2)\ ,$$

so that

$$M(3,1) = 3\ M(1,1/3) = (3\pi/2)/K'(1/9)\ ,$$

$$M(3,2\sqrt{2}) = 3\ M(1,2\sqrt{2}/3) = (3\pi/2)/K'(8/9)\ .$$

2. THE BORCHARDT ALGORITHM - TRIGONOMETRIC CASE

We take the original trigonometric case, when $x_0 = \cos\varphi$, $y_0 = 1$, and where, for $n \geq 0$, x_n, y_n are defined by (2), with the positive square root always taken. We assume $x_0 \neq 1$ and $0 < \varphi < \frac{1}{2}\pi$. It is easy to show that $x_n \uparrow$, $y_n \downarrow$; and since

$$y_{n+1}^2 - x_{n+1}^2 = \frac{1}{4}(y_n^2 - x_n^2)\ ,$$

it follows that

$$\lim x_n = \lim y_n = B(\cos\varphi,1)\ .$$

Also if $x_0 = 1$, we have $x_n \equiv y_n \equiv 1$ and $B(1,1) = 1$.

To find $B(0,\varphi)$, we show by induction that, for $n \geq 1$,

$$\begin{aligned} x_n &= \cos(2^{-1}\varphi)\cos(2^{-2}\varphi)\cdots\cos(2^{-n+1}\varphi)\cos^2(2^{-n}\varphi)\ ,\\ y_n &= \cos(2^{-1}\varphi)\cos(2^{-2}\varphi)\cdots\cos(2^{-n+1}\varphi)\cos(2^{-n}\varphi)\ . \end{aligned} \tag{5}$$

By repeated use of the relation $2\sin x\cos x = \sin 2x$, we find

$$\begin{aligned} 2^n\sin(2^{-n}\varphi)\cdot y_n &= 2^{n-1}\{[\cos(2^{-1}\varphi)\cdots\cos(2^{-n+1}\varphi)][2\cos(2^{-n}\varphi)\sin(2^{-n}\varphi)]\}\\ &= 2^{n-2}\{[\cos(2^{-1}\varphi)\cdots\cos(2^{-n+2}\varphi)]\cdot[2\cos(2^{-n+1}\varphi)\sin(2^{-n+1}\varphi)]\}\\ &= 2^{n-2}\{[\cdots]\cdot\sin(2^{-n+2}\varphi)\}\\ &= \cdots\\ &= \sin\varphi\ . \end{aligned} \tag{6}$$

Hence, letting $n \to \infty$, we have

$$\lim y_n = B(\cos\varphi,\ 1) = (\sin\varphi)/\varphi\ . \tag{7}$$

Following Gauss, we call this the "simplest" (Borchardt) algorithm. We now turn to the general case in which we replace (2) by

(8) $$x_{n+1} = \tfrac{1}{2}(x_n + y_n) , \quad y_{n+1} = \varepsilon_n \operatorname{sign}(x_{n+1})(x_{n+1}y_n)^{\frac{1}{2}} ,$$

where $\varepsilon_n = \pm 1$. It will be convenient to write $\varepsilon_n = (-1)^{\alpha_n}$, where $\alpha_n = 0,1$. Then the general algorithm is specified by a binary sequence $\{ \cdots \alpha_n \cdots \alpha_2\alpha_1\alpha_0\}$. The "simplest" algorithm corresponds to the sequence $\{ \ldots 0 \ldots 000\}$. It will appear that the more interesting cases occur when there are only a finite number of 1's. The algorithm is then specified by the binary number

(9) $$a = \alpha_N 2^N + \alpha_{N-1}2^{N-1} + \cdots + \alpha_1 2 + \alpha_0 ,$$

where $\alpha_N \neq 0$ and $\varepsilon_r = 1$, $\alpha_r = 0$ for $r > N$. We shall call a the <u>index</u> of the algorithm.

We shall show that the a-algorithm has a limit, and we denote it by $B_a(\cos \varphi, 1)$.

We first note that all the quantities defined by (8) are real and that $|y_n| > |x_n|$. This is proved by induction. The case $n = 0$ is trivial. Assume that $x_0,\ldots,x_r$, $y_0,\ldots,y_r$ are real and that $|y_r| > |x_r|$. Then from

$$x_{r+1} = \tfrac{1}{2}(x_r + y_r)$$

it follows that x_{r+1} is real and nonzero and that

$$\operatorname{sign} x_{r+1} = \operatorname{sign} y_r .$$

Hence y_{r+1} is real and

(10) $$\begin{aligned} |y_{r+1}|^2 - |x_{r+1}|^2 &= |x_{r+1}|\,|y_r| - |x_{r+1}|^2 \\ &= \tfrac{1}{2}|x_{r+1}|[2|y_r| - |(x_r + y_r)|] \\ &\geq \tfrac{1}{2}|x_{r+1}|[2|y_r| - |x_r| - |y_r|] \\ &= \tfrac{1}{2}|x_{r+1}|[\,|y_r| - |x_r|] > 0 , \end{aligned}$$

so that $|y_{r+1}| > |x_{r+1}|$. This completes the induction proof.

It is heuristically obvious that there will be convergence to zero unless all but a finite number of the α_n are 1, because of cancellation in the determination of the arithmetic mean. If there are only a finite number of $\alpha_n = 1$, then we can apply the result of the simplest case to the tails of the sequences to conclude that there is convergence to a nonzero limit.

To establish the first statement, consider a step when a negative determination of the square root is taken. Then we have either

$$x_{n+1} > 0, y_{n+1} < 0 \quad \text{or} \quad x_{n+1} < 0, y_{n+1} > 0 .$$

In the first case,

$$x_{n+2} = -\tfrac{1}{2}(|y_{n+1}| - x_{n+1}), \quad |y_{n+2}| = \{-\tfrac{1}{2}(|y_{n+1}| - x_{n+1})|y_{n+1}|\}^{\frac{1}{2}} ,$$

so that

$$\frac{|x_{n+2}|}{|y_{n+1}|} < \tfrac{1}{2} , \quad \frac{|y_{n+2}|}{|y_{n+1}|} < \frac{1}{\sqrt{2}} ;$$

i.e., we have a contraction with ratio at least $\sqrt{2}$:

$$\max(|x_{n+2}|, |y_{n+2}|) < 2^{-\frac{1}{2}} \max(|x_{n+1}|, |y_{n+1}|) .$$

The second case is dealt with similarly.

Hence when there is an infinite number of negative square roots, we have convergence to zero.

THEOREM 1. Let

$$a = 2^{N_1} + 2^{N_2} + \dots + 2^{N_m} , \tag{11}$$

where $0 \leq N_1 < N_2 < \dots < N_m$, and where the N's are integral. If $\sin\theta \neq 0$, then

$$\frac{1}{B_a(\cos\theta, 1)} = \frac{\theta + 2a^*\pi}{\sin\theta} , \tag{12}$$

where

$$a^* = -2^{N_1} + 2^{N_2} - \dots + (-1)^m 2^{N_m} . \tag{13}$$

<u>Proof</u>. This is accomplished by a careful reworking of the proof already given in the "simplest" case. Note that the "limiting" cases $a = 0$, $a = \infty$ are covered by (12).

Let us look at the sequences defined by (5) when we now allow φ to be any angle $\not\equiv \pi \pmod{2\pi}$. In this excepted case, we have $x_0 = -1$, $y_0 = 1$, and $x_n \equiv y_n = 0$ for $n \geq 1$. The results of (6) and (7) still apply.

We now assert that the "simplest" algorithm, applied in the case of $\psi = \varphi + 2a^*\pi$, is precisely the algorithm determined by a, applied in the φ case. We use induction. The initial values are the same in both cases for $\cos(\theta + 2a^*\pi) = \cos\theta$.

Suppose that the algorithms coincide up to the k-th step. Then

$$x_k = \cos(2^{-1}\psi) \cdots \cos(2^{-k+1}\psi)\cos^2(2^{-k}\psi) ,$$
$$y_k = \cos(2^{-1}\psi) \ldots \cos(2^{-k+1}\psi)\cos(2^{-k}\psi) .$$

By definition of the a-algorithm (with $\varepsilon_r = 1$ if $r > N$),

$$(14) \qquad \begin{aligned} x_{k+1} &= \tfrac{1}{2}(x_k + y_k) = \cos(2^{-1}\psi) \ldots \cos(2^{-k})\psi \cos^2(2^{-k-1}\psi) , \\ y_{k+1} &= \varepsilon_k \operatorname{sign}(x_{k+1})(x_{k+1}y_k)^{1/2} \\ &= (-1)^{\alpha_k} \operatorname{sign}(x_{k+1}) |\cos(2^{-1}\psi) \ldots \cos(2^{-k}\psi)\cos(2^{-k-1}\psi)| . \end{aligned}$$

If we can show that

$$(15) \qquad \operatorname{sign}[\cos(2^{-1}\psi) \ldots \cos(2^{-k-1}\psi)] = (-1)^{\alpha_k} \operatorname{sign}(x_{k+1}) ,$$

then we have

$$y_{k+1} = \cos(2^{-1}\psi) \ldots \cos(2^{-k-1}\psi) ,$$

and we have shown that the ψ-formulas are correct at the $(k+1)$th step, so that the proof of Theorem 1 will be complete.

Referring back to (14), we see that (15) follows if we show that

$$(16) \qquad \operatorname{sign}(\cos 2^{-k-1}\psi) = (-1)^{\alpha_k} .$$

In order to clarify the proof of (16), we discuss first a special case. Take $a = 26 = 2^4 + 2^3 + 2$, so that $a^* = -2 + 2^3 - 2^4 = -10$.

We have to consider, in the case $x_0 = .5$, $y_0 = 1$, the simplest algorithm beginning with

$$x_0 = \cos(\tfrac{1}{3}\pi - (2 - 2^3 + 2^4)2\pi) , \; y_0 = 1 .$$

Let us examine (16) in this case first when $k = 2$. We have $\alpha_k = 0$ and

$$2^{-3}(\tfrac{1}{3}\pi - 2^2\pi + 2^4\pi + 2^5\pi) \equiv 2^{-3}(\tfrac{1}{3}\pi - 2^2\pi) \pmod{2\pi} = -(11/24)\pi$$

so that (16) is satisfied. It is also satisfied when $k = 3$, for the $\alpha_k = 1$ and

$$2^{-4}(\tfrac{1}{3}\pi - 2^2\pi + 2^4\pi - 2^5\pi) \equiv 2^{-4}(\tfrac{1}{3}\pi - 2^2\pi) + \pi \pmod{2\pi} = -(11/48)\pi + \pi .$$

We now take up the general case

$$2^{-k-1}\psi = 2^{-k-1}\varphi + 2^{-k-1}\left[-2^{N_1} + 2^{N_2} - \ldots + (-1)^m 2^{N_m}\right](2\pi) .$$

In order to determine sign $(\cos 2^{-k-1}\psi)$, we can discard all the powers of 2 in $[\ldots]$ which are greater than k, and combine all those less than k

with the outside term $2^{-k-1}\varphi$, to get a sum φ_k, say. Hence

$$2^{-k-1}\psi = \varphi_k + \theta_k ,$$

where

$$\theta_k = 2^{-k-1}(-1)^k 2^k(2\pi) = (-1)^k\pi \text{ if there is an } N_k = k, \text{ i.e., if } \alpha_k = 1 ,$$

and θ_k is otherwise zero. It is clear that φ_k is of the opposite sign to θ_k when $\alpha_k = 1$, and that $|\varphi_k| < \frac{1}{2}\pi$ in all cases.

Hence we have

$$\alpha_k = 1 , \quad \operatorname{sign}(\cos 2^{-k-1}\psi) = -1 ,$$
$$\alpha_k = 0 , \quad \operatorname{sign}(\cos 2^{-k-1}\psi) = 1 ,$$

which establishes (16). The proof of Theorem 1 is complete.

The values of a^* for $a = 1,2,\dots$ are

-1, -2, 1, -4, 3; 2, -3, -8, 7, 6;
-7, 4, 5, 6, 5; -16, 15, 16, -15, 12;
-13, -14, 13, 8, -9; -10, 9, -12, 11, 10;
-11, -32, 31, ...

It is easy to see that $a \to a^*$ is a one-one mapping of positive integers onto the set of all nonzero integers. We illustrate this by two examples.

(i)
$$\begin{aligned} a^* = 26 &= 2 + 2^3 + 2^4 \\ &= -2 + 2^2 + (-2^3 + 2^4) + 2^4 \\ &= -2 + 2^2 - 2^3 + 2^5 , \end{aligned}$$
$$a = 2 + 2^2 + 2^3 + 2^5 = 46 .$$

(ii)
$$\begin{aligned} a^* = -26 &= -2 - 2^3 - 2^4 \\ &= -2 + 2^3 - 2^4 - 2^4 \\ &= -2 + 2^3 - 2^5 , \end{aligned}$$
$$a = 2 + 2^3 + 2^5 = 42 .$$

It is also easy to prove that

$$\frac{1}{3} a \le |a^*| \le a .$$

We give here the early terms of the Borchardt sequences in the case $x_0 = \frac{1}{2}$, $y_0 = 1$ for the algorithms corresponding to $a = 0(1)3$ together with the limits and the reciprocals of the limits multiplied by $(3\sqrt{3}/2\pi)$.

a = 0		a = 1		a = 2		a = 3	
.5000	1.0000	.5000	1.0000	.5000	1.0000	.5000	1.0000
.7500	.8660	.7500	-.8660	.7500	.8660	.7500	-.8660
⋮		-.0580	-.1778	.8080	-.8365	-.0580	.2241
		-.1411	-.1684	-.0143	-.1092	.0830	.1364
		⋮		-.0617	-.0821	.1098	.1224
				⋮		⋮	

$$B_0 = .82699 \qquad (3\sqrt{3}/2\pi)B_0^{-1} = 1$$

$$B_1 = .16540 \qquad (3\sqrt{3}/2\pi)B_1^{-1} = -5$$

$$B_2 = .07518 \qquad (3\sqrt{3}/2\pi)B_2^{-1} = -11$$

$$B_3 = .11814 \qquad (3\sqrt{3}/2\pi)B_3^{-1} = 7$$

Theoretically, since in this case $\theta = \frac{1}{3}\pi$, the limit in the simplest case is $(\sin\theta)/\theta = 3\sqrt{3}/2\pi = .826993$.

We conclude our discussion of the Borchardt case with a brief sketch of the invariant-integral approach in the simplest case. This will serve as a motivation for the representation (5) and as a model for some later developments.

Take $y > x > 0$, and define

$$I(x,y) = \frac{1}{2(y^2 - x^2)^{\frac{1}{2}}} \int_{2x^2y^{-2}-1}^{1} \frac{dX}{(1 - X^2)^{\frac{1}{2}}} . \tag{17}$$

If we change the variable by writing $X = 2t^2 - 1$, we get

$$I(x_0,y_0) = \frac{1}{(y_0^2 - x_0^2)^{\frac{1}{2}}} \int_{x_0y_0^{-1}}^{1} \frac{dt}{(1 - t^2)^{\frac{1}{2}}} , \tag{18}$$

and elementary algebra shows that the right-hand side is $I(x_1,y_1)$. In fact, since

$$x_1 = \tfrac{1}{2}(x_0 + y_0) , \quad y_1 = \left[\tfrac{1}{2}(x_0 + y_0)y_0\right]^{\frac{1}{2}} ,$$

we have

$$2x_1^2y_1^{-2} - 1 = 2 \cdot \frac{(x_0 + y_0)^2}{4} \frac{2}{y_0(x_0 + y_0)} - 1 = \frac{x_0 + y_0}{y_0} - 1 = \frac{x_0}{y_0}$$

and

$$2(y_1^2 - x_1^2)^{\frac{1}{2}} = 2\left[\frac{(x_0 + y_0)y_0}{2} - \frac{(x_0 + y_0)^2}{4}\right]^{\frac{1}{2}} = (y_0^2 - x_0^2)^{\frac{1}{2}} .$$

We have therefore established

$$(19) \qquad I(x_0,y_0) = I(x_1,y_1) = \cdots = I(x_n,y_n) = \cdots .$$

We determine the common limit $B_0 = B_0(x_0,y_0)$ as follows. Evaluating $I(x,y)$ directly gives

$$2(y^2 - x^2)^{\frac{1}{2}} I(x,y) = \tfrac{1}{2}\pi - \arcsin(2x^2y^{-2} - 1) = \arccos(2x^2y^{-2} - 1)$$
$$= 2 \arccos xy^{-1} .$$

Hence, if we write $\theta_n = \arccos x_ny_n^{-1}$, we have

$$I(x_n,y_n) = \frac{\theta_n}{(y_n^2 - x_n^2)^{\frac{1}{2}}} = \frac{\theta_n}{y_n \sin \theta_n} .$$

Since, as $n \to \infty$, x_n,y_n have a common limit, it follows that $\lim x_ny_n^{-1} = 1$ as $\lim \theta_n = 0$. Hence

$$I(x_0,y_0) = \lim I(x_n,y_n) = B_0^{-1} ,$$

so that, as before,

$$B_0 = \frac{(y_0^2 - x_0^2)^{\frac{1}{2}}}{\arccos x_0y_0^{-1}} .$$

3. THE BORCHARDT ALGORITHM - HYPERBOLIC CASE

We now discuss the sequences (2) when $0 \leq y_0 < x_0$. This can be done in several ways. We proceed in an elementary manner, relying on the analogy between trigonometric and hyperbolic functions.

We may assume $x_0 = \cosh u$, $y_0 = 1$, and take $u > 0$. Using the facts that

$$1 + \cosh t = 2 \cosh^2 \tfrac{1}{2}t , \quad \sinh t = 2 \sinh \tfrac{1}{2}t ,$$

we see, just as in Section 2, that the sequences

$$(20) \qquad \begin{aligned} x_n &= \cosh(2^{-1}u)\cosh(2^{-2}u) \cdots \cosh(2^{-n+1}u)\cosh^2(2^{-n}u) , \\ y_n &= \cosh(2^{-1}u)\cosh(2^{-2}u) \cdots \cosh(2^{-n+1}u)\cosh(2^{-n}u) \end{aligned}$$

have a common limit

$$\lim x_n = \lim y_n = B_0(\cosh u, 1) = u^{-1} \sinh u .$$

This deals with the "simplest" case.

We next note that the result just established does not depend on the reality of u, and to discuss the general case we observe what happens when we apply (20) when u is replaced by $u + 2a^*i\pi$. We have to find the appropriate replacement for (8). This turns out to be:

The positive square root y_{n+1} of $x_{n+1}y_n$ is that for which

$$R(y_{n+1}/x_{n+1}) > 0 \ , \tag{21}$$

or, when $R(y_{n+1}/x_{n+1}) = 0$, that for which

$$I(y_{n+1}/x_{n+1}) > 0 \ , \tag{22}$$

and (8) is to be replaced by

$$\begin{aligned} x_{n+1} &= \tfrac{1}{2}(x_n + y_n) \ , \\ y_{n+1} &= (-1)^{\alpha_n}(\text{positive square root of } x_{n+1}y_n) \ . \end{aligned} \tag{23}$$

We have assumed that $x_0 \neq y_0$; this implies that no x_n vanishes, so that (21) and (22) are meaningful.

Geometrically, the condition (21) means that the positive square root is the one nearer to x_{n+1}. Referring to the first diagram, we note that x_{n+1} is the middle-point of the segment $x_n y_n$, and that $\pm y_{n+1}$ are on the bisector of the angle $x_{n+1}0y_n$. To justify the position of y_{n+1} outside the triangle, we use elementary geometry. Referring to the second diagram, if AD_1 is the bisector of angle BAC, then equating expressions (in terms of the sides) for $\cos \measuredangle BAD_1$ and $\cos \measuredangle CAD_1$, we get

$$AD_1^2 = AB \cdot BC - BD_1 \cdot D_1C \ ,$$

so that AD_1 is less than the geometric mean of AB and BC.

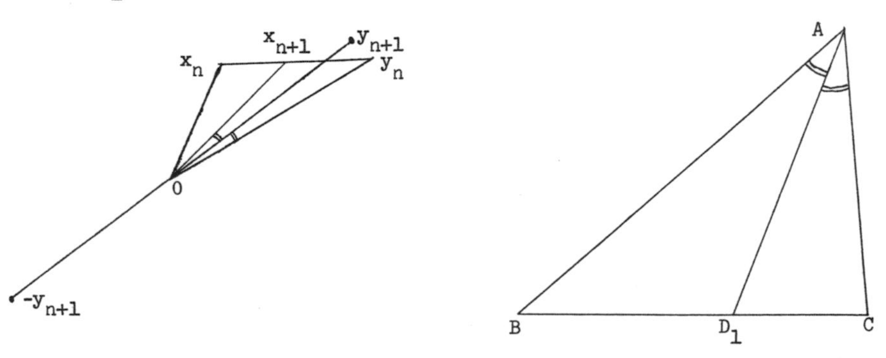

Arithmetically, if $z = x + iy$, $Z = X + iY$, then

$$R(zZ^{-1}) = (xX + yY)(X^2 + Y^2)^{-1}$$

and

$$|z \mp Z|^2 = x^2 + y^2 + X^2 + Y^2 \mp 2(xX + yY) ,$$

so that $|z - Z| < |z + Z|$ is equivalent to $(xX + yY) > 0$.

Note that

$$\text{sign } R(Z^{-1}) = \text{sign } R(Z) , \quad \text{sign } I(Z^{-1}) = -\text{sign } I(Z) .$$

The progress of the algorithms defined by (21)-(23) is illustrated by the cases when the index is 6 and when it is 11. We take $u > 0$ and use the fact that

$$\cosh(a \pm ib) = \cosh a \cos b \pm i \sinh a \sin b .$$

$\underline{a = 6}$, $\underline{a^* = 2}$

$k = 0$ $\cosh(u + 4i\pi) = \cosh u$

$k = 1$ $\cosh(\frac{1}{2}u + 2i\pi) = \cosh(\frac{1}{2}u)$

$k = 2$ $\cosh(\frac{1}{4}u + i\pi) = -\cosh(\frac{1}{4}u)$

$k = 3$ $\cosh(\frac{1}{8}u + \frac{1}{2}i\pi) = i \sinh(\frac{1}{8}u)$

$k = 4$ $\cosh(\frac{1}{16}u + \frac{1}{4}i\pi)$ has positive real part

$\underline{a = 11}$, $\underline{a^* = -7}$

$k = 0$ $\cosh(u - 14i\pi) = \cosh u$

$k = 1$ $\cosh(\frac{1}{2}u - 7i\pi) = -\cosh(\frac{1}{2}u)$

$k = 2$ $\cosh(\frac{1}{4}u - \frac{7}{2}i\pi) = i \sinh(\frac{1}{4}u)$

$k = 3$ $\cosh(\frac{1}{8}u - \frac{7}{4}i\pi)$ has positive real part

$k = 4$ $\cosh(\frac{1}{16}u - \frac{7}{8}i\pi)$ has negative real part

$k = 5$ $\cosh(\frac{1}{32}u - \frac{7}{16}i\pi)$ has positive real part

In order to get an analogue of Theorem 1, it is sufficient to rework the proof carefully. We make use of the conformal mapping given by $w = \cos z$.

With the "finite" case disposed of in this way, there remains the infinite case, e.g., when negative square roots are taken at the stages 2, 3, 5, 7, ..., i.e., when the algorithm is specified by the infinite string

$$\ldots 100010101100 .$$

One approach is direct and is an appropriate modification of the discussion of v. Dávid [17]. Another is to make use of the results of the finite case and certain monotony properties of $\max(|x_n|, |y_n|) = \theta_n$. We take a finite segment of the infinite string so long that its limit is less than ε in absolute value, choose a term in that sequence which is less than 2ε in absolute value, and compare it with the corresponding term in the infinite sequence.

We indicate the progress of the algorithm of index 1 with initial values $x_0 = 3$, $y_0 = 1$.

$$\begin{array}{ll} 3 & 1 \\ 2 & -\sqrt{2} \\ \frac{1}{2}(2-\sqrt{2}) = .2929 & i\sqrt{\sqrt{2}-1} = .6436\,i \\ .1464 + .3218i & .1011 + .4661i \\ .1238 + .3940i & \\ \vdots & \vdots \end{array}$$

$$\text{limit} = \frac{2\sqrt{2}}{\operatorname{arccosh} 3 - 2i\pi} = .1171 + .4173i$$

SUMMARY. This completes our outline of a discussion of the Borchardt case when x_0, y_0 are real and positive. No essentially new problems arise if we allow arbitrary x_0, y_0.

Observe that we have found the limits of all algorithms; and, given a possible limit, we can construct the algorithm which gives that limit.

Gauss discussed in detail the backward continuation of the arithmetic-geometric mean algorithm. This can be done also for the Borchardt algorithm.

Finally, the "Borchardt" algorithm was known to Gauss and to Pfaff long before the publication of Borchardt's paper. (See Carlson [2].)

4. THE LOGARITHMIC ALGORITHM

We discuss only the "simplest" case. Additional care is needed in the general case because we have two ambiguities at each stage. See Todd [14]. Suppose that $y_0 > x_0 > 0$ and that x_n, y_n are defined by (3), where the positive square root is always taken. It is easy to show that $x_n \uparrow, y_n \downarrow$; and since, from (3),

$$(y_{n+1}^2 - x_{n+1}^2) = \frac{1}{2}(y_n^2 - x_n^2) , \tag{24}$$

it follows that the sequences have a common limit $\ell_0(x_0, y_0) = \ell$ between x_0 and y_0.

We next observe that

$$(25)\qquad \frac{(y_n/x_n) - 1}{y_n^2 - x_n^2} = \frac{y_n - x_n}{x_n(y_n^2 - x_n^2)} = \frac{1}{x_n(y_n + x_n)} \to \frac{1}{2\ell^2} \ .$$

It follows from (3) that

$$(26)\qquad (y_{n+1}/x_{n+1}) = (y_n/x_n)^{\frac{1}{2}} \ .$$

We now require the following result, which was used by Hurwitz [9] in his development of the theory of the elementary transcendental functions by means of square roots.

If $x > 0$ then

$$(27)\qquad \lim 2^n(x^{2^{-n}} - 1) = \log x \ .$$

This is a special case of: If $x > 0$ then

$$(28)\qquad \lim n(x^{1/n} - 1) = \log x \ .$$

Using (24) and (26) repeatedly, we find

$$(29)\qquad \frac{(y_n/x_n) - 1}{y_n^2 - x_n^2} = \frac{(y_{n-1}/x_{n-1})^{\frac{1}{2}} - 1}{2^{-1}(y_{n-1}^2 - x_{n-1}^2)} = \cdots = \frac{(y_0/x_0)^{2^{-n}} - 1}{2^{-n}(y_0^2 - x_0^2)} \ .$$

Hence, by (25) and (27),

$$(30)\qquad \frac{1}{2\ell^2} = \frac{\log(y_0/x_0)}{y_0^2 - x_0^2} \ , \text{ i.e., } \ \ell = \left[\frac{2\log(y_0/x_0)}{y_0^2 - x_0^2}\right]^{-\frac{1}{2}} \ .$$

We give another proof of this by means of an invariant integral (see Carlson [2]).

If

$$I(x,y) = \int_0^\infty (t + x^2)^{-1}\,(t + y^2)^{-1}\,dt \ ,$$

then by changing the variable from t to s, where

$$t = s(s + xy)[s + \{(x + y)/2\}^2]^{-1} \ ,$$

we find

$$I(x_{n+1}, y_{n+1}) = I(x_n, y_n) \ .$$

Hence,

$$\int_0^\infty (t + \ell^2)^{-2}\,dt = \int_0^\infty (t + x_0^2)^{-1}\,(t + y_0^2)^{-1}\,dt \ ;$$

i.e., by elementary integrals,

$$\ell^{-2} = [2 \log(y_0/x_0)]/(y_0^2 - x_0^2) .$$

We now give a third proof which will serve as a pattern for later work. For simplicity, take $x_0 = 1$, $y_0 = t > 1$. Then we have

$$x_1 : y_1 = 1 : \sqrt{t} \text{ with multiplier } [\tfrac{1}{2}(1 + t)]^{\frac{1}{2}} ,$$

$$x_2 : y_2 = 1 : \sqrt[4]{t} \text{ with multiplier } [\tfrac{1}{2}(1 + \sqrt{t})]^{\frac{1}{2}} .$$

Since $t^{2^{-n}} \to 1$, we have $x_n/y_n \to 1$. The square of the cumulated multiplier is

$$\left[\frac{1 + t}{2} \cdot \frac{1 + t^{2^{-1}}}{2} \cdot \ldots \cdot \frac{1 + t^{2^{-n}}}{2}\right] .$$

Multiplying this by $(1 - t^{2^{-n}})/(1 - t^{2^{-n}})$, we find that the factors in square brackets telescope and hence

$$[\ \ldots \ ] = \frac{1}{1 - t^{2^{-n}}} \cdot \frac{1 - t^2}{2^n} .$$

Proceeding to the limit, using (27), we recover (30).

We conclude our discussion with an example: the case when $x_0 = \sqrt{2}$, $y_0 = 1$. We find

$$x_1 = 1.3065, \quad x_2 = 1.2535, \quad x_3 = 1.2272, \ \ldots ,$$

$$y_1 = 1.0987, \quad y_2 = 1.1495, \quad y_3 = 1.1752, \ \ldots ,$$

giving

$$\ell(\sqrt{2}, 1) = [\log_e 2]^{-\frac{1}{2}} = 1.201122409 .$$

5. THE LEMNISCATE ALGORITHM

We shall only outline a discussion of the "simplest" case; the general case will be discussed in a later paper (Todd [14]). We assume $x_0 > y_0 > 0$. Then, although the intervals (x_n, y_n) are 'nested,' since

$$x_{n+1}^2 - y_{n+1}^2 = -\tfrac{1}{4}(x_n^2 - y_n^2) ,$$

the sequences are not monotonic, but the even and odd sequences are monotonic. A common limit $L = L(x_0, y_0)$ therefore exists.

To evaluate L, we use the approach of Carlson [2]. Define

$$I(x,y) = \frac{1}{4}\int_0^\infty (t + x^2)^{-3/4} (t + y^2)^{-1/2} \, dt .$$

Using the change of variable

$$t = \frac{s(s + xy)}{s + [\frac{1}{2}(x + y)]^2} ,$$

we find

$$I(x_n, y_n) = I(x_{n+1}, y_{n+1}) .$$

It follows that

$$I(x_0, y_0) = \int_0^{\infty} (t + L^2)^{-5/4}\, dt = \left[-(t + L^2)^{-1/4}\right]_0^{\infty} = L^{-1/2} .$$

We get a more convenient form for $I(x,y)$ by a change of variable,

$$t + x^2 = (x^2 - y^2)T^{-4} .$$

A little algebra gives

$$I(x,y) = \frac{1}{(x^2 - y^2)^{1/4}} \int_0^{(1-x^{-2}y^2)^{1/4}} \frac{dT}{(1 - T^4)^{1/2}} .$$

The lemniscate sine and its inverse are defined by

$$\operatorname{arcsl} z = \int_0^z \frac{dT}{(1 - T^4)^{1/2}} , \quad z = \int_0^{\operatorname{sl} z} \frac{dT}{(1 - T^4)^{1/2}} .$$

This nomenclature is justified because if we compute the arclength of the lemniscate of Bernoulli,

$$r^2 = \cos 2\theta ,$$

we get

$$s = \int_r^1 (1 + r^2(d\theta/dr)^2)^{1/2}\, dr$$

$$= \int_r^1 (1 - t^4)^{-1/2}\, dt .$$

Note that $r = 0$ gives the quarter-perimeter

$$A = \int_0^1 (1 - t^4)^{-1/2}\, dt = 1.3110287771 \ldots = 2^{-1/2} K(1/2) .$$

In this notation, we have

$$[L(x_0, y_0)]^{-1/2} = (x_0^2 - y_0^2)^{-1/4} \operatorname{arcsl}(1 - x_0^{-2} y_0^2)^{1/4} . \tag{31}$$

In order to prepare for the discussion of the general case, we have to parameterize the algorithm (4). For this purpose, we require some of the properties of the lemniscate functions. We now summarize the relevant parts of the theory. The theory goes back to Gauss; there has been an excellent exposition by Markushevich [11].

It seems more convenient to use the lemniscate cosine than the sine in our development. We define

$$\text{arccl } x = A - \text{arcsl } x . \tag{32}$$

The following relations hold:

$$\text{sl } A = 1 , \quad \text{cl } A = 0 ,$$

$$\text{cl}^2 x + \text{sl}^2 x + \text{cl}^2 x \,\text{sl}^2 x = 1 . \tag{33}$$

Both $\text{sl } x$ and $\text{cl } x$ are elliptic functions with equal quarter periods A. In terms of the usual Jacobi functions,

$$\begin{aligned} \text{sl } x &= 2^{-1/2} \,\text{sn}(2^{1/2}x)/\text{dn}(2^{1/2}x) , \\ \text{cl } x &= \text{cn}(2^{1/2}x) , \end{aligned} \tag{34}$$

where the modulus $k^2 = 1/2$. The addition formula

$$\text{cl}(\alpha + \beta) = \frac{\text{cl}^2 \beta - \text{sl}^2 \alpha}{\text{cl } \alpha \text{ cl } \beta \,(1 + \text{sl}^2\alpha) + \text{sl } \alpha \text{ sl } \beta \,(1 + \text{cl}^2\beta)} \tag{35}$$

gives the following results which are crucial later:

$$\text{cl}(x + 2A) = \text{cl}(x + 2iA) = -\text{cl } x, \quad \text{cl}(x + 2A + 2iA) = \text{cl } x . \tag{36}$$

Since $k'^2 = 1 - k^2 = 1/2$, the Jacobi imaginary transformation expresses the lemniscate functions with argument ix algebraically in terms of those with argument x. In particular,

$$\text{cl}(ix) = [\text{cl } x]^{-1} , \quad \text{sl}(ix) = i \text{ sl } x . \tag{37}$$

This is the simplest example of "complex multiplication."

From the addition formula (35) we find, writing c for $\text{cl } \tau$, s for $\text{sl } \tau$, and using (33),

$$\begin{aligned} \text{cl}(1 + i)\tau &= \frac{c^2(1 + s^2) - is^2(1 + c^2)}{1 - s^2c^2} \cdot \frac{1 + c^2}{1 + c^2} \\ &= \frac{c^2[1 - c^2 + 1 + c^2] - i[(1 - c^2)(1 + c^2)]}{1 + c^2 - c^2(1 - c^2)} \\ &= \frac{i(c^2 - i)}{c^2 + i} . \end{aligned}$$

Write $u_1 = u_0/(1 + i)$. Then we have just proved

$$\text{cl } u_0 = i \,\frac{\text{cl}^2 u_1 - i}{\text{cl}^2 u_1 + i} . \tag{38}$$

(The corresponding result for the sl-function is given by Gauss [3, p. 409].)

Suppose that we have

$$x_0 = 1 + \mathrm{cl}^2\, u_0 \;, \; y_0 = 2\, \mathrm{cl}\, u_0 \;. \tag{39}$$

Substitute from (38) in (39) to get, where now $c_1 = \mathrm{cl}\, u_1$,

$$x_0 = \frac{(c_1^2 + i)^2 - (c_1^2) - i)^2}{(c_1^2 + i)^2} = \frac{4ic_1^2}{(c_1^2 + i)^2} \;, \quad y_0 = 2i\, \frac{(c_1^2 - i)}{(c_1^2 + i)} \;.$$

The recurrence relation (4) now gives

$$x_1 = \frac{i(1 + c_1^2)^2}{(c_1^2 + i)^2} \;, \quad y_1 = \frac{2ic_1(1 + c_1^2)}{(c_1^2 + i)^2} \;,$$

so that

$$x_1 : y_1 = (1 + c_1^2) : 2c_1 \;, \tag{40}$$

where the multiplier is

$$\mu_1 = \frac{i(1 + c_1^2)}{(c_1^2 + i)^2} \;. \tag{41}$$

Comparison of (35) and (40) shows that we have accomplished the parameterization announced in Section 1, and this will enable us to get an analogue of Theorem 1.

The results (39), (40), together with the fact that $\mathrm{cl}\, x \to 1$ as $x \to 0$, establish the existence of the limit. However, to evaluate it, we need to evaluate the infinite product of the successive multipliers. This appears awkward using the parameterization (39); but if instead we use

$$x_0 = \vartheta_2^2(u_0) + \vartheta_4^2(u_0) \;, \quad y_0 = 2\vartheta_2(u_0)\, \vartheta_4(u_0) \;, \tag{42}$$

where the parameter of the ϑ-functions is $q = e^{-\pi}$ or $\tau = i$, we can represent the product in a telescoping form and recover the limit by use of the relation

$$\lim z^{-1}\, \vartheta_1(z) = \vartheta_1' = \vartheta_2\vartheta_3\vartheta_4 = 2^{1/4}\, \vartheta_2^2$$

(where the ϑ's have zero argument) instead of

$$\lim \theta^{-1} \sin \theta = 1 \;.$$

ACKNOWLEDGMENT. I am indebted to my students Frank Liang and Douglas Tyler for computations carried out on a PDP-10.

REFERENCES

1. C. W. Borchardt, Gesammelte Werke, Reimer, Berlin, 1888.

2. B. C. Carlson, Algorithms involving arithmetic and geometric means, Amer. Math. Monthly 78 (1971), 696-705.

3. A. Erdélyi et al., eds., Higher transcendental functions, II, McGraw-Hill, New York, 1953.

4. L. Fejér, Gesammelte Arbeiten, II, Birkhäser, Basel, 1970.

5. C. F. Gauss, Werke, I - XII, B. C. Teubner, Leipzig, 1870-1929.

6. H. Geppert, ed., Ostwalds Klassiker, #225, C. F. Gauss, Anziehung eines elliptischen Ringes, Akademische Verlagsgesellschaft, Leipzig, 1925.

7. H. Geppert, Zur Theorie des arithmetisch-geometrischen Mittels, Math. Ann. 99 (1928), 162-180.

8. H. Geppert, Wie Gauss zur elliptischen Modulfunktion kam, Deutsche Math. 5 (1940-41), 158-175.

9. A. Hurwitz, Über die Einführung der elementaren transzendenten Funktionen in der algebraische Analysis, Math. Ann. 70 (1911), 33-47 ≡ Mathematische Werke, I, 706-721, Birkhäuser, Basel, 1932.

10. A. I. Markuschevich, Die Arbeiten von C. F. Gauss über Funktionentheorie, 151-182 in Reichardt [12].

11. A. I. Markuschevich, The remarkable sine functions, Elsevier, New York, 1966.

12. H. Reichardt, ed., Gauss 1777-1855, Gedenkband, B. G. Teubner, Leipzig, 1957.

13. John Todd, The lemniscate constants, Comm. ACM 18 (1975), 16-19.

14. John Todd, The many values of mixed means, II, to appear.

15. J. V. Uspensky, On the arithmetic-geometric means of Gauss, Math. Notae 5 (1945), 1-28, 57-88, 129-161.

16. L. von Dávid, Zur Gaussischen Theorie der Modulfunktion, Rend. Circ. Mat. Palermo, 35 (1913), 82-89.

17. L. von Dávid, Arithmetisch-geometrisches Mittel und Modulfunktion, J. Reine Angew. Math. 159 (1928), 154-170.

18. E. T. Whittaker and G. N. Watson, A course of modern analysis, 4th ed., Cambridge University Press, London, 1962.

INEQUALITIES AND FUNCTIONAL EQUATIONS IN THE THEORY OF THE PRICE INDEX

Wolfgang Eichhorn
Institut für Wirtschaftstheorie
und Operations Research
Universität Karlsruhe
D-75 Karlsruhe
WEST GERMANY

ABSTRACT. Five axioms which play a role in the theory of the price index are introduced in the form of inequalities and functional equations. It is shown that there exists a subset of four of these axioms which are independent and which imply the fifth one.

1. INTRODUCTION

In this note, a price index is regarded as a positive-valued measure of the prices of n commodities of a base year and of the current year. We introduce the price indices by a system of five natural properties which we call axioms. One of these axioms is the statement that a price index is a strictly increasing function of the current prices. In our approach, this condition can be called natural, since we think of a price index as a measure of the prices which is sensitive with respect to any change of any single price.

The five axioms are expressed in the form of inequalities and functional equations. It will be shown that there exists an independent quadruple of the axioms which implies the fifth one. Several examples of functions satisfying the axioms, that is, of price indices, will be given. The general solution of the system of axioms is not known.

A more detailed version of this note, where also price indices depending on both the prices and the quantities are considered, will be part of a book of the author [5]; see also [4].

Note that we do not consider here the so-called economic theoretic approach, where the preferences of an individual consumer (or consumer group) are involved in such a way that the prices and quantities constituting the price index become functions of each others (see, in this connection, S. N. Afriat [1], C. Blackorby and R. R. Russell [2], W. E. Diewert [3], F. M. Fisher and K. Shell [6], J. Muellbauer [8], F. A. Pollak [9], and P. A.

Samuelson and S. Swamy [10]).

2. NOTATIONS, AXIOMS, AND EXAMPLES

The following notations will be used:

$\mathbb{R}_{++} = \{r \mid r \text{ a positive real number}\}$

$\underline{x} = (x_1,\dots,x_n)$, $\underline{y} = (y_1,\dots,y_n)$, $\underline{xy} = x_1y_1 + \dots + x_ny_n$

$\underline{x} > \underline{y}$ if and only if $x_1 > y_1,\dots,x_n > y_n$

$\underline{x} \geqq \underline{y}$ if and only if $x_1 \geq y_1,\dots,x_n \geq y_n$

$\underline{x} \geq \underline{y}$ if and only if $\underline{x} \geqq \underline{y}$, $\underline{x} \neq \underline{y}$

Let

$$\underline{p}^o = (p_1^o,\dots,p_n^o) \in \mathbb{R}^n_{++}, \quad \underline{p} = (p_1,\dots,p_n) \in \mathbb{R}^n_{++}$$

be the vectors of the prices of n commodities of a base year and of the current year, respectively.

We regard a price index as a function

$$P : \mathbb{R}^{2n}_{++} \to \mathbb{R}_{++}, \quad (\underline{p}^o,\underline{p}) \mapsto P(\underline{p}^o,\underline{p})$$

which satisfies the following five natural properties which we call axioms. The value $P(\underline{p}^o,\underline{p})$ is called the <u>value of the price index at the price situation</u> $(\underline{p}^o,\underline{p})$.

(A.0) MEAN-VALUE AXIOM. For all $\underline{p}^o \in \mathbb{R}^n_{++}$, $\underline{p} \in \mathbb{R}^n_{++}$,

$$\min\left\{\frac{p_1}{p_1^o},\dots,\frac{p_n}{p_n^o}\right\} \leq P(\underline{p}^o,\underline{p}) \leq \max\left\{\frac{p_1}{p_1^o},\dots,\frac{p_n}{p_n^o}\right\}.$$

In other words, the value of the price index P is a mean value between the minimum and the maximum of the ratios of the corresponding prices of the two periods under consideration.

(A.1) MONOTONICITY AXIOM. The function P is strictly increasing with respect to $\underline{p}$ and strictly decreasing with respect to $\underline{p}^o$; that is, for every quadruple $\underline{p}^o, \bar{\underline{p}}^o, \underline{p}, \bar{\underline{p}}$ of vectors of $\mathbb{R}^n_{++}$,

$$P(\underline{p}^o,\underline{p}) > P(\underline{p}^o,\bar{\underline{p}}) \quad \text{if} \quad \underline{p} \geq \bar{\underline{p}}$$

and

$$P(\underline{p}^o,\underline{p}) < P(\bar{\underline{p}}^o,\underline{p}) \quad \text{if} \quad \underline{p}^o \geq \bar{\underline{p}}^o.$$

(A.2) LINEAR-HOMOGENEITY AXIOM. The function P is linearly homogeneous with respect to $\underline{p}$; that is, for all $\underline{p}^o \in \mathbb{R}^n_{++}$, $\underline{p} \in \mathbb{R}^n_{++}$, $\lambda \in \mathbb{R}_{++}$,

$$P(\underline{p}^o,\lambda\underline{p}) = \lambda P(\underline{p}^o,\underline{p}).$$

In other words, if the current prices change with the same percentage, then the value of the price index changes with this percentage.

(A.3) IDENTITY AXIOM. For all $\underline{p}^o \in \mathbb{R}^n_{++}$, $P(\underline{p}^o,\underline{p}^o) = 1$. In other words, if the prices of the base year do not change, then the value of the price index is equal to one.

(A.4) DIMENSIONALITY AXIOM. For all $\underline{p}^o \in \mathbb{R}^n_{++}$, $\underline{p} \in \mathbb{R}^n_{++}$, $\lambda \in \mathbb{R}_{++}$,

$$P(\lambda\underline{p}^o,\lambda\underline{p}) = P(\underline{p}^o,\underline{p}).$$

In other words, if two economies are identical except for the definition of the unit of money, then the values of the price indices are equal.

Examples of functions P satisfying axioms (A.0)-(A.4), i.e., of price indices, include the following:

$$(1) \qquad P(\underline{p}^o,\underline{p}) = \frac{\underline{c}\underline{p}}{\underline{c}\underline{p}^o} \qquad (\underline{c} = (c_1,\ldots,c_n) \in \mathbb{R}^n_{++})$$

$$(2) \qquad P(\underline{p}^o,\underline{p}) = \frac{\left[\beta_1 p_1^{-\rho} + \cdots + \beta_n p_n^{-\rho}\right]^{-1/\rho}}{\left[\beta_1 (p_1^o)^{-\rho} + \cdots + \beta_n (p_n^o)^{-\rho}\right]^{-1/\rho}}$$

$$\begin{cases} \rho \in \mathbb{R}, \neq 0, \ \Sigma\beta_\nu = 1, \\ \beta_1 \in \mathbb{R}_{++},\ldots,\beta_n \in \mathbb{R}_{++} \end{cases}$$

$$(3) \qquad P(\underline{p}^o,\underline{p}) = \left[\beta_1\left(\frac{p_1}{p_1^o}\right)^{-\rho} + \cdots + \beta_n\left(\frac{p_n}{p_n^o}\right)^{-\rho}\right]^{-1/\rho}$$

$$(4) \qquad P(\underline{p}^o,\underline{p}) = \left(\frac{p_1}{p_1^o}\right)^{\alpha_1}\left(\frac{p_2}{p_2^o}\right)^{\alpha_2} \cdots \left(\frac{p_n}{p_n^o}\right)^{\alpha_n} \qquad \begin{cases} \alpha_1 \in \mathbb{R}_{++},\ldots,\alpha_n \in \mathbb{R}_{++} \\ \Sigma\alpha_\nu = 1 \end{cases}$$

$$(5) \qquad P(\underline{p}^o,\underline{p}) = \left[\frac{\underline{a}\underline{p}}{\underline{a}\underline{p}^o}\,\frac{\underline{b}\underline{p}}{\underline{b}\underline{p}^o}\right]^{-1/2} \qquad \begin{cases} \underline{a} = (a_1,\ldots,a_n) \in \mathbb{R}^n_{++} \\ \underline{b} = (b_1,\ldots,b_n) \in \mathbb{R}^n_{++} \end{cases}$$

3. MAIN RESULTS

THEOREM 1. The Mean-Value Axiom (A.0) is a consequence of axioms (A.1), (A.2), (A.3) in the following sense: Every function $F : \mathbb{R}^{2n}_{++} \to \mathbb{R}_{++}$ which satisfies (A.1), (A.2), (A.3) also satisfies (A.0).

Hence, we may define a price index to be a function $P : \mathbb{R}^{2n}_{++} \to \mathbb{R}_{++}$ which satisfies the <u>four</u> axioms (A.1)-(A.4).

<u>Proof of Theorem 1.</u> By definition,

$$\min\left\{\frac{p_1}{p_1^o},\dots,\frac{p_n}{p_n^o}\right\} \underline{p}^o \leqq \underline{p} \leqq \max\left\{\frac{p_1}{p_1^o},\dots,\frac{p_n}{p_n^o}\right\} \underline{p}^o. \tag{6}$$

Now, on the one hand,

$$\begin{aligned}\min\left\{\frac{p_1}{p_1^o},\dots,\frac{p_n}{p_n^o}\right\} &=: \mu(\underline{p}^o,\underline{p}) \\ &= \mu(\underline{p}^o,\underline{p})F(\underline{p}^o,\underline{p}^o) && \text{(by (A.3))} \\ &= F(\underline{p}^o,\mu(\underline{p}^o,\underline{p})\underline{p}^o) && \text{(by (A.2))} \\ &\leq F(\underline{p}^o,\underline{p}) && \text{(by (A.1) and (6));}\end{aligned}$$

and on the other hand,

$$\begin{aligned}\max\left\{\frac{p_1}{p_1^o},\dots,\frac{p_n}{p_n^o}\right\} &=: M(\underline{p}^o,\underline{p}) \\ &= M(\underline{p}^o,\underline{p})\, F(\underline{p}^o,\underline{p}^o) && \text{(by (A.3))} \\ &= F(\underline{p}^o,M(\underline{p}^o,\underline{p})\underline{p}^o) && \text{(by (A.2))} \\ &\geq F(\underline{p}^o,\underline{p}) && \text{(by (A.1) and (6)).}\end{aligned}$$

This completes the proof.

THEOREM 2. Axioms (A.1)-(A.4) are independent in the following sense: Any three of them can be satisfied by a function $F : \mathbb{R}^{2n}_{++} \to \mathbb{R}_{++}$ which does not satisfy the remaining axiom.

<u>Proof.</u> The function F given by

$$F(\underline{p}^o,\underline{p}) = \left(\frac{p_1}{p_1^o}\right)^{\alpha_1}\left(\frac{p_2}{p_2^o}\right)^{\alpha_2}\cdots\left(\frac{p_n}{p_n^o}\right)^{\alpha_n} \qquad \begin{cases}\alpha_1 \in \mathbb{R}_{++},\dots,\alpha_{n-1} \in \mathbb{R}_{++} \\ -\alpha_n \in \mathbb{R}_{++},\ \Sigma\alpha_\nu = 1\end{cases}$$

satisfies (A.2), (A.3), (A.4), but not (A.1) (since $\alpha_n < 0$). The function F given by

$$F(\underline{p}^o,\underline{p}) = \left[\frac{\underline{a}\underline{p}}{\underline{a}\underline{p}^o}\right]^{1/2} \qquad (\underline{a} = (a_1,\ldots,a_n) \in \mathbb{R}^n_{++})$$

satisfies (A.1), (A.3), (A.4), but not (A.2). The function F given by

$$F(\underline{p}^o,\underline{p}) = \frac{\underline{a}\underline{p}}{\underline{b}\underline{p}^o} \qquad \begin{cases} \underline{a} = (a_1,\ldots,a_n) \in \mathbb{R}^n_{++} \\ \underline{b} = (b_1,\ldots,b_n) \in \mathbb{R}^n_{++} \end{cases}$$

satisfies (A.1), (A.2), (A.4), but not (A.3). Finally, the function F given by

$$F(\underline{p}^o,\underline{p}) = \frac{\Sigma p^o_\nu}{\Sigma p^o_\nu+1}\,\frac{1}{n}\,\Sigma\,\frac{p_\nu}{p^o_\nu} + \frac{1}{\Sigma p^o_\nu+1}\max\left\{\frac{p_1}{p^o_1},\ldots,\frac{p_n}{p^o_n}\right\}$$

satisfies (A.1), (A.2), (A.3), but not (A.4). This completes the proof.

4. PROBLEM

The set of all price indices, i.e., the set of all functions P which satisfy axioms (A.1)-(A.4), is not known. As one easily sees, the following is true.

THEOREM 3. If $P_1,\ldots,P_k$ are an arbitrary number k of price indices, then

$$G := (\beta_1 P_1^\delta + \cdots + \beta_k P_k^\delta)^{1/\delta} \qquad \begin{cases} \delta \neq 0,\ \beta_1 \geq 0,\ldots,\beta_k \geq 0 \\ \text{real constants},\ \Sigma\beta_k = 1 \end{cases} \tag{7}$$

and

$$H := P_1^{\delta_1} P_2^{\delta_2} \cdots P_k^{\delta_k} \qquad \begin{cases} \delta_1 \geq 0,\ \delta_2 \geq 0,\ldots,\delta_k \geq 0 \\ \text{real constants},\ \Sigma\delta_k = 1 \end{cases} \tag{8}$$

are also price indices.

Here P^δ is defined by

$$(p^o,p) \mapsto \left[P(\underline{p}^o,\underline{p})\right]^\delta .$$

We note that (8) with $\delta_k = \beta_k$ follows from (7) for $\delta \to 0$.

As one sees from (7) or (8), the set of all price indices is a convex set C. The problem of determining this set C seems to be difficult. In order to solve this problem, the theorem of M. Krein and D. Milman [7] may be helpful.

ACKNOWLEDGMENT. I am indebted to H. Funke and J. Voeller for valuable remarks and suggestions concerning this paper.

REFERENCES

1. S. N. Afriat, The theory of international comparisons of real income and prices, in: D. J. Daly, ed., International comparisons of prices and outputs, National Bureau of Economic Research, New York, 1972, 13-69.

2. C. Blackorby and R. R. Russell, Indices and subindices of the cost of living and the standard of living, International Economic Review, forthcoming.

3. W. E. Diewert, Exact and superlative index numbers, J. of Econometrics 4 (1976), 115-145.

4. W. Eichhorn, Fisher's tests revisited, Econometrica 44, (1976), 247-256.

5. W. Eichhorn, Functional equations in economics, Reading, forthcoming.

6. F. M. Fisher and K. Shell, The economic theory of price indices, two essays on the effect of taste, quality and technological change, New York - London, 1972.

7. M. Krein and D. Milman, On extreme points of regular convex sets, Studia Mathematica, 9 (1940), 133-138.

8. J. Muellbauer, The cost of living and taste and quality change, J. of Economic Theory 10, (1975), 269-283.

9. R. A. Pollak, Subindices in the cost of living index, International Economic Review 16, (1975), 135-150.

10. P. A. Samuelson and S. Swamy, Invariant economic index numbers and canonical duality: survey and synthesis, American Economic Review 64, (1974), 566-593.

INDUCTION WITH INEQUALITIES INVOLVING NONQUASIARITHMETIC MEANS

Georg Aumann
Institut für Mathematik
Technische Universität München
WEST GERMANY

ABSTRACT. For the proof of inequalities involving quasiarithmetic mean values, the method of induction in the number of variables is very common. But if the mean values are not quasiarithmetic, then the question of what a coherent family of means with a variable number of variables should be has no unique answer. A practicable answer is given by the elevation algorithm ([1], [2]), which produces a mean of $n+1$ variables from one of n variables in a natural way. As this method is not mentioned in current monographs on inequalities (e.g., [3], [4]), a short report is given here incidentally extending the old results to not necessarily symmetric means.

1. THE ELEVATION ALGORITHM

DEFINITION 1. Let $[a,b] =: J$, a real interval, and let $n \in \mathbb{N}$. A continuous function $m : J^n \to J$ is called an n-mean on J if it is isotonic in each variable and if there is a constant r ("contraction factor") with $0 < r \leq 1/2$ such that for all $(x_1,\ldots,x_n) \in J^n$ we have

$$\underline{x} + r(\overline{x} - \underline{x}) \leq m(x_1,\ldots,x_n) \leq \overline{x} - r(\overline{x} - \underline{x}),$$

where $\underline{x} := \min_\nu x_\nu$ and $\overline{x} := \max_\nu x_\nu$.

REMARKS. (i) There is no condition of symmetry on the mean m.

(ii) For fixed positive values $a_1,\ldots,a_n$ with $\sum_\nu a_\nu = 1$, the function $A_{a_1 \cdots a_n} : (x_1,\ldots,x_n) \mapsto \sum_\nu a_\nu x_\nu$ is the arithmetic n-mean with weights a_ν; any finite interval can serve as domain.

(iii) If $f : [a',b'] \to [a,b]$ is bijective and in both directions Lipschitzian, then $f^{-1} \circ m \circ f$, or $(x_1',\ldots,x_n') \mapsto f^{-1}(m(f(x_1'),\ldots,f(x_n')))$, is an n-mean on $[a',b']$, the f-map of m; if m is arithmetic, then $f^{-1} \circ m \circ f$ is called quasiarithmetic.

DEFINITION 2. Given an n-mean m on J, the following algorithmic limit process, called <u>elevation</u>, furnishes an $(n+1)$-mean m' on J, called an <u>elevated mean</u> of m. It is determined by choosing an <u>arrangement</u> R consisting of a permutation P of $\{1,2,\ldots,n+1\}$ and $n+1$ orderings $(p_{u1},\ldots,p_{un})$ of the set $\{1,\ldots,n+1\}\setminus\{P(u)\}$, $u=1,\ldots,n+1$:

$$R := (P,(p_{11},\ldots,p_{1n}),\ldots,(p_{n+1\,1},\ldots,p_{n+1\,n})) .$$

Now given $(x_1,\ldots,x_{n+1}) \in J^{n+1}$, the iterative construction of a sequence $(x_1^{(q)},\ldots,x_{n+1}^{(q)})$, $q = 0,1,2,\ldots$, is the following: $x_u^{(0)} := x_u$, and for $q \geq 0$,

$$x_u^{(q+1)} := m(x_{p_{u1}}^{(q)},\ldots,x_{p_{un}}^{(q)}), \quad u = 1,\ldots,n+1 .$$

One can prove ([1]) that

$$\lim_{q\to\infty} x_u^{(q)} = m_R'(x_1,\ldots,x_{n+1})$$

exists, is independent of u, and is an $(n+1)$-mean on J with a contraction factor $r' \geq r^2$; furthermore, m_R' satisfies the functional equation

$$\text{(F)}\qquad m_R'(x_1,\ldots,x_{n+1}) = m_R'(m(x_{p_{11}},\ldots,x_{p_{1n}}),\ldots,m(x_{p_{n+1\,1}},\ldots,x_{p_{n+1\,n}}))$$

for all $(x_1,\ldots,x_{n+1}) \in J^{n+1}$, and m_R' is the only $(n+1)$-mean on J satisfying (F).

REMARKS. (i) The f-map of an elevated mean m' is the elevated mean of the f-map of m (with the same arrangement R).

(ii) If m is quasiarithmetic, then m_R' is also quasiarithmetic with the same mapping function f and with weight coefficients satisfying algebraic equations corresponding to (F).

(iii) In other cases, m_R' may be of a quite unexpected nature. For example, if $m(x_1,x_2) := (x_1^2+x_2^2)/(x_1+x_2)$ for, say, $x_1,x_2 \in [1,2]$, then $m'(x_1,x_2,x_3)$ (independent of R because of the symmetry of m) is a transcendental function and is different from $(x_1^2+x_2^2+x_3^2)/(x_1+x_2+x_3)$.

2. APPLICATION TO INEQUALITIES OF CONVEXITY TYPE

DEFINITION 3. Let $k,n \in \mathbb{N}$, $\Phi : J^k \to J^0$, $J^0 := [a_0,b_0]$, $m_1,\ldots,m_k$ n-means on J, and m_0 an n-mean on J_0. (The argument $(x_1,\ldots,x_k)$ of Φ is considered to form a "column vector," the argument $(x_1,\ldots,x_n)$ of an m_i to form a "row vector.") Φ is called <u>convex with respect to</u> $(m_1,\ldots,m_k;\, m_0)$ if for every $n \times k$-matrix

$$X := \begin{pmatrix} x_{11} & \cdots & x_{1n} \\ \cdots & \cdots & \cdots \\ x_{k1} & \cdots & x_{kn} \end{pmatrix}, \quad x_{st} \in J,$$

the inequality

$$(K_n) \qquad \Phi(m_1(x_{1.}),\dots,m_k(x_{k.})) \leq m_0(\Phi(x_{.1}),\dots,\Phi(x_{.n}))$$

is true ($x_{s.}$ is the s-th row vector, $x_{.t}$ the t-th column vector, of X). For brevity, let us write instead of (K_n),

$$\Phi MX \leq m_0 \Phi X,$$

where M is an abbreviation for $(m_1,\dots,m_k)$.

Clearly the classical inequalities of Cauchy, Jensen, Hölder, and Minkowski are specimens of (K_n).

THEOREM OF INDUCTION. If Φ is convex with respect to $(m_1,\dots,m_k;\ m_0)$, then it is continuous and also convex with respect to $(m'_{1R},\dots,m'_{kR};\ m'_{0R})$; that is,

$$(K_{n+1}) \qquad \Phi M'_R X' \leq m'_{0R} \Phi X'$$

holds for every $(n+1) \times k$-matrix and $M'_R := (m'_{1R},\dots,m'_{kR})$.

Proof. (i) The continuity of Φ is shown as in [2].

(ii) For the proof of $(K_n) \Longrightarrow (K_{n+1})$, we need a more elaborate notation. Let x^0_{su} denote the elements of X'; furthermore, for $q \geq 0$, let

$$x^{q+1}_{su} := m_s(x^q_{sp_{u1}},\dots,x^q_{sp_{un}})$$

for $s = 1,\dots,k$, $u = 1,\dots,n+1$, and let

$$y^q_u := \Phi(x^q_{1u},\dots,x^q_{ku}).$$

Then from (K_n), it follows that

$$\Phi(m_1(x^q_{1p_{u1}},\dots,x^q_{1p_{un}}),\dots,m_k(x^q_{kp_{u1}},\dots,x^q_{kp_{un}}))$$
$$\leq m_0(\Phi(x^q_{1p_{u1}},\dots,x^q_{kp_{u1}}),\dots,\Phi(x^q_{1p_{un}},\dots,x^q_{kp_{un}})),$$

or, for $q \geq 0$,

$$y^{q+1}_u = \Phi(x^{q+1}_{1u},\dots,x^{q+1}_{ku}) \leq m_0(y^q_{p_{u1}},\dots,y^q_{p_{un}}) =: \tilde{y}^{q+1}_u.$$

Introducing now $z_u^0 := y_u^0$, $z_u^{q+1} := m_0(z^q_{p_{u1}},\dots,z^q_{p_{un}})$, $q \geq 0$, we have $\lim_q z_u^q = m_0'(y_1^0,\dots,y_{n+1}^0)$. Furthermore,

$$z_u^1 = m_0(y^0_{p_{u1}},\dots,y^0_{p_{un}}) = \tilde{y}_u^1 \geq y_u^1.$$

Assuming that

$$z_u^q \geq y_u^q \quad \text{for} \quad u = 1,\dots,n+1,$$

from the isotonicity of m_0 we get

$$z_u^{q+1} = m_0(z^q_{p_{u1}},\dots,z^q_{p_{un}}) \geq m_0(y^q_{p_{u1}},\dots,y^q_{p_{un}}) = \tilde{y}_u^{q+1} \geq y_u^{q+1},$$

and for $q \to \infty$, by the continuity of Φ, we have

$$m'_{0R}\,\Phi X' \geq \Phi M'_R X'.$$

REMARKS. (i) As to the equality sign in (K_n) and in (K_{n+1}), statements may be formulated analogous to those in [2].

(ii) A special (classical) case: If $m_1 = \cdots = m_k = m_0 =: m_-$, and Φ is a k-mean $m_|$ on $J_0 = J$ (the subscripts $-$ and $|$ indicate, respectively, that we have to do with a row or column operation), then the inequality

$$(M_{k,n}) \qquad m_|\, m_-\, X \leq m_-\, m_|\, X$$

can be understood as "concavity" of m_- with respect to $m_|$, such that the induction theorem with reversed inequality sign is applicable in the number k. So from $(M_{k,n})$ we can deduce not only $(M_{k,n+1})$ but also $(M_{k+1,n})$ involving the corresponding elevated means m'_- and $m'_|$.

3. PROBLEM

The nonsymmetry of the mean m evolves a question concerning the elevation algorithm: If m is a (nonsymmetric) n-mean, and if a sequence $R_1, R_2, \dots, R_q, \dots$ of arrangements is given, one can carry through the algorithm in a generalized way by using for the q-th iterative step the arrangement R_q. There is again convergence to an (n+1)-mean $m'_{R_1R_2\cdots}$ which in general does not satisfy a functional equation. Nevertheless, one can ask for a characterization of the family F_m of all those (n+1)-means $m'_{R_1R_2\cdots}$ descending from one m by generalized elevation.

REFERENCES

1. G. Aumann, Aufbau von Mittelwerten mehrerer Argumente, I, Math. Ann. 109 (1933), 235-253.

2. G. Aumann, Konvexe Funktionen und die Induktion bei Ungleichungen zwischen Mittelwerten, Sitz. Ber. Bayer. Ak. Wiss. Math. Kl. 1933, 403-415.

3. E. F. Beckenbach and R. Bellman, Inequalities, Springer Verlag, Berlin, 1971 (3rd ed.).

4. D. S. Mitrinovic, Analytic inequalities, Springer Verlag, Berlin, 1970.

REFERENCES

1. G. Aumann, Aufbau von Mittelwerten mehrerer Argumente, I, Math. Ann. 109 (1933), 235-253.

2. G. Aumann, Konvexe Funktionen und die Induktion bei Ungleichungen zwischen Mittelwerten, Sitz. Ber. Bayer. Ak. Wiss. Math. Kl. 1933, 403-415.

3. E. F. Beckenbach and R. Bellman, Inequalities, Springer Verlag, Berlin, 1971 (3rd ed.).

4. D. S. Mitrinovic, Analytic inequalities, Springer Verlag, Berlin, 1970.

ON TURÁN-TYPE INEQUALITIES

Dieter K. Ross
Department of Mathematics
La Trobe University
Bundoora, Victoria 3083
AUSTRALIA

ABSTRACT. Many of the special functions y_n which occur in mathematical physics satisfy an inequality of the type $y_{n+1}^2 - y_n y_{n+2} \geq 0$, for $n = 0,1,2,\ldots$. This paper deals primarily with a relationship between this inequality and a three-term recurrence relation which can be considered as the generator of the y_n. Only a few examples are given here, but many more can be derived just as easily.

1. INTRODUCTION

There are many theorems about special functions which can be used to prove inequalities of the form

$$W_n(x) \equiv y_{n+1}^2(x) - y_n(x)y_{n+2}(x) \geq 0, \quad \text{for} \quad n = 0,1,2,\ldots,$$

where the $y_n(x)$ are particular sequences of functions defined for some real values of x. An extensive literature survey on this topic is to be found in Karlin and Szegö [1]. Amongst the simplest of sequences satisfying $W_n(x) \geq 0$ are the sequences of functions 1, x^n, e^{nx}, $(a \sin nx + b \cos nx)$, and $(a \sinh nx + b \cosh nx)$, where a and b are real constants.

A rather less obvious such sequence of functions is that given by the Legendre polynomials $P_n(x)$ defined on $-1 \leq x \leq 1$. This was first proved by Turán [5], who was concerned with estimating the distribution of their zeros. His proof was based on the recurrence relation

$$(n + 1)P_{n+1}(x) = (2n + 1)xP_n(x) - nP_{n-1}(x)$$

and on the differential relation

$$(1 - x^2)P_n'(x) = nP_{n-1}(x) - nxP_n(x).$$

Thus, he proves that

$$W_n(x) \equiv P_{n+1}^2(x) - P_n(x)P_{n+2}(x) \geq 0 \quad \text{for} \quad x^2 \leq 1.$$

Another example is afforded by the Bessel functions of the first kind, denoted by $J_n(x)$. For these functions it is known that

$$J_{n+1}(x) = (2n/x)J_n(x) - J_{n-1}(x)$$

and

$$J_n'(x) = J_{n-1}(x) - (n/x)J_n(x),$$

from which it follows that

$$W_n(x) \equiv J_{n+1}^2(x) - J_n(x)J_{n+2}(x) = 2\int_0^1 t\,J_{n+1}^2(xt)dt.$$

That is, for these functions we have $W_n(x) \geq 0$, $x \in \mathbb{R}$.

In the two examples cited above, both the three-term recurrence relation and expressions for the derivative are used. However, there are simple instances where the recurrence relation alone, sometimes without reference to the values of $y_0(x)$ and $y_1(x)$, is sufficient to establish a result like $W_n(x) \geq 0$.

2. THEOREMS AND EXAMPLES

Let $y_n(x)$, with $n = 0,1,2,\ldots$, be a sequence of functions defined for some real values of x by the recurrence relation

$$y_n(x) = B_n y_{n+1}(x) + C_n y_{n-1}(x), \quad n = 1,2,3,\ldots, \tag{1}$$

where the numbers B_n and C_n may be functions of x. Define

$$W_n(x) \equiv y_{n+1}^2(x) - y_n(x)y_{n+2}(x). \tag{2}$$

Then the following simple identities exist:

$$W_n(x) = [y_{n+1}(x) + y_n(x)/(2B_{n+1})]^2 + (4B_{n+1}C_{n+1} - 1)y_n^2(x)/(4B_{n+1}^2) \tag{3}$$

and

$$B_{n+2}W_{n+1}(x) = C_{n+1}W_n(x) + \Delta C_{n+1}y_{n+1}^2(x) + \Delta B_{n+1}y_{n+1}^2(x), \tag{4}$$

where Δ is the forward-difference operator: $\Delta U_n \equiv U_{n+1} - U_n$.

THEOREM 1. Let $y_n(x)$ be a sequence of functions defined as in (1), where the numbers B_n and C_n have either the property

(i) $4B_nC_n \geq 1$,

or the property

(ii) $B_{n+1} \geq B_n > 0$ and $C_{n+1} \geq C_n > 0$.

Then

$$W_n(x) \equiv y_{n+1}^2 - y_n(x)y_{n+2}(x) \geq 0.$$

Proof. This follows in a trivial manner from (3) and (4).

EXAMPLE 1a. Consider the recurrence relation satisfied by the ultra-spherical polynomials $P_n^{(\lambda)}(x)$, namely

$$(5) \qquad (n+1)y_{n+1}(x) = 2(n+\lambda)xy_n(x) - (n+2\lambda-1)y_{n-1}(x).$$

Then the condition $4B_nC_n \geq 1$ leads to the inequality

$$W_n(x) \equiv [P_{n+1}^{(\lambda)}(x)]^2 - P_n^{(\lambda)}(x)P_{n+2}^{(\lambda)}(x) \geq 0 \quad \text{for} \quad x^2 \leq 1 - (1-\lambda)^2/(n+\lambda)^2.$$

This inequality is valid for any of the solutions of (5) and is independent of the values of $y_0(x)$ and of $y_1(x)$, supposing only that these should be defined. This result can be strengthened if we know that

$$|P_n^{(\lambda)}(x)/P_n^{(\lambda)}(1)| \leq 1 \quad \text{for} \quad 0 \leq \lambda \leq 1 \quad \text{and} \quad x^2 \leq 1; \quad \text{see Example 2.}$$

The case $\lambda = 1/2$ is interesting, for then (5) is also the recurrence relation for the Legendre functions of the second kind, usually denoted by $Q_n(x)$, so that

$$Q_{n+1}^2(x) - Q_n(x)Q_{n+2}(x) \geq 0 \quad \text{for} \quad x^2 < 1 - \frac{1}{(2n+1)^2}.$$

EXAMPLE 1b. The Hermite polynomials $H_n(x)$ satisfy the relation

$$y_{n+1}(x) = xy_n(x) - ny_{n-1}(x).$$

This corresponds to the case $B_n = 1/x$ and $C_n = n/x$ in (4), so that

$$W_n(x) \equiv H_{n+1}^2(x) - H_n(x)H_{n+2}(x) \geq 0 \quad \text{for} \quad x \in \mathbb{R}^+.$$

In fact, the inequality is valid for all $x \in \mathbb{R}$. Once again the inequality is valid no matter what is the value of $H_0(x)$, $H_1(x)$ then being $xH_0(x)$, supposing only that $H_0(x)$ should be defined.

THEOREM 2. Let $y_n(x)$ be a sequence of functions defined as in (1), where the B_n and C_n have the properties

$$C_0 = 0, \quad C_i \text{ and } B_{i-1} \neq 0 \quad \text{for} \quad i = 1,2,3,\dots .$$

Then

$$W_n(x) = (C_1/C_{n+1}) \prod_{i=1}^{n+1} (C_i/B_i)y_0^2(x) + (\Delta B_n/B_{n+1})y_{n+1}^2(x)$$
$$+ \sum_{i=1}^{n} \frac{C_{i+1}C_{i+2}\cdots C_n}{B_iB_{i+1}B_{i+2}\cdots B_nB_{n+1}} \Delta(B_{i-1}C_i)y_i^2(x). \tag{6}$$

Proof. This follows directly from the identity in (4) by carrying out the appropriate summations and rearrangements.

EXAMPLE 2. Let $F_n(x) \equiv P_n^{(\lambda)}(x)/P_n^{(\lambda)}(1)$ denote a standardized ultraspherical polynomial, so that $F_n(x)$ satisfies

$$(n + 2\lambda)y_{n+1}(x) = 2(n + \lambda)xy_n(x) - ny_{n-1}(x).$$

It follows from Theorem 5, and from the inequality $|F_n(x)| \leq 1$ for $x^2 < 1$ (see Szegö [4]), that

$$W_n(x) \equiv F_{n+1}^2(x) - F_n(x)F_{n+2}(x) \geq \lambda \frac{[1 - F_{n+1}^2(x)]}{(n + \lambda)(n + 2\lambda + 1)} \geq 0$$

for $0 \leq \lambda \leq 1$. This result was known to Szász [3]. In the same way, it can be shown that $W_n(x) \geq 0$ for $\lambda \geq 1$ and $x^2 \geq 1$. This appears to be a new result.

Several other extensions are possible, but these will not be given here.

Many new inequalities and identities can be found for special functions by exploring in more detail the identities (3) and (4). Nevertheless, the above examples illustrate some of the relationships between three-term recurrence relations and Turán-type inequalities. Further results of this kind, some of which use more sophisticated methods, are given in Ross [2].

REFERENCES

1. S. Karlin and G. Szegö, On certain determinants whose elements are orthogonal polynomials, J. d'Analyse Math. 8, (1960 - 1961), 1-157.

2. D. K. Ross, Inequalities and identities for $y_n^2 - y_{n-1}y_{n+1}$, to appear.

3. O. Szász, Inequalities concerning ultraspherical polynomials and Bessel functions, Amer. Math. Soc. Proc. 1 (1950), 256-267.

4. G. Szegö, Orthogonal polynomials, Amer. Math. Soc. Colloquium Publications 23, 1959.

5. P. Turán, On the zeros of the polynomials of Legendre, Časopis pro Pěstování Matematiky a Fysiky, 75 (1950), 113-122.

ON THE POSITIVITY OF CIRCULANT AND SKEW-CIRCULANT DETERMINANTS

E. F. Beckenbach
Department of Mathematics
University of California
Los Angeles, California 90024
U.S.A.

Richard Bellman
Department of Mathematics
University of Southern California
Los Angeles, California 90007
U.S.A.

ABSTRACT. For a given vector $A := (a_0, a_1, \ldots, a_{n-1})$ in $\mathbb{R}^n$, simple necessary and sufficient conditions on $a_0, a_1, \ldots, a_{n-1}$ are established for the determinant of the circulant matrix of A to be positive, or negative, or zero. There is a striking difference between the conditions for n odd and the conditions for n even. The determinant of the skew-circulant matrix of A is similarly discussed.

1. DEFINITIONS AND STATEMENT OF RESULTS

For a given vector $A := (a_0, a_1, \ldots, a_{n-1})$ in $\mathbb{R}^n$, let $C(A)$ denote the <u>circulant matrix</u> of A,

$$C(A) := \begin{bmatrix} a_0 & a_1 & \cdots & a_{n-1} \\ a_{n-1} & a_0 & \cdots & a_{n-2} \\ \cdot & \cdot & \cdots & \cdot \\ a_1 & a_2 & \cdots & a_0 \end{bmatrix},$$

and let $\det C(A)$ denote the determinant of $C(A)$.

We shall prove the following results:

THEOREM 1. For the circulant matrix $C(A)$ of a given vector

$$A := (a_0, a_1, \ldots, a_{n-1}) \quad \text{in } \mathbb{R}^n,$$

if n is odd, $n = 2r + 1$, and if

$$\sum_{j=0}^{n-1} a_j \geq 0, \tag{1}$$

then

$$\det C(A) \geq 0. \tag{2}$$

THEOREM 2. For the circulant matrix $C(A)$ of a given vector

$$A := (a_0, a_1, \ldots, a_{n-1}) \quad \text{in } \mathbb{R}^n,$$

if n is even, $n = 2r + 2$, and if

$$\left| \sum_{j=0}^{r} a_{2j} \right| \geq \left| \sum_{j=0}^{r} a_{2j+1} \right| , \tag{3}$$

then

$$\det C(A) \geq 0 . \tag{4}$$

If the inequality sign is reversed in (1) or (3), then it is reversed also in (2) or (4), respectively. Accordingly, if the sign of equality holds in (1) or (3), then it holds also in (2) or (4), respectively. But the sign of equality can hold in (2) or (4) without holding in (1) or (3); this is the case, for example, when all the a_j are positive and equal. In Section 4, necessary and sufficient conditions will be given for equality, and for strict inequality, to hold in (2) and (4).

2. LEMMA

For a given vector $A := (a_0, a_1, \ldots, a_{n-1})$ in $\mathbb{R}^n$, let

$$f(x) := a_0 + a_1 x + \cdots + a_{n-1} x^{n-1} , \tag{5}$$

let ω denote the n-th root of unity of least positive argument,

$$\omega := \cos \frac{2\pi}{n} + i \sin \frac{2\pi}{n} , \tag{6}$$

and let

$$\omega_j := \omega^j, \quad j = 0, 1, \ldots, n . \tag{7}$$

We shall use the following known result as a principal tool.

LEMMA 1. With the notation given above, we have

$$\det C(A) = \prod_{j=0}^{n-1} f(\omega_j) . \tag{8}$$

Proof of Lemma 1. The following simple (cf. [2, pp. 106, 107], [4, pp. 444, 445]) proof is given for completeness.

For

$$M := \begin{bmatrix} \omega_0^0 & \omega_1^0 & \cdots & \omega_{n-1}^0 \\ \omega_0^1 & \omega_1^1 & \cdots & \omega_{n-1}^1 \\ \cdots & \cdots & \cdots & \cdots \\ \omega_0^{n-1} & \omega_1^{n-1} & \cdots & \omega_{n-1}^{n-1} \end{bmatrix} ,$$

matrix multiplication gives

$$(9) \qquad C(A) \times M = \begin{bmatrix} \omega_0^0 f(\omega_0) & \omega_1^0 f(\omega_1) & \cdots & \omega_{n-1}^0 f(\omega_{n-1}) \\ \omega_0^1 f(\omega_0) & \omega_1^1 f(\omega_1) & \cdots & \omega_{n-1}^1 f(\omega_{n-1}) \\ \cdots & \cdots & \cdots & \cdots \\ \omega_0^{n-1} f(\omega_0) & \omega_1^{n-1} f(\omega_1) & \cdots & \omega_{n-1}^{n-1} f(\omega_{n-1}) \end{bmatrix},$$

whence

$$(10) \qquad [\det C(A)] \det M = \left[\prod_{j=0}^{n-1} f(\omega_j)\right] \det M .$$

Since

$$\omega_j \neq \omega_k \quad \text{for} \quad j \neq k \quad (j,k = 0,1,\dots,n-1),$$

the Vandermonde determinant

$$\det M = \prod_{0 \leq j < k \leq n-1} (\omega_k - \omega_j)$$

does not vanish, and therefore (8) follows from (10).

We note, incidentally (see [1, pp. 242, 243], [3, p. 66]), that (9) is an expression of the fact that the vectors

$$\begin{bmatrix} \omega_j^0 \\ \omega_j^1 \\ \vdots \\ \omega_j^{n-1} \end{bmatrix}, \quad j = 0,1,\dots,n-1 ,$$

are the eigenvectors of $C(A)$, with corresponding eigenvalues of $f(\omega_j)$.

3. PROOF OF THEOREMS 1 AND 2

Since

$$\omega_{n-j} = \overline{\omega}_j \quad (j = 0,1,\dots,n) ,$$

and since $a_0, a_1, \dots, a_{n-1}$ are real numbers, we have

$$(11) \qquad f(\omega_{n-j}) = \overline{f(\omega_j)} .$$

Proof of Theorem 1. For n odd, $n = 2r + 1$, (8) and (11) give

$$\det C(A) = f(\omega_0) \prod_{j=1}^{r} |f(\omega_j)|^2 . \tag{12}$$

Since

$$f(\omega_0) = f(0) = \sum_{j=0}^{n} a_j , \tag{13}$$

and since $\prod_{j=1}^{r} |f(\omega_j)|^2 \geq 0$, it follows from (12) that (1) implies (2).

Proof of Theorem 2. For n even, $n = 2r + 2$, (8) and (11) give

$$\det C(A) = f(\omega_0)f(\omega_{r+1}) \prod_{j=1}^{r} |f(\omega_j)|^2 . \tag{14}$$

Since

$$\begin{aligned} f(\omega_0)f(\omega_{r+1}) &= \left(\sum_{j=0}^{2r+1} a_j\right)\left[\sum_{j=0}^{2r+1} (-1)^j a_j\right] \\ &= \left(\sum_{j=0}^{r} a_{2j} + \sum_{j=0}^{r} a_{2j+1}\right)\left(\sum_{j=0}^{r} a_j - \sum_{j=0}^{r} a_{2j+1}\right) \\ &= \left(\sum_{j=0}^{r} a_{2j}\right)^2 - \left(\sum_{j=0}^{r} a_{2j+1}\right)^2 , \end{aligned} \tag{15}$$

and since again $\prod_{j=1}^{r} |f(\omega_j)|^2 \geq 0$, it follows from (14) that (3) implies (4).

4. NECESSARY AND SUFFICIENT CONDITIONS

By logical exclusion, equations (12) - (15) yield the following results:

THEOREM 3. With the notation of Section 2, if n is odd, $n = 2r + 1$, then:

(i) $$\det C(A) = 0$$

if and only if at least one of the equations

$$f(\omega_j) = 0 , \quad j = 0,1,\ldots,r ,$$

is satisfied;

(ii) $\det C(A) > 0$ [(iii) $\det C(A) < 0$]

if and only if each of the inequalities

$$f(\omega_j) \neq 0 , \quad j = 1,2,\ldots,r ,$$

and

$$\sum_{j=0}^{n-1} a_j > 0 \qquad \left[\sum_{j=0}^{n-1} a_j < 0\right]$$

is satisfied.

THEOREM 4. With the notation of Section 2, if n is even, $n = 2r + 2$, then:

(i)
$$\det C(A) = 0$$

if and only if at least one of the equations

$$f(\omega_j) = 0 \ , \quad j = 0,1,\ldots,r+1 \ ,$$

is satisfied;

(ii) $\det C(A) > 0$ [(iii) $\det C(A) < 0$]

if and only if each of the inequalities

$$f(\omega_j) \neq 0 \ , \quad j = 1,2,\ldots,r \ ,$$

and

$$\left| \sum_{j=0}^{r} a_{2j} \right| > \left| \sum_{j=0}^{r} a_{2j+1} \right| \qquad \left[\left| \sum_{j=0}^{r} a_{2j} \right| < \left| \sum_{j=0}^{r} a_{2j+1} \right| \right]$$

is satisfied.

5. EXAMPLES

Perhaps a consideration of some simple examples of one-parameter families of vectors in $\mathbb{R}^5$ will be illuminating.

EXAMPLE 1. Consider the one-parameter family of vectors in $\mathbb{R}^5$ defined by

$$A(x) := (x,1,0,0,1) \ , \quad x \in \mathbb{R} \ .$$

By (5), (6), (7), and (12), we have

$$\det C(A(x)) = f(0)|f(\omega)|^2|f(\omega^2)|^2$$

$$= (x + 2)(x + 2 \cos \tfrac{2\pi}{5})^2(x + 2 \cos \tfrac{4\pi}{5})^2 \ ,$$

so that, with the known trigionometric-function values,

$$\det C(A(x)) = (x + 2)\left(x + \frac{\sqrt{5} - 1}{2}\right)^2\left(x - \frac{\sqrt{5} + 1}{2}\right)^2.$$

Thus, $\det C(A(x))$ has a zero of order one at $x = -2$, and has zeros of order two at $x = -(\sqrt{5} - 1)/2$ and $x = (\sqrt{5} + 1)/2$. Since, further, $f(0) < 0$ for $x < -2$, and $f(0) > 0$ for $x > -2$, we have

$$\det C(A(x)) = 0 \quad \text{for} \quad x = -2 \quad \text{or} \quad x = \frac{\pm\sqrt{5} + 1}{2} \ ;$$

$$\det C(A(x)) < 0 \quad \text{for} \quad x < -2 \; ;$$

$$\det C(A(x)) > 0 \quad \text{for} \quad x > -2 \; , \quad x \neq \frac{\pm\sqrt{5}+1}{2}$$

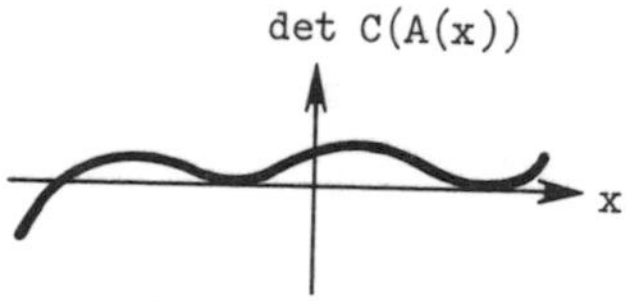

EXAMPLE 2. Next consider the family of vectors in $\mathbb{R}^5$ defined by

$$A(x) := (x,1,-1,-1,1) \; , \quad x \in \mathbb{R} \, .$$

For members of this family, we have

$$\det C(A(x)) = f(0)\,|f(\omega)|^2\,|f(\omega^2)|^2$$

$$= x\left(x + 2\cos\frac{2\pi}{5} - 2\cos\frac{4\pi}{5}\right)^2 \left(x - 2\cos\frac{2\pi}{5} + 2\cos\frac{4\pi}{5}\right)^2$$

$$= x(x+\sqrt{5})^2(x-\sqrt{5})^2 \; .$$

Now

$$\det C(A(x)) = 0 \quad \text{for} \quad x = -\sqrt{5}, \quad x = 0, \quad \text{or} \quad x = \sqrt{5} \; ;$$

$$\det C(A(x)) < 0 \quad \text{for} \quad x < 0 \; , \quad x \neq -\sqrt{5}$$

$$\det C(A(x)) > 0 \quad \text{for} \quad x > 0 \; , \quad x \neq \sqrt{5} \; .$$

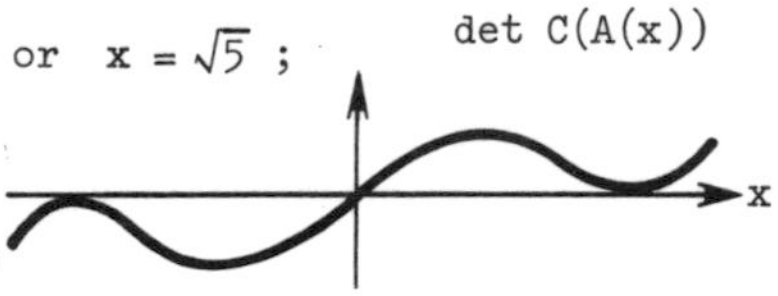

EXAMPLE 3. Finally, consider the family of vectors in $\mathbb{R}^5$ defined by

$$A(x) := (x,1,1,1,1) \; , \quad x \in \mathbb{R} \, .$$

For members of this family, we have

$$\det C(A(x)) = f(0)\,|f(\omega)|^4$$

$$= (x+4)\left(x + 2\cos\frac{2\pi}{5} + 2\cos\frac{4\pi}{5}\right)^4 ,$$

$$= (x+4)(x-1)^4 \; .$$

Now

$$\det C(A(x)) = 0 \quad \text{for} \quad x = -4 \quad \text{or} \quad x = 1 \; ;$$

$$\det C(A(x)) < 0 \quad \text{for} \quad x < -4 \; ;$$

$$\det C(A(x)) > 0 \quad \text{for} \quad x > -4 \; , \quad x \neq 1 \, .$$

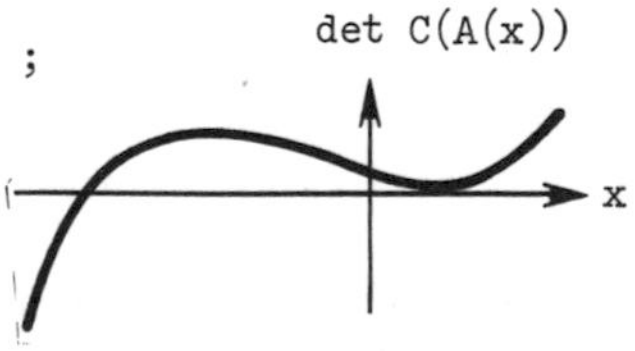

6. ON SKEW-CIRCULANT DETERMINANTS

A matrix that can be obtained by changing the sign of each element on one side of the principal diagonal of a circulant matrix is called a <u>skew-circulant matrix</u>.

For a given vector $A := (a_0, a_1, \ldots, a_{n-1})$ in $\mathbb{R}^n$, let $S(A)$ denote the skew-circulant matrix of A,

$$S(A) := \begin{bmatrix} a_0 , & a_1 , & \ldots , & a_{n-1} \\ -a_{n-1} , & a_0 , & \ldots , & a_{n-2} \\ \cdot & \cdot & \cdot & \cdot \\ -a_1 , & -a_2 , & \ldots , & a_0 \end{bmatrix} ,$$

and let $\det S(A)$ denote the determinant of $S(A)$.

We shall prove the following results:

THEOREM 5. For the skew-circulant matrix $S(A)$ of a given vector

$$A := (a_0, a_1, \ldots, a_{n-1}) \quad \text{in } \mathbb{R}^n ,$$

if n is odd, $n = 2r + 1$, and if

$$\sum_{j=0}^{r} a_{2j} \geq \sum_{j=0}^{r-1} a_{2j+1} , \tag{16}$$

then

$$\det S(A) \geq 0 . \tag{17}$$

THEOREM 6. For the skew-circulant matrix $S(A)$ of a given vector

$$A := (a_0, a_1, \ldots, a_{n-1}) \quad \text{in } \mathbb{R}^n ,$$

if n is even, $n = 2r + 2$, then

$$\det S(A) \geq 0 . \tag{18}$$

Let σ denote the n-th root of -1 of least positive argument,

$$\sigma := \cos \frac{\pi}{n} + i \sin \frac{\pi}{n} ,$$

and let

$$\tau_j := \sigma\omega^j , \qquad j = 0,1,\ldots,n .$$

We shall use the following result as a principal tool in proving Theorems 5 and 6.

LEMMA 2. With the notation given above, we have

$$\det S(A) = \prod_{j=0}^{n-1} f(\tau_j) . \tag{19}$$

Proof of Lemma 2. For

$$N := \begin{bmatrix} \tau_0^0 & \tau_1^0 & \cdots & \tau_{n-1}^0 \\ \tau_0^1 & \tau_1^1 & \cdots & \tau_{n-1}^1 \\ \cdots & \cdots & \cdots & \cdots \\ \tau_0^{n-1} & \tau_1^{n-1} & \cdots & \tau_{n-1}^{n-1} \end{bmatrix},$$

matrix multiplication gives

$$(20) \qquad S(A) \times N = \begin{bmatrix} \tau_0^0 f(\tau_0) & \tau_1^0 f(\tau_1) & \cdots & \tau_{n-1}^0 f(\tau_{n-1}) \\ \tau_0^1 f(\tau_0) & \tau_1^1 f(\tau_1) & \cdots & \tau_{n-1}^1 f(\tau_{n-1}) \\ \cdots & \cdots & \cdots & \cdots \\ \tau_0^{n-1} f(\tau_0) & \tau_1^{n-1} f(\tau_1) & \cdots & \tau_{n-1}^{n-1} f(\tau_{n-1}) \end{bmatrix},$$

whence

$$(21) \qquad [\det S(A)] \det N = \left| \prod_{j=0}^{n-1} f(\tau_j) \right| \det N .$$

Since

$$\tau_j \neq \tau_k \quad \text{for} \quad j \neq k \quad (j,k = 0,1,\ldots,n-1) ,$$

the Vandermonde determinant

$$\det N = \prod_{0 \le j < k \le n-1} (\tau_k - \tau_j)$$

does not vanish, and therefore (19) follows from (21).

We note, incidentally, that (20) is an expression of the fact that the vectors

$$\begin{bmatrix} \tau_j^0 \\ \tau_j^1 \\ \vdots \\ \tau_j^{n-1} \end{bmatrix}, \quad j = 0,1,\ldots,n-1 ,$$

are the eigenvectors of $S(A)$, with corresponding eigenvalues $f(\tau_j)$.

Proof of Theorems 5 and 6. Since

$$\tau_{n-1-j} = \overline{\tau_j} \quad (j = 0,1,\ldots,n-1) ,$$

and since $a_0, a_1, \ldots, a_{n-1}$ are real numbers, we have

(22) $$f(\tau_{n-1-j}) = \overline{f(\tau_j)} .$$

For Theorem 5, with n odd, $n = 2r + 1$, (19) and (22) give

(23) $$\det S(A) = f(\tau_r) \prod_{j=0}^{r-1} |f(\tau_j)|^2 .$$

Since

$$f(\tau_r) = f(-1) = \sum_{j=0}^{r} a_{2j} - \sum_{j=0}^{r-1} a_{2j+1} ,$$

and since $\Pi_{j=0}^{r-1} |f(\tau_j)|^2$ is nonnegative, it follows from (23) that (16) implies (17).

For Theorem 6, with n even, $n = 2r + 2$, (19) and (22) give

(24) $$\det S(A) = \prod_{j=0}^{r} |f(\tau_j)|^2 ,$$

which implies (18).

Alternative proofs of Theorems 5 and 6 can be given by means of the algebra of circulant and skew-circulant determinants.

Thus, for n odd [4, p. 466], changing the sign of each element in each even-numbered row of $S(A)$, and then changing the sign of each element in each even-numbered column of the resulting matrix, transforms $S(A)$ into $C(B)$, where

$$B := (b_0, b_1, \ldots, b_{n-1}) ,$$

with

(25) $$b_j := (-1)^j a_j , \quad j = 0,1,\ldots,n-1 .$$

Since signs have been changed in a total of an even number of rows and columns of $S(A)$ to obtain $C(B)$, we have

$$\det S(A) = \det C(B) .$$

Therefore, by (25), Theorem 5 follows from Theorem 1.

For n even [4, pp. 445, 446], a more involved argument shows that $\det S(A)$ can be expressed as a sum of two squares of real numbers. This establishes Theorem 6.

Conditions under which the sign of equality holds in (17) and in (18) can be read off from the eigenvalue products in (23) and (24), respectively:

THEOREM 7. With the foregoing notation, if n is odd, $n = 2r + 1$, then:

(i) $$\det S(A) = 0$$

if and only if at least one of the equations

$$f(\tau_j) = 0 \ , \quad j = 0,1,\ldots,r \ ,$$

is satisfied;

(ii) $\det S(A) > 0$ [(iii) $\det S(A) < 0$]

if and only if each of the inequalities

$$f(\tau_j) \neq 0 \ , \quad j = 0,1,\ldots,r-1 \ ,$$

and

$$\sum_{j=0}^{r} a_{2j} > \sum_{j=0}^{r-1} a_{2j+1} \qquad \left[\sum_{j=0}^{r} a_{2j} < \sum_{j=0}^{r-1} a_{2j+1} \right]$$

is satisfied.

THEOREM 8. With the foregoing notation, if n is even, $n = 2r + 2$, then:

(i) $$\det S(A) = 0$$

if and only if at least one of the equations

$$f(\tau_j) = 0 \ , \quad j = 0,1,\ldots,r \ ,$$

is satisfied;

(ii) $$\det S(A) > 0$$

if and only if each of the inequalities

$$f(\tau_j) \neq 0 \ , \quad j = 0,1,\ldots,r \ ,$$

is satisfied.

REFERENCES

1. Richard Bellman, Introduction to matrix analysis, 2nd edition, McGraw-Hill Book Company, New York, 1970.

2. Gerhard Kowalewski, Einführung in die Determinantentheorie, Walter de Gruyter, Berlin, 1942.

3. Marvin Marcus and Henryk Minc, A survey of matrix theory and matrix inequalities, Allyn and Bacon, Boston, 1964.

4. Thomas Muir and William H. Metzler, A treatise on the theory of determinants, Dover Publications, New York, 1960.

CONVEXITY IN THE THEORY OF THE GAMMA FUNCTION

H.-H. Kairies
Mathematisches Institut der Technischen Universität
3392 Clausthal-Zellerfeld
WEST GERMANY

ABSTRACT. In analogy with Artin's axiomatic treatment of the gamma function, it is here investigated to what extent $\Gamma : \mathbb{R}_+ \to \mathbb{R}_+$ can be characterized as a convex solution of suitable combinations of the functional equations

$$f(x+1) = xf(x), \quad f(\tfrac{x}{2})\, f(\tfrac{x+1}{2}) = 2\sqrt{\pi}\; 2^{-x} f(x), \quad f(x)\, f(1-x) = \pi/\sin \pi x.$$

1. INTRODUCTION

The subject of our considerations is the function

$$\text{(1)} \qquad \Gamma : \mathbb{R}_+ \to \mathbb{R}_+, \quad \Gamma(x) := \lim_{n\to\infty} \frac{n!\; n^x}{x(x+1)\cdots(x+n)} .$$

The function Γ is analytic and log-convex on $\mathbb{R}_+$; further, Γ satisfies the functional equations

$$\text{(F)} \qquad \forall x \in \mathbb{R}_+ : f(x+1) = xf(x),$$

$$\text{(M)} \qquad \forall p \in \mathbb{N}\ \forall x \in \mathbb{R}_+ : \prod_{k=0}^{p-1} f(\tfrac{x+k}{p}) = (2\pi)^{(p-1)/2}\, p^{1/2-x} f(x),$$

$$\text{(R)} \qquad \forall x \in (0,1) : f(x)\, f(1-x) = \pi/\sin \pi x.$$

We shall treat the following problem: Is it possible to characterize Γ as a convex solution of certain combinations of the functional equations above, together with some regularity conditions?

The main topic in Section 2 will be (F), and in Section 3 it will be the Gauss multiplication theorem (M) in its simplest nontrivial form $(p = 2)$:

$$(M_2) \qquad \forall x \in \mathbb{R}_+ : f(\tfrac{x}{2})\, f(\tfrac{x+1}{2}) = 2\sqrt{\pi}\, 2^{-x} f(x).$$

Equation (M_2) is known as Legendre's duplication formula. The reflection formula (R) is not well adopted for a global characterization of Γ, because it connects arguments only in the open interval $(0,1)$.

Our discussion of the axiomatic theory of the gamma function is motivated by the famous monograph of E. Artin [2]. Artin considered Γ to be a very elementary function, and in this spirit he used only elementary (natural) assumptions in his characterization theorems. We try to follow this line.

We use a standardized notation in the theorems to make it easy to compare analogous characterizing properties. Some minor improvements (which would distroy the simple structure of a theorem) are added in subsequent remarks.

2. CONVEX SOLUTIONS OF (F)

The main result in this section is due to H. Bohr/J. Mollerup [3]:

THEOREM. The conditions

(a) $f : \mathbb{R}_+ \to \mathbb{R}_+$ log-convex on $\mathbb{R}_+$,

(b) $\forall x \in \mathbb{R}_+ : f(x + 1) = xf(x)$,

(c) $f(1) = 1$

imply $f = \Gamma$.

In Artin's elegant proof [2], the product representation (1) for f is deduced directly from the inequality

$$(2) \quad \frac{\log f(n-1) - \log f(n)}{(n-1) - n} \le \frac{\log f(n+x) - \log f(n)}{(n+x) - n} \le \frac{\log f(n+1) - \log f(n)}{(n+1) - n},$$

for $x \in (0,1]$, $n \in \mathbb{N}$, $n \ge 2$, which holds because of (a).

The theorem remains true if "f log-convex on $\mathbb{R}_+$" is replaced by "f log-convex on (α,∞) for some $\alpha \in \mathbb{R}_+$." In this case, (2) has to be considered for $n \ge \alpha + 2$.

It is well known that "$f : \mathbb{R}_+ \to \mathbb{R}_+$ convex on R_+" instead of (a) is not sufficient for a characterization. For instance, A. E. Mayer [7] has shown: The function $w : \mathbb{R}_+ \to \mathbb{R}_+$, defined by

$$w(x) = 1 \quad \text{for} \quad x \in [1,2),\ w(x + 1) = xw(x), \tag{3}$$

is convex on $\mathbb{R}_+$. Now $w \notin C^1(\mathbb{R}_+)$; we constructed in [4] a convex C^∞-function $f \neq \Gamma$, satisfying (b) and (c). The construction in [4] was unnecessarily complicated; we obtain a stronger and simpler result in the following:

THEOREM 1. The conditions

(a) $f : \mathbb{R}_+ \to \mathbb{R}_+$ convex on $\mathbb{R}_+$ and log-convex on $(0,\alpha]$ for $\alpha \in \mathbb{R}_+$,

(b) f analytic on $\mathbb{R}_+$,

(c) $\forall x \in \mathbb{R}_+ : f(x + 1) = xf(x)$,

(d) $\forall x \in (0,1) : f(x)\, f(1 - x) = \pi/\sin \pi x$,

(e) $f(1) = 1$

do not imply $f = \Gamma$.

<u>Proof</u>. For any given $\alpha \in \mathbb{R}_+$, choose $\gamma \in \mathbb{R}_+$ such that

$$\gamma < \min\{\psi'(\alpha)/4\pi^2,\ \psi(2)/4\pi\}. \tag{4}$$

For numerical details concerning Γ, ψ $(\psi(x) = \Gamma'(x)/\Gamma(x))$, and their derivatives, consult [1]. Now define

$$w^*(x) := \Gamma(x) \cdot \exp v(x) \quad \text{for} \quad x \in \mathbb{R}_+$$

and let $v(x) := \gamma \sin 2\pi x$; then

$$\forall x \in \mathbb{R}_+ : v(x + 1) = v(x) \quad \text{and} \quad \forall x \in (0,1) : v(x) = -v(1 - x).$$

Consequently w^* satisfies conditions (b), (c), (d), and (e). To show (a), we first take any $x \in (0,\alpha]$. Then, by (4),

$$(\log \circ w^*)''(x) = -4\pi^2\gamma \sin 2\pi x + \psi'(x) \geq -4\pi^2\gamma + \psi'(\alpha) \geq 0.$$

Thus w^* is log-convex on $(0,\alpha]$. There is no harm in supposing $\alpha \geq 3$, and we do so. Then w^* is convex on $(0,3]$. For $x \in [2,3]$, we get

$$\begin{aligned} w^{*\prime}(x) &= \exp(\gamma \sin 2\pi x)[2\pi\gamma \cos 2\pi x \cdot \Gamma(x) + \Gamma'(x)] \\ &\geq \exp(\gamma \sin 2\pi x)[-2\pi\gamma \cdot \Gamma(3) + \Gamma'(2)] > 0, \end{aligned}$$

if $\Gamma'(2) > 4\pi\gamma$, which is true again because of (4). Now take any $n \in \mathbb{N}$, $n \geq 3$ and $x \in [n, n+1)$. Then

$$w^*(x) = (x-1)(x-2) \cdots (x-n+2)w^*(x-n+2) \quad \text{and} \quad x-n+2 \in [2,3)$$

implies: w^* is convex on $[n, n+1)$ as a product of positive convex functions with a positive derivative. Finally, $w^* \in C^1(\mathbb{R}_+)$ assures: w^* convex on $\mathbb{R}_+$.

REMARKS. (i) The preceding theorems show in a striking manner that the concept of log-convexity is fully adequate to characterize a solution of (F), whereas pure convexity is not. The reason for the different behaviour may be seen in the following facts:

There exists a $\beta \in \mathbb{R}_+$ such that for all $x \in \mathbb{R}_+ : \Gamma''(x) \geq \beta > 0$.

On the other hand, there is no $\beta \in R_+$ with: $\forall x \in \mathbb{R}_+ : (\log \circ \Gamma)''(x) \geq \beta \quad 0$; we have just $\forall x \in \mathbb{R}_+ : (\log \circ \Gamma)''(x) \geq 0$.

(ii) An interesting open problem related to the Bohr/Mollerup theorem may be formulated in this way: Find all functions $\varphi : \mathbb{R}_+ \to \mathbb{R}$ such that $f : \mathbb{R}_+ \to \mathbb{R}_+$, $\varphi \circ f$ convex on $\mathbb{R}_+$, (F), and $f(1) = 1$ imply $f = \Gamma$.

(iii) M. Muldoon [8] gave a characterization of Γ by means of the concept of complete monotonicity (a much stronger condition than log-convexity), where (F) is used only on $\mathbb{N}$:

THEOREM. The conditions

(a) $f : \mathbb{R}_+ \to \mathbb{R}_+$,

$\forall x \in \mathbb{R}_+ \forall k \in \mathbb{N} \cup \{0\} : (-1)^k (\log \circ f)^{(k+2)}(x) \geq 0$,

(b) $\forall n \in \mathbb{N} : f(n) = \Gamma(n)$

imply $f = \Gamma$.

3. CONVEX SOLUTIONS OF (M_2)

To start the discussion of the functional equation

$$(M_2) \qquad \forall x \in \mathbb{R}_+ : f(\tfrac{x}{2})\, f(\tfrac{x+1}{2}) = 2\sqrt{\pi}\, 2^{-x} f(x),$$

we refer to some results from [2] and [4], where the notion of convexity is not explicitly used. In Artin's monograph, Γ is characterized as C^1-solution of (M_2) and (F), while [4] contains a characterization of Γ by means of (M_2) alone. The limit condition used there can be weakened and simplified:

THEOREM 2. The conditions

(a) $f : \mathbb{R}_+ \to \mathbb{R}_+$, $f \in C^1(\mathbb{R}_+)$,

(b) $\forall x \in \mathbb{R}_+ : f(\tfrac{x}{2})\, f(\tfrac{x+1}{2}) = 2\sqrt{\pi}\, 2^{-x} f(x)$,

(c) $\lim_{x \to 0+} x f(x) \in [0, \infty]$,

(d) $f(1) = 1$

imply $f = \Gamma$.

Proof. Define $g(x) := \log f(x) - \log \Gamma(x)$ for $x \in \mathbb{R}_+$. Then $g \in C^1(\mathbb{R}_+)$ and $\forall x \in \mathbb{R}_+ : g(x) = g(x/2) + g((x+1)/2)$. Moreover, define $h(x) := g'(x)$ for $x \in \mathbb{R}_+$. Then $h \in C(\mathbb{R}_+)$ and

$$(5) \qquad \forall x \in \mathbb{R}_+ : h(x) = \tfrac{1}{2}[h(\tfrac{x}{2}) + h(\tfrac{x+1}{2})].$$

The limit condition (c) may be written in the form

$$\lim_{x\to 0+} xf(x) = e^{\beta} \quad \text{with some} \quad \beta \in [-\infty,\infty].$$

This is equivalent to

$$\lim_{x\to 0+} f(x)/\Gamma(x) = e^{\beta},$$

hence also equivalent to

$$\lim_{x\to 0+} g(x) = \beta, \quad \beta \in [-\infty,\infty]. \tag{6}$$

In the sequel we make no use of $h \in C(\mathbb{R}_+)$; all we need is the existence of the Riemann integral $\int_1^x h(t)dt$ for $x \in (0,1+\delta]$, where $\delta \in \mathbb{R}_+$ may be choosen arbitrarily small. On account of (d), we have

$$g(x) = \int_1^x h(t)dt \quad \text{for} \quad x \in (0,1+\delta]$$

and

$$\lim_{x\to 0+} \int_1^x h(t)dt = \beta.$$

Iteration of (5) yields

$$\forall x \in (0,\delta]\ \forall m \in \mathbb{N} : h(x\cdot 2^m) = \frac{1}{2^m} \sum_{k=0}^{2^m-1} h(x + \frac{k}{2^m}).$$

Taking the limit as $m \to \infty$, we obtain

$$\forall x \in (0,\delta] : \lim_{m\to\infty} h(x\cdot 2^m) = \int_x^{x+1} h(t)dt =: \alpha_x. \tag{7}$$

Note that $\alpha_x \in \mathbb{R}$. Now

$$\lim_{m\to\infty} h(x\cdot 2^m) = \lim_{m\to\infty} h(\frac{x}{2^s}\cdot 2^m)$$

for any $s \in \mathbb{N}$. Therefore,

$$\forall s \in \mathbb{N}\ \forall x \in (0,\delta] : g(\frac{x}{2^s} + 1) - g(\frac{x}{2^s}) = \alpha_x. \tag{8}$$

Passing to the limit as $s \to \infty$ in (8) we get: $\alpha_x = g(1) - \beta = -\beta$. That means: α_x, defined by (7), does not depend on x. Furthermore, $\beta \in \mathbb{R}$

because of $\alpha_x \in \mathbb{R}$. Now (7) becomes

$$\forall x \in (0,\delta] : g(x+1) - g(x) = -\beta, \quad \text{hence} \quad \forall x \in (0,\delta] : h(x+1) - h(x) = 0.$$

Extend h by $h(0) := h(1)$. Then h is Riemann integrable on $[0,1+\delta]$. Consequently (see [5]) we infer from (5):

$$(9) \qquad \forall x \in \mathbb{R}_+ : h(x) = \lim_{m\to\infty} \frac{1}{2^m} \sum_{k=0}^{2^m-1} h\left(\frac{x+k}{2^m}\right) = \int_0^1 h(t)dt = -\beta.$$

Thus $\forall x \in \mathbb{R}_+ : g(x) = -\beta x + \gamma$ and $\beta = \gamma = 0$ because of $g(1) = g(1/2) = 0$. This proves our theorem.

REMARKS. (i) The crucial step was to ensure that (5) has only one solution, namely (9). This is guaranteed by the existence of the Riemann integral $\int_0^x h(t)dt$ for $x \in [0,1+\delta]$. In [5] we have shown that this condition is optimally adapted to the uniqueness problem; e.g., $h \in \mathfrak{L}^1(\mathbb{R}_+)$ is not sufficient to get (9).

(ii) The limit condition (c) is equivalent to (6), and (6) is fulfilled if $\lim_{x\to 0+} h(x) \in [-\infty,\infty]$ or if h is bounded on one side for $x \to 0+$.

(iii) In connection with (F), the limit condition (c) is in fact very natural. If we assume (F) to be satisfied and if f is continuous at 1, then $\lim_{x\to 0+} xf(x) = f(1)$.

Now we investigate what happens if the limit condition (c) of Theorem 2 is deleted.

THEOREM 3. The conditions

(a) $f : \mathbb{R}_+ \to \mathbb{R}_+$, $f \in C^n(\mathbb{R}_+)$ for $n \in \mathbb{N}$,

(b) $\forall x \in \mathbb{R}_+ : f(\frac{x}{2})\, f(\frac{x+1}{2}) = 2\sqrt{\pi}\, 2^{-x} f(x)$,

(c) $f(1) = 1$

do not imply $f = \Gamma$.

<u>Proof</u>. We shall construct a function $f \neq \Gamma$ satisfying (a), (b), and

(c), and shall represent f in the form:

$$f(x) = \Gamma(x) \exp \int_1^x u(t)dt .$$

In fact, we have some freedom in the choice of u. Namely, $n \in \mathbb{N}$ being given arbitrarily, there clearly exists a function $u \in C^{n-1}([\frac{1}{2},1])$, different from the zero function, such that

(10) $$u(\tfrac{1}{2}) = u(1) = u_+^{(m)}(\tfrac{1}{2}) = u_-^{(m)}(1) = 0, \quad 1 \leq m \leq n-1,$$

(11) $$\|u\| \leq 1, \ \|u^{(m)}\| \leq 1, \ 1 \leq m \leq n-1,$$

(12) $$\int_{1/2}^{1} u(t)dt = 0.$$

We are going to extend u by means of the functional equation (5) to a function defined on $\mathbb{R}_+$. The extension onto $(0,1]$ is described below in A, the extension onto $[1,2]$ in B, and the extension onto $[2,\infty)$ in C. To avoid unnecessary complications in the notation, we denote the extended function after every extension step again by u; we hope no confusion will arise by doing that.

A. Define u_1 on $[\frac{1}{4},\frac{1}{2}]$ by

$$u_1(\tfrac{x}{2}) := 2u(x) - u(\tfrac{x+1}{2}), \ x \in [\tfrac{1}{2},1].$$

Clearly $u_1 \in C^{n-1}((\frac{1}{4},\frac{1}{2}))$. Furthermore, $u_1(\frac{1}{2}) = 2u(1) - u(1) = 0 = u(\frac{1}{2})$ because of (10). Taking left derivatives, we get

$$u_{1_-}^{(m)}(\tfrac{x}{2}) = 2^{m+1} u_-^{(m)}(x) - u_-^{(m)}(\tfrac{x+1}{2}) \quad \text{for } x \in (\tfrac{1}{2},1], \ 1 \leq m \leq n-1.$$

Thus

$$u_{1_-}^{(m)}(\tfrac{1}{2}) = 2^{m+1} u_-^{(m)}(1) - u_-^{(m)}(1) = 0 = u_+^{(m)}(\tfrac{1}{2}),$$

again because of (10). Next we define $u(x) := u_1(x)$ on $[\frac{1}{4},\frac{1}{2}]$, and the preceding considerations show:

$$u \in C^{n-1}([\tfrac{1}{4},1]) \quad \text{and} \quad \forall x \in [\tfrac{1}{2},1] : u(x) = \tfrac{1}{2}[u(\tfrac{x}{2}) + u(\tfrac{x+1}{2})].$$

Now the induction procedure is clear. If

$$u \in C^{n-1}([\tfrac{1}{2^k},1]) \quad \text{and} \quad \forall x \in [\tfrac{1}{2^{k-1}},1] : u(x) = \tfrac{1}{2}[u(\tfrac{x}{2}) + u(\tfrac{x+1}{2})],$$

define u_k on $[\frac{1}{2^{k-1}},\frac{1}{2^k}]$ by

$$u_k(\tfrac{x}{2}) := 2u(x) - u(\tfrac{x+1}{2}), \quad x \in [\tfrac{1}{2^k},\tfrac{1}{2^{k-1}}].$$

Clearly $u_k \in C^{n-1}((\frac{1}{2^{k+1}},\frac{1}{2^k}))$. Furthermore,

$$u_k(\tfrac{1}{2^k}) = 2u(\tfrac{1}{2^{k-1}}) - u(\tfrac{2^{k-1}+1}{2^k}) = u(\tfrac{1}{2^k})$$

and

$$\begin{aligned} u_{k\,-}^{(m)}(\tfrac{1}{2^k}) &= 2^{m+1}\, u_-^{(m)}(\tfrac{1}{2^{k-1}}) - u_-^{(m)}(\tfrac{2^{k-1}+1}{2^k}) \\ &= 2^{m+1}\, u_+^{(m)}(\tfrac{1}{2^{k-1}}) - u_+^{(m)}(\tfrac{2^{k-1}+1}{2^k}) \\ &= u_+^{(m)}(\tfrac{1}{2^k}). \end{aligned}$$

Now we define

$$u(x) := u_k(x) \quad \text{on} \quad [\tfrac{1}{2^{k+1}},\tfrac{1}{2^k}]$$

and have

$$u \in C^{n-1}([\tfrac{1}{2^{k+1}},1]) \quad \text{and} \quad \forall x \in [\tfrac{1}{2^k},1] : u(x) = \tfrac{1}{2}[u(\tfrac{x}{2}) + u(\tfrac{x+1}{2})].$$

B. An extension $U : [1,2] \to \mathbb{R}$ of $u : (0,1] \to \mathbb{R}$ by means of (5) must satisfy

$$\forall x \in [1,2] : U(\tfrac{x+1}{2}) - 2U(x) = -u(\tfrac{x}{2}). \tag{13}$$

Equation (13) may be interpreted as a linear inhomogeneous functional equation for U. Equations of type (13) are discussed in great detail in M. Kuczma's monograph [6] on that subject. Iteration of (13) shows:

$$\forall x \in [1,2]\ \forall k \in \mathbb{N} : U(x) = \frac{1}{2^k} U\left(\frac{x + 2^k - 1}{2^k}\right) + \frac{1}{2} \sum_{\kappa=0}^{k-1} \frac{1}{2^\kappa} u\left(\frac{x + 2^\kappa - 1}{2^{\kappa+1}}\right).$$

If there is a continuous solution U of (13), satisfying $U(1) = 0$, it must be given by

$$\text{(14)} \qquad \forall x \in [1,2] : U(x) = \frac{1}{2} \sum_{\kappa=0}^{\infty} \frac{1}{2^\kappa} u\left(\frac{x + 2^\kappa - 1}{2^{\kappa+1}}\right).$$

On the other hand, it is easily checked that the function U defined by (14) is a solution of (13) with $U(1) = 0$. Because of (11), we have $U \in C^{n-1}([1,2])$, and the derivatives may be obtained by termwise differentiation of (14). By (10), we have finally

$$U_+^{(m)}(1) = u_-^{(m)}(1) = 0, \qquad 1 \leq m \leq n - 1.$$

Hence, defining

$$u(x) := \frac{1}{2} \sum_{\kappa=0}^{\infty} \frac{1}{2^\kappa} u\left(\frac{x + 2^\kappa - 1}{2^{\kappa+1}}\right) \quad \text{for } x \in [1,2],$$

we obtain

$$u \in C^{n-1}((0,2]) \quad \text{and} \quad \forall x \in (0,2] : u(x) = \tfrac{1}{2}[u(\tfrac{x}{2}) + u(\tfrac{x+1}{2})].$$

C. The extension procedure onto $[2,\infty)$ is essentially the same as in A. Therefore, we write down just one step. Let

$$u \in C^{n-1}((0,2^k + 1]) \quad \text{and} \quad \forall x \in (0,2^k + 1] : u(x) = \tfrac{1}{2}[u(\tfrac{x}{2}) + u(\tfrac{x+1}{2})].$$

Define

$$U_k \text{ on } [2^k + 1, 2^{k+1} + 1] \text{ by } U_k(x) := \tfrac{1}{2}[u(\tfrac{x}{2}) + u(\tfrac{x+1}{2})].$$

Observe that

$$x \in [2^k + 1, 2^{k+1} + 1] \text{ implies } \frac{x+1}{2} \in [2^{k-1} + 1, 2^k + 1].$$

We have

$$U_k(2^k + 1) = \tfrac{1}{2}[u(\frac{2^k + 1}{2}) + u(2^{k-1} + 1)] = u(2^k + 1)$$

and

$$\begin{aligned} U_{k_+}^{(m)}(2^k + 1) &= \frac{1}{2^{m+1}}[u_+^{(m)}(\frac{2^k + 1}{2}) + u_+^{(m)}(2^{k-1} + 1)] \\ &= \frac{1}{2^{m+1}}[u_-^{(m)}(\frac{2^k + 1}{2}) + u_-^{(m)}(2^{k-1} + 1)] \\ &= u_-^{(m)}(2^k + 1). \end{aligned}$$

Hence, defining

$$u(x) := U_k(x) \quad \text{for} \quad x \in [2^k + 1, 2^{k+1} + 1],$$

we obtain

$$u \in C^{n-1}((0, 2^{k+1} + 1])$$

and

$$\forall x \in (0, 2^{k+1} + 1) : u(x) = \tfrac{1}{2}[u(\tfrac{x}{2}) + u(\frac{x+1}{2})].$$

Now for $x \in \mathbb{R}_+$, let

$$v(x) := \int_1^x u(t)dt.$$

Clearly, $v \in C^n(\mathbb{R}_+)$, $v(1) = 0$, and v is not the zero function. Finally, by (12),

$$v(x) = \int_1^x \tfrac{1}{2}[u(\tfrac{t}{2}) + u(\tfrac{t+1}{2})]dt = \int_{1/2}^{x/2} u(\tau)d\tau + \int_1^{(x+1)/2} u(\tau)d\tau$$

$$= v(\tfrac{x}{2}) + v(\tfrac{x+1}{2}) + \int_{1/2}^{1} u(\tau)d\tau = v(\tfrac{x}{2}) + v(\tfrac{x+1}{2}).$$

Thus $f := \Gamma \cdot \exp \circ v$ meets all the conditions indicated in the theorem.

In the following theorems, we state simple characterizations of Γ as a convex solution of (M_2) satisfying some additional conditions. The appearance of these additional conditions is partly justified by the preceding theorems.

THEOREM 4. The conditions

(a) $f : \mathbb{R}_+ \to \mathbb{R}_+$ convex on $\mathbb{R}_+$,

(b) $\forall x \in \mathbb{R}_+ : f(\tfrac{x}{2})\, f(\tfrac{x+1}{2}) = 2\sqrt{\pi}\, 2^{-x} f(x)$,

(c) $\lim_{x \to 0+} xf(x) \in [0,\infty]$,

(d) $f(1) = 1$

imply $f = \Gamma$.

THEOREM 5. The conditions

(a) $f : \mathbb{R}_+ \to \mathbb{R}_+$ convex on $\mathbb{R}_+$,

(b) $\forall x \in \mathbb{R}_+ : f(\tfrac{x}{2})\, f(\tfrac{x+1}{2}) = 2\sqrt{\pi}\, 2^{-x} f(x)$,

(c) $\forall x \in \mathbb{R}_+ : f(x+1) = xf(x)$

imply $f = \Gamma$.

THEOREM 6. The conditions

(a) $f : \mathbb{R}_+ \to \mathbb{R}_+$ convex on $\mathbb{R}_+$,

(b) $\forall x \in R_+ : f(\tfrac{x}{2})\, f(\tfrac{x+1}{2}) = 2\sqrt{\pi}\, 2^{-x} f(x)$

(c) $\forall x \in (0,1) : f(x)\, f(1-x) = \pi/\sin \pi x$

(d) $f(1) = 1$

imply $f = \Gamma$.

Proof. f convex on $\mathbb{R}_+$ implies: f'_+ is a monotone function on $\mathbb{R}_+$. Define

$$g := \log \circ (f/\Gamma) \quad \text{and} \quad h := g'_+.$$

Then $g \in C(\mathbb{R}_+)$, and for every $x \in \mathbb{R}_+$ the Riemann integral $\int_1^x h(t)dt$ exists. Furthermore, g' is defined on $\mathbb{R}_+ \setminus A$, where $A \subset \mathbb{R}_+$ is countable. On $\mathbb{R}_+ \setminus A$, we have $g' = g'_+$, hence

$$g(x) - g(1) = \int_1^x h(t)dt.$$

Finally, as a consequence of (b), we have

$$\forall x \in R_+ : g(x) = g(\tfrac{x}{2}) + g(\tfrac{x+1}{2}) \quad \text{and} \quad h(x) = \tfrac{1}{2}[h(\tfrac{x}{2}) + h(\tfrac{x+1}{2})].$$

The proof of Theorem 4 now is verbally the same as the proof of Theorem 2.

According to (c) of Theorem 5, we have

$$\lim_{x \to 0+} g(x) = \lim_{x \to 0+} g(x+1) = g(1).$$

Thus g satisfies the limit condition (6).

The proof of Theorem 2 shows g to be of the form $x \mapsto \beta x + \gamma$. Taking into account $g(1/2) = 0$ and $g(1) = \lim_{x \to 0+} g(x)$, we get $\beta = \gamma = 0$, which ends the proof of Theorem 5. Note that in this case $f(1) = 1$ is a consequence of (a), (b), and (c).

Finally, condition (c) of Theorem 6 gives

$$\lim_{x \to 0+} g(x) = -\lim_{x \to 0+} g(1-x) = g(1) = 0.$$

Thus (6) is fulfilled, and again the proof of Theorem 2 tells us: g is the zero function, which completes the proof of Theorem 6.

REMARKS. (i) The functional equations (F) in Theorem 5 and (R) in Theorem 6 have been used only in a local form. Replacing these conditions by

$\forall x \in (0,\delta] : f(x + 1) = xf(x)$ resp. $\forall x \in (0,\delta] : f(x)\, f(1 - x) = \pi/\sin \pi x$

for an arbitrarily small $\delta \in \mathbb{R}_+$, we get the same implications.

(ii) If we replace (M_2) in Theorem 4 or in Theorem 6 by (F), we no longer get a characterization of Γ; this fact is emphatically expressed in Theorem 1. In this sense, (M_2) owns a higher characterizing power with respect to the set of convex functions $f : \mathbb{R}_+ \to \mathbb{R}_+$ than the functional equation (F) does.

REFERENCES

1. M. Abramowitz and I. A. Stegun (Editors), Handbook of mathematical functions, Dover, New York, 1965.

2. E. Artin, The gamma function, Holt, Rinehart and Winston, New York, 1964.

3. H. Bohr and J. Mollerup, Laerebog i matematisk analyse, III, Jul. Gjellerups Forlag, Kopenhagen, 1922.

4. H.-H. Kairies, Zur axiomatischen Charakterisierung der Gammafunktion, J. Reine Angew. Math. 236 (1969), 103-111.

5. H.-H. Kairies, Definitionen der Bernoulli-Polynome mit Hilfe ihrer Multiplikationstheoreme, Manuscripta Math. 8 (1973), 363-369.

6. M. Kuczma, Functional equations in a single variable, Polish Scientific Publishers, Warszawa, 1968.

7. A. E. Mayer, Konvexe Lösung der Funktionalgleichung $1/f(x + 1) = xf(x)$, Acta Math. 70 (1939), 57-62.

8. M. E. Muldoon, Some characterizations of the gamma function involving the notion of complete monotonicity, Aequationes Math. 8 (1972), 212-215.

ON THE HAUSDORFF MOMENT PROBLEM

A. Jakimovski
Department of Mathematics
Tel-Aviv University
Tel-Aviv
ISRAEL

D. C. Russell
Department of Mathematics
York University
Downsview, Ontario M3J 1P3
CANADA

ABSTRACT. New necessary and sufficient conditions for solutions of the Hausdorff moment problem are presented.

1. INTRODUCTION AND MAIN RESULTS

The classical moment problem for a finite interval concerns necessary and sufficient conditions under which a (real or complex) sequence

$$u = \{u_n\}_{n \geq 0}$$

is representable in the form

$$u_n = \int_0^1 t^n \, d\chi(t) \qquad (n = 0,1,\ldots)$$

for some $\chi \in BV[0,1]$. For an extensive survey of moment problems, see [17].

The foregoing form of the problem was solved by Hausdorff [4], who extended it by replacing

$$t^n \text{ with } t^{\lambda_n},$$

where $\lambda = \{\lambda_n\}$ is an arbitrarily given fixed sequence satisfying

$$(1) \qquad 0 = \lambda_0 < \lambda_1 < \cdots < \lambda_n \to \infty .$$

By a change of variable, the extended problem is equivalent to a representation

$$(2) \qquad u_n = \int_0^\infty e^{-\lambda_n \sigma} d\psi(\sigma) \quad (n = 0,1,\cdots), \qquad \int_0^\infty |d\psi(\sigma)| < \infty .$$

Hausdorff [5] gave necessary and sufficient conditions (involving only u and λ) for such a representation (2), in the case where

$$\sum_1^\infty \frac{1}{\lambda_n} = +\infty .$$

The case

$$\sum_1^\infty \frac{1}{\lambda_n} < \infty$$

was considered by Hallenbach [3], who showed that Hausdorff's conditions are then necessary but not sufficient, and the choice of ψ in (2) is not unique.

Korenblyum [11] considered the subclass of Hallenbach's case in which

$$\frac{\lambda_{n+1}}{\lambda_n} \geq K > 1 ,$$

and in this subclass he showed that the simple condition

$$\sum_{1}^{\infty} |\Delta u_n| < \infty$$

is both necessary and sufficient for the representation (2) (we write, throughout,

$$\Delta s_n = s_n - s_{n+1}$$

for any sequence $\{s_n\}$). In a second theorem, Korenblyum imposed the weaker assumption

$$\frac{\lambda_{n+2}}{\lambda_n} \geq K > 1 .$$

In the present paper, we first extend Korenblyum's results by considering, for a nonnegative integer p, the property

$$(H_p) : \inf_{n\geq 1} (\lambda_{n+p+1}/\lambda_n) > 1 ,$$

and we obtain sufficient conditions (involving only u and λ) for the representation (2) when, for some p, (H_p) is assumed. Our conditions are also necessary, with no assumption on λ other than (1). In addition, we consider the problem for ψ belonging to spaces other than $BV[0,\infty)$ (in this connection, see Endl [2]).

Let $\lambda = \{\lambda_n\}$ be a fixed sequence satisfying (1), and p be a non-negative integer. Then the difference operator

$$D^p = D^p_\lambda$$

is defined, for any sequence $u = \{u_n\}$, by

$$D^p u_n = (-1)^p \sum_{i=n}^{n+p} \pi_i u_i, \quad \text{where} \quad \pi_i = \prod_{n\leq j\leq n+p,\ i\neq j} (\lambda_i - \lambda_j)^{-1};$$

as usual, empty products are defined throughout as 1, and empty sums as 0. This difference operator represents a divided difference of the sequence $\{u_n\}$, and, in fact, if

$$u_i = f(\lambda_i) ,$$

then

$$D^p u_n = D^p f(\lambda_n) = (-1)^p [\lambda_n, \ldots, \lambda_{n+p}]_f$$

(e.g., see [15, p. 423]); the factor $(-1)^p$ is inserted for convenience in avoiding alternating signs for differences of certain decreasing functions. The inductive definition of a divided difference gives

$$\Delta D^p u_n = D^p u_n - D^p u_{n+1} = (\lambda_{n+p+1} - \lambda_n) D^{p+1} u_n ,$$

and we also denote

$$E_n^0 = 1, \quad E_n^p = \lambda_{n+1} \cdots \lambda_{n+p} \quad (p \geq 1) .$$

We shall prove the following results:

THEOREM 1. Let λ satisfy (1). If, for a sequence $u = \{u_n\}_{n\geq 0}$,

$$(3) \quad \exists\, \psi(\cdot) \in BV[0,\infty) \text{ such that } u_n = \int_0^\infty e^{-\lambda_n \sigma} d\psi(\sigma) \quad (n = 0,1,2,\cdots) ,$$

then $\{u_n\}$ is bounded and, for each $p = 0,1,2,\cdots$, we have

$$(4) \quad \text{(i)} \sum_{n=0}^{\infty} E_n^p |\Delta D^p u_n| < \infty \quad \text{and} \quad \text{(ii)} \ E_n^p D^p u_n = o(1) \quad (n \to \infty) ;$$

moreover,

$$(5) \begin{cases} \text{(i)} \ \sup_{p\geq 0} \sum_{n=0}^{\infty} E_n^p |\Delta D^p u_n| \leq \|\psi\|_{BV} \equiv \int_0^\infty |d\psi(\sigma)| \\ \text{(ii)} \ \sup_{n,p} |E_n^p D^p u_n| \leq \|\psi\|_{BV} . \end{cases}$$

Conversely, let p be a fixed nonnegative integer, and suppose that $\{u_n\}$ is bounded, that the condition

$$(H_p) : \inf_{n\geq 1} (\lambda_{n+p+1}/\lambda_n) > 1$$

is satisfied, and that (4) holds; then u has the representation (3).

THEOREM 2. Let λ satisfy (1). If, for a sequence $u = \{u_n\}_{n\geq 1}$,

$$(6) \quad \exists\, m(\cdot) \in L^\infty[0,\infty) \text{ such that } u_n = \int_0^\infty e^{-\lambda_n \sigma} m(\sigma) d\sigma \quad (n = 1,2,\cdots) ,$$

then $\{\lambda_n u_n\}$ is bounded and, for each $p = 0,1,2,\cdots$, we have

$$(7) \quad \text{(i)} \sum_{i=1}^{n} E_i^p \Delta D^p(\lambda_i u_i) = O(1) \quad \text{and} \quad \text{(ii)} \ E_n^p D^p(\lambda_n u_n) = O(1), \quad (n \to \infty) ;$$

moreover,

$$(8) \begin{cases} \text{(i)} \ \sup_{n,p} \left| \sum_{i=1}^{n} E_i^p \Delta D(\lambda_i u_i) \right| \leq \|m\|_{L^\infty} = \operatorname{ess\,sup}_{\sigma \geq 0} |m(\sigma)| , \\ \text{and} \\ \text{(ii)} \ \sup_{n,p} |E_n^p D^p(\lambda_n u_n)| \leq 2\|m\|_{L^\infty} . \end{cases}$$

Conversely, let p be a fixed nonnegative integer, and suppose that $\{\lambda_n u_n\}$

is bounded, that (H_p) is satisfied, and that (7) holds; then u has the representation (6).

The necessity of the conditions (5) and (8) we prove by direct computation. However, the sufficiency parts of our proofs depend on the integral representation of a sequence u being equivalent to the existence of a continuous linear functional $f(\cdot)$ over a certain Banach sequence space, such that

$$u_n = f(e^n)$$

for each n, where

$$e^n = (0,\ldots,0,1,0,\ldots) \text{ with } 1 \text{ in the } n^{th} \text{ place.}$$

The condition (H_p) then allows us to identify this space with the field of a matrix transformation, for which the continuous linear functional representation is more tractable.

REMARKS. (i) In the case $p = 0$ of Theorem 1, (4)(i) (namely $\Sigma\,|\Delta u_n| < \infty$) implies (4)(ii) (namely $u_n = O(1)$), and we obtain the result of Korenblyum [11, Theorem 2].

(ii) When $p \geq 1$ (so that $E_n^p \nearrow$ as $n \nearrow \infty$), (4)(ii) implies

$$D^p u_n = o(1) .$$

However, if

$$D^p u_n = o(1)$$

and (4)(i) holds, then

$$|E_n^p D^p u_n| = \left|E_n^p \sum_{k=n}^{\infty} \Delta D^p u_k\right| \leq \sum_{k=n}^{\infty} E_k^p |\Delta D^p u_k| \to 0 \quad \text{as} \quad n \to \infty .$$

Consequently, for $p = 1,2,\cdots$, (4) is equivalent to

$$\sum_{n=0}^{\infty} E_n^p |\Delta D^p u_n| < \infty \quad \text{and} \quad D^p u_n = o(1) \quad (n \to \infty) . \tag{9}$$

If (4) is replaced with (9) in our Theorem 1, the case $p = 1$ then gives the result of Korenblyum [11, Theorem 3].

(iii) A partial summation shows that, for any sequence $\{u_n\}$,

$$\sum_{n=0}^{N} E_n^{p-1} \Delta D^{p-1} u_n = E_N^p D^p u_N + \sum_{n=0}^{N-1} E_n^p \Delta D^p u_n \quad (p \geq 1) . \tag{10}$$

Consequently, if (4)(i) holds for $p = p_1$ and for $p = p_1 - 1$, then (4)(ii) holds for $p = p_1$; and a similar remark applies to (7).

A companion theorem for $m \in L^q$ $(1 \leq q < \infty)$ is as follows:

THEOREM 3. Let λ satisfy (1), $1 \leq q < \infty$, and let

$$u_n = \int_0^\infty e^{-\lambda_n \sigma} m(\sigma)d\sigma, \qquad \|m\|_{L^q} \equiv \left(\int_0^\infty |m(\sigma)|^q d\sigma\right)^{1/q} < \infty . \tag{11}$$

Then

$$\begin{cases} \text{(i)} \quad \sup_{p\geq 0} \sum_{n=1}^{\infty} (\lambda_n^{-1} - \lambda_{n+p+1}^{-1})^{-q+1} \, |E_n^p \, \Delta D^p u_n|^q \leq \|m\|_{L^q}^q , \\ \text{and} \\ \text{(ii)} \quad \sup_{n,p} \lambda_n^{1-(1/q)} \, E_n^p |D^p u_n| \leq \|m\|_{L^q} . \end{cases} \tag{12}$$

We are not at present able to prove a converse of Theorem 3 corresponding to the second (sufficiency) parts of Theorems 1 and 2. In fact, by comparison with Theorem 1, the condition (12) in the case $q = 1$ (even with the imposition of one of the conditions (H_p)) cannot be sufficient for the representation (11) with $q = 1$, since this would imply that every function of bounded variation is the integral of an L^1-function, which of course is not the case.

2. AUXILIARY RESULTS

A <u>sequence space</u> is a vector subspace of ω, the space of all complex-valued sequences

$$x = \{x_k\}_{k\geq 0} .$$

A <u>BK-space</u> is a Banach sequence space in which all the coordinate functionals P_k,

$$P_k(x) = x_k ,$$

are continuous. Examples are the spaces ℓ_∞, c, c^0 of bounded, convergent, convergent to zero, sequences, respectively, all with $\|x\| = \sup_k |x_k|$;

$$v^0 = \{x : x \in c^0 \text{ and } \|x\| \equiv \sum_0^\infty |\Delta x_k| < \infty\} ;$$

$$bs = \{x : \|x\| \equiv \sup_n \left| \sum_0^n x_k \right| < \infty\} ;$$

and

$$\ell_q = \{x : \|x\| \equiv \left(\sum_0^\infty |x_k|^q\right)^{1/q} < \infty\}, \quad 1 \leq q < \infty; \text{ we write } \ell_1 = \ell .$$

A countable collection of points, $\{a^k\}$, of a BK-space F, is a (<u>Schauder</u>) <u>basis</u> for F if there are unique functionals f_k $(k = 0,1,\cdots)$ such that

$$x = \sum_{k=0}^\infty f_k(x)a^k$$

for each $x \in F$ (in the topology of F). If a BK-space has the basis $\{e^k\}$, where

$$e^k = (0,\dots,0,1,0,\dots) \text{ with } 1 \text{ in rank } k ,$$

then necessarily

$$f_k(x) = x_k .$$

The spaces c^0, v^0, ℓ_q $(1 \le q < \infty)$ have $\{e^k\}$ as a basis (throughout this paper, "basis" alwans means "Schauder basis"). Some dual spaces of F are denoted as follows:

F^* = <u>continuous dual</u> of F = space of all continuous linear functionals on F;

F^β = β-<u>dual</u> of $F = \{y : \Sigma_0^\infty x_k y_k$ converges for each $x \in F\}$;

F^γ = γ-<u>dual</u> of $F = \{y : \sup_n |\Sigma_0^n x_k y_k| < \infty$ for each $x \in F\}$.

The β-dual of F is the space of all <u>convergence factors</u> of F, and when F has basis $\{e^k\}$, then $F^\beta = F^\gamma$ (see Jakomovski and Livne [6, p. 356, Remark]). For example,

$$(c^0)^\beta = (c^0)^\gamma = \ell_1, \quad \ell_1^\beta = \ell_1^\gamma = \ell_\infty, \quad (v^0)^\beta = (v^0)^\gamma = bs .$$

For an infinite matrix

$$A = (a_{nk})_{n,k\ge 0}$$

and an infinite sequence

$$x = \{x_k\}_{k\ge 0}$$

of complex numbers, the sequences Ax and xA have general terms

$$(Ax)_n = \sum_k a_{nk}x_k, \quad (xA)_k = \sum_n x_n a_{nk} ,$$

where each series (if it is infinite) is supposed convergent. If F is a sequence space, we write

$$F_A = \{x : Ax \in F\} .$$

The matrix A is called F-<u>reversible</u> when, for each $y \in F$, $y = Ax$ has a unique solution $x \in \omega$; if $\{e^k\} \subset F$, there then exists a unique right inverse matrix A' $(AA' = I$, the unit matrix). A <u>normal</u> matrix

$$A \quad (a_{nk} = 0, \quad k > n; \quad a_{nn} \ne 0)$$

is ω-reversible. It is known (see Zeller [18, Satz 4.10], and [8, Lemma 1(b)]) that if F is a BK-space and A is F-reversible, then F_A is a BK-space with

$$\|x\|_{F_A} = \|Ax\|_F .$$

We write

$$J = (j_{nk}), \quad j_{nk} = 1 \quad \text{for} \quad 0 \leq k \leq n, \quad j_{nk} = 0 \quad \text{for} \quad k > n ,$$

with inverse

$$J' = (j'_{nk}), \quad j'_{nn} = 1, \quad j'_{n,n-1} = -1, \quad j'_{nk} = 0 \quad \text{otherwise} .$$

Thus a series

$$\Sigma a_k, \quad a = \{a_k\} ,$$

is related to its sequence of partial sums $x = \{x_k\}$ by $x = Ja$, $a = J'x$.

LEMMA 1. Let E be a BK-space with basis $\{e^j\}$, and let A be E-reversible, with right inverse A'. Let

$$z^{[n]} \equiv (z_0, \dots, z_n, 0, 0, \dots) .$$

Then

$$(13) \qquad z \in (E_A)^\beta \iff [zA' \text{ exists and } \sup_n \|z^{[n]}A'\|_{E^\gamma} < \infty] .$$

<u>Proof</u>. Write

$$t_{nr} = (z^{[n]}A')_r = \sum_{k=0}^{n} z_k a'_{kr} .$$

Then with the given hypotheses (see Jakimovski and Livne [6, Theorem 5.1] or Jakimovski and Russell [7, Corollary 3.1]) the conditions on the right of (13) are equivalent to $T \in (E \to c)$ (i.e., the matrix $T = (t_{nr})$ defines a map from E to c). Now if $\{e^j\}$ is a basis for E, and A is E-reversible, then by results of Banach [1, pp. 47, 111], $x = A'y$ gives a bijective map from E to E_A, and then

$$\sum_{k=0}^{n} z_k x_k = \sum_{k=0}^{n} z_k \sum_{r=0}^{\infty} a'_{kr} y_r = \sum_{r=0}^{\infty} t_{nr} y_r ;$$

because of the bijection, this equation may be quantified either $\forall x \in E_A$ or $\forall y \in E$. On the one hand,

$$\lim_n \sum_{k=0}^{n} z_k x_k$$

exists $\forall x \in E_A$, if and only if $z \in (E_A)^\beta$; and on the other hand,

$$\lim_n \sum_{r=0}^{\infty} t_{nr} y_r$$

exists $\forall y \in E$, if and only if $T \in (E \to c)$. The result now follows.

The series-to-sequence transformation matrix $\bar{T}^p$ [15, p. 419], and the corresponding sequence-to-sequence matrix T^p, are defined by

$$(14) \qquad C^p_n = E^p_n(T^p x)_n = E^p_n(\bar{T}^p a)_n = \sum_{i=0}^{n} (\lambda_{n+1} - \lambda_i) \cdots (\lambda_{n+p} - \lambda_i) a_i ,$$

where

$$E_n^0 = 1, \quad E_n^p = \lambda_{n+1} \cdots \lambda_{n+p} \quad (p \geq 1) ,$$

and

$$x = Ja, \quad \bar{T}^p = T^p J .$$

The spaces

$$(C,\lambda,p)_0 = \{x\colon T^p x \in c^0\} \quad \text{with norm} \quad \|x\|_1 = \sup_{n\geq 0} |(T^p x)_n| ,$$

$$|C,\lambda,p|_0 = \{x\colon T^p x \in v^0\} \quad \text{with norm} \quad \|x\|_1' = \sum_{n=0}^{\infty} |\Delta(T^p x)_n|$$

are BK-spaces (see Jakimovski and Russell [8, Lemma 2(b)]). Since T^p and $\bar{T}^p$ are normal, they have inverses $(T^p)'$, $(\bar{T}^p)'$, related to each other by

$$(\bar{T}^p)' = J'(T^p)' ,$$

and to the divided differences in Section 1 above (see [16, pp. 297-298]) by

$$\bar{u}(T^p)' = u(\bar{T}^p)' = \{E_n^p \,\Delta D^p u_n\}_{n\geq 0} , \tag{15}$$

where here, and throughout, we write, for a sequence u,

$$\bar{u} = uJ' = \{\Delta u_n\}_{n\geq 0} .$$

Note that

$$T^0 = I \quad \text{and} \quad \bar{T}^0 = J .$$

Successive T or $\bar{T}$ matrices are related to each other by

$$T^p = HT^{p-1}, \quad \bar{T}^p = H\bar{T}^{p-1} \quad (p \geq 1) ,$$

where $H = (h_{nk})$ is given by

$$h_{nk} = (E_k^p - E_{k-1}^p)/E_n^p \quad \text{for} \quad 0 \leq k \leq n, \quad h_{nk} = 0 \quad \text{for} \quad k > n \quad (E_{-1}^p = 0) ; \tag{16}$$

since H is normal, it has a unique right inverse H' for $p \geq 1$, namely

$$h'_{rr} = \lambda_{r+p}/(\lambda_{r+p} - \lambda_r), \quad h'_{r+1,r} = -\lambda_{r+1}/(\lambda_{r+p+1} - \lambda_{r+1}), \quad h'_{kr} = 0 \quad \text{otherwise.}$$

Thus zH' exists for any sequence z and, in the notation of Lemma 1 (with $A = H$),

$$t_{nr} = \sum_{k=0}^{n} z_k h'_{kr} = z_r h'_{rr} + z_{r+1} h'_{r+1,r} = (zH')_r \quad \text{for} \quad r < n, \quad t_{nn} = z_n h'_{rr} .$$

If we take $E = c^0$, so that $E^\gamma = \ell$, the condition

$$\sup_n \|z^{[n]}H'\|_{E^\gamma} < \infty \quad \text{becomes} \quad \sup_n \sum_{r=0}^{n} |t_{nr}| < \infty ,$$

which is equivalent to the two conditions

$$zH' \in \ell, \quad z_n h'_{nn} = O(1) \ ;$$

while if we take $E = v^0$, so that $E^\gamma = bs$, the condition

$$\sup_n \|z^{[n]}H'\|_{E^\gamma} < \infty \text{ is equivalent to } zH' \in bs, \quad z_n h'_{nn} = O(1) \ .$$

In particular, choose

$$z = u(\bar{T}^{p-1})' = \bar{u}(T^{p-1})' \ ;$$

then (see (15)) we have

$$zH' = u(\bar{T}^{p-1})'H' = u(\bar{T}^p)' = \{E_n^p \, \Delta D^p u_n\}_{n \geq 0} \ ,$$

and

$$z_n h'_{nn} = (E_n^{p-1} \, \Delta D^{p-1} u_n)\lambda_{n+p}/(\lambda_{n+p} - \lambda_n) = E_n^p \, D^p u_n \ .$$

Hence from Lemma 1 we obtain:

COROLLARY 1.1. Let $p \geq 1$. Then (4) holds if and only if

$$\bar{u}(T^{p-1})' \in (c_H^0)^\beta \ . \tag{17}$$

COROLLARY 1.2. Let $p \geq 1$. In order that

$$\bar{u}(T^{p-1})' \in (v_H^0)^\beta \ , \tag{18}$$

it is necessary and sufficient that

$$\sum_{i=0}^{n} E_i^p \, \Delta D^p u_i = O(1) \quad \text{and} \quad E_n^p \, D^p u_n = O(1), \quad \text{as} \quad n \to \infty \ .$$

LEMMA 2. Let $p \geq 0$. If $D^{p+1}u_n = 0$ $(n = 0,1,\ldots)$, then

$$u_n = k_0 + \sum_{r=1}^{p} k_r \lambda_n^r$$

for some constants $k_0,\ldots,k_p$.

<u>Proof</u>. From the formula (e.g., see [15, Lemma 1])

$$\sum_{i=0}^{n} u_i a_i = \sum_{i=0}^{n} C_i^p \, \Delta D^p u_i + \sum_{r=0}^{p} C_n^r D^r u_{n+1} \ , \tag{19}$$

choose

$$\{a_i\}_{i \geq 0} = e^j$$

for an arbitrary fixed $j \geq 0$; then we have from (14) and (19), since

$$\Delta D^p u_i = (\lambda_{i+p+1} - \lambda_i) D^{p+1} u_i = 0$$

(by hypothesis),

$$u_j = \sum_{r=0}^{p} (\lambda_{n+1} - \lambda_j) \cdots (\lambda_{n+r} - \lambda_j) D^r u_{n+1} = k_{n0} + \sum_{r=1}^{p} k_{nr} \lambda_j^r ,$$

say, for any $n \geq j$. Thus

$$D^p u_j = (-1)^p \, k_{np} \quad (n \geq j) ,$$

and hence

$$k_{np} = [(-1)^p \, D^p u_j]_{j=0} = k_p ,$$

for every $n \geq 0$. If now $k_{nr} = k_r$ for $q < r \leq p$, then

$$k_{nq} = [(-1)^q \, D^q (u_j - k_p \lambda_j^p - \cdots - k_{q+1} \lambda_j^{q+1})]_{j=0} = k_q ,$$

and the required result follows by induction.

COROLLARY 2. Let $p \geq 1$. If $w \in bs$ and $w(T^{p-1})' = 0$, then $w = 0$.

Proof. The case $p = 1$ requires no proof (and $w \in bs$ is not needed), so let $p \geq 2$. Writing

$$w = \bar{u} = \{\Delta u_n\}_{n \geq 0} ,$$

we have, by (15),

$$\bar{u}(T^{p-1})' = \{E_n^{p-1} \, \Delta D^{p-1} u_n\}_{n \geq 0} ,$$

and so, by hypothesis,

$$D^p u_n = 0 .$$

Thus, by Lemma 2,

$$u_i = k_0 + \sum_{r=1}^{p-1} k_r \lambda_i^r ,$$

whence

$$\sum_0^n w_i = u_0 - u_{n+1} = - \sum_{r=1}^{p-1} k_r \lambda_{n+1}^r ;$$

since $\{\lambda_n\}$ is unbounded, the hypothesis $w \in bs$ ensures that $k_1 = \cdots = k_{p-1} = 0$. Hence $\Sigma_0^n w_i = 0$, and so $w_n = 0$, for every $n \geq 0$.

LEMMA 3. Let F be a BK-space with basis $\{e^j\}$, let the matrix B be

F-reversible with a column-finite right inverse B', and let $\{e^j\} \subset F_B$. Then

(a) $d^j = B'e^j$ $(j \geq 0)$ [equivalent to $Bd^j = e^j$] is a basis for F_B, and

$$\forall x \in F_B, \quad x = \sum_{j=0}^{\infty} (Bx)_j d^j ;$$

(b) $$[f \in F_B^*, \quad w_i = f(e^i) \quad (i = 0,1,\ldots)] \Rightarrow wB' \in F^\beta .$$

Suppose, in addition, that $w \in bs$ and $wB' = 0$ together imply $w = 0$, and that

$$\left\| B\left(\sum_0^n e^i\right)\right\|_F \leq M$$

for all $n \geq 0$. Then

(c) $$[w \in bx, \quad wB' \in F^\beta] \Rightarrow [\exists\, f \in F_B^*, \quad w_i = f(e^i) \quad (i = 0,1,\ldots)] .$$

If $wB' = 0$ alone implies $w = 0$, then the hypothesis $w \in bs$ is not required in (c).

Proof. (a) See Jakimovski and Livne [6, Theorem 2.7].

(b) $(wB')_j = \Sigma_i w_i b'_{ij} = \Sigma_i f(e^i)b'_{ij} = f(\Sigma_i e^i b'_{ij}) = f(d^j)$, since the sums are infinite and since

$$(\sum_i e^i b'_{ij})_n = b'_{nj} = \sum_i b'_{ni} e^j_i = (B'e^j)_n = d^j_n .$$

By part (a), it now follows that

$$\forall x \in F_B, \quad x = \sum_{j=0}^{\infty} (Bx)_j d^j \quad \text{and} \quad f(x) = \sum_{j=0}^{\infty} (Bx)_j f(d^j) = \sum_{j=0}^{\infty} (Bx)_j (wB')_j ,$$

a convergent series; hence, since B is F-reversible, $wB' \in F^\beta$.

(c) If $wB' \in F^\beta$, then

$$\forall y \in F, \quad g(y) = \sum_{k=0}^{\infty} (wB')_k y_k$$

satisfies

$$g \in F^*, \quad g(e^j) = (wB')_j .$$

Now $\forall x \in F_B$, we have $y = Bx \in F$, and $f(x) \equiv g(Bx)$ satisfies $f \in F_B^*$; moreover,

$$(wB')_j = g(e^j) = g(Bd^j) = f(d^j) = f\Big(\sum_i e^i b'_{ij}\Big) = \sum_i f(e^i) b'_{ij} ,$$

and hence $zB' = 0$, where

$$z_i = w_i - f(e^i).$$

Since, by hypothesis, $w \in bs$, and also

$$\left|\sum_0^n f(e^i)\right| = \left|f\left(\sum_0^n e^i\right)\right| \le \|f\| \left\|\sum_0^n e^i\right\|_{F_B} \le M\|f\| ,$$

we have $z \in bs$. Thus the additional hypothesis allows us to conclude that $z = 0$; that is, $w_i = f(e^i)$.

In particular, in Lemma 3 take

$$F = c_H^0 \text{ (where } H \text{ is given by (16)),} \quad B = T^{p-1}, \quad w = \bar{u} .$$

Then

$$F_B = (c_H^0)_{T^{p-1}} = c_{T^p}^0 = (C,\lambda,p)_0 ,$$

and

$$\left\|B\left(\sum_0^n e^i\right)\right\|_F = \left\|T^p\left(\sum_0^n e^i\right)\right\|_{c^0} \le 1 ,$$

by (14). Also all the other requirements of Lemma 3 hold, including (by Corollary 2) the additional condition required for the proof of part (c). Hence we have:

COROLLARY 3.1. Let $p \geq 1$. Then

(a) $[f \in (C,\lambda,p)_0^*,\ \bar{u}^i = f(e^i)\ (i = 0,1,\ldots)] \Rightarrow \bar{u}(T^{p-1})' \in (c_H^0)^\beta$;

(b) $[u \in \ell_\infty,\ \bar{u}(T^{p-1})' \in (c_H^0)^\beta] \Rightarrow [\exists\, f \in (C,\lambda,p)_0^*,\ \bar{u}_i = f(e^i)$ $(i = 0,1,\ldots)]$.

If we take $F = c^0$, $B = I$, $w = \bar{u}$ in Lemma 3, we get the well-known result:

COROLLARY 3.2. $\bar{u} \in \ell \iff [\exists\, f \in (c^0)^*,\ \bar{u}_i = f(e^i)\ (i = 0,1,\ldots)]$.

There are immediate analogues for absolute summability fields, obtained by taking $F = v_H^0$ or $F = v^0$ in Lemma 3, namely:

COROLLARY 3.3. Let $p \geq 1$. Then

(a) $[f \in |C,\lambda,p|_0^*,\ \bar{u}_i = f(e^i)\ (i = 0,1,\ldots)] \Rightarrow \bar{u}(T^{p-1})' \in (v_H^0)^\beta$;

(b) $[u \in \ell_\infty,\ \bar{u}(T^{p-1})' \in (v_H^0)^\beta] \Rightarrow [\exists\, f \in |C,\lambda,p|_0^*,\ \bar{u}_i = f(e^i)$ $(i = 0,1,\ldots)]$.

COROLLARY 3.4. $u \in \ell_\infty \iff [\exists\, f \in (v^0)^*,\ \bar{u}_i = f(e^i)\ (i = 0,1,\ldots)]$.

Given λ satisfying (1) and a nonnegative integer p, we define the <u>Riesz sums</u> of a sequence $x = \{x_n\}$, with $a_n = x_n - x_{n-1}$ $(n = 0,1,\ldots;$ $x_{-1} = 0)$, by

$$A^p(x,t) = \sum_{\lambda_j < t} (t - \lambda_j)^p a_j \quad (t > 0), \quad A^p(x,0) = 0 ,$$

and the Abel means by

$$F^p(x,\sigma) = \frac{\sigma^p}{p!} \int_0^\infty e^{-t\sigma} dA^p(x,t) \quad (\sigma > 0) .$$

Denote

$$(A,\lambda,p)_0 = \{x : \lim_{\sigma \to 0+} F^p(x,\sigma) = 0\} ,$$

$$|A,\lambda,p|_0 = (A,\lambda,p)_0 \cap \{x : F^p(x,\sigma) \in BV(0,\infty)\} .$$

LEMMA 4. Let p be a nonnegative integer. Then

(a) $(H^p) \Longleftrightarrow [(A,\lambda,p)_0 = (C,\lambda,p)_0]$;

(b) $(H^p) \Longleftrightarrow [|A,\lambda,p|_0 = |C,\lambda,p|_0]$;

(c) $(H_p) \Rightarrow [(A,\lambda,p)_0$ is a BK-space with norm $\|x\|_2 = \sup_{\sigma>0} |F^p(x,\sigma)|]$;

(d) $(H_p) \Rightarrow [|A,\lambda,p|_0$ is a BK-space with norm $\|x\|_2' = \int_0^\infty |dF^p(x,\sigma)|]$.

Proof. If $(R,\lambda,p)_0$ is the space of sequences Riesz-summable to zero, the equivalence

$$(R,\lambda,p)_0 = (C,\lambda,p)_0$$

follows from Russell [15, Theorem 4] and Meir [13]; combining this with

$$(H^p) \Longleftrightarrow [(A,\lambda,p)_0 = (R,\lambda,p)_0] ,$$

due to Jakimovski and Russell [9, Theorem I], we obtain (a). For (b) we similarly use

$$|R,\lambda,p|_0 = |C,\lambda,p|_0 \quad \text{(Körle [12])}$$

and

$$(H_p) \Longleftrightarrow [|A,\lambda,p|_0 = |R,\lambda,p|_0] \quad \text{(Jakimovski and Russell [9, Theorem I])}.$$

Now (H_p) implies, by (a) and [9, Lemma 2 (38) and Lemma 7 (63)], that

$$\forall x \in \omega, \sup_{\sigma > 0} |F^p(x,\sigma)| \le \sup_{n \ge 0} |(T^p x)_n| \le K_p \sup_{\sigma \ge 0} |F^p(x,\sigma)| .$$

Since $(C,\lambda,p)_0$ is a BK-space with the norm

$$\|x\|_1 = \sup_{n\geq 0} |(T^p x)_n| ,$$

(c) follows. For (d), we similarly use (b) and [9; Lemma 2 (40) and Lemma 7 (64)].

LEMMA 5. Let p be a nonnegative integer. Suppose that (H_p) holds. Then

(a) $f \in (A,\lambda,p)_0^*$ if and only if $\exists\, \psi(\cdot) \in BV[0,\infty)$ such that

$$\forall x \in (A,\lambda,p)_0, \quad f(x) = \int_0^\infty F^p(x,\sigma)\, d\psi(\sigma); \tag{20}$$

(b) $f \in |A,\lambda,p|_0^*$ if and only if $\exists\, m(\cdot) \in L^\infty[0,\infty)$ such that

$$\forall x \in |A,\lambda,p|_0, \quad f(x) = \int_0^\infty m(\sigma)\, dF^p(x,\sigma) . \tag{21}$$

Proof. By Lemma 4(c), $(A,\lambda,p)_0$ is, with the norm $\|\cdot\|_2$, a BK-space; for sequences in $(A,\lambda,p)_0$, the range-space of the transformation is a vector subspace of the Banach space $C_0[0,\infty)$ of functions continuous and bounded on $[0,\infty)$ with the value 0 at 0, and with the sup-norm. Therefore it follows by the Hahn-Banach theorem and the representation theorem for continuous linear functionals on $C_0[0,\infty)$ (see Banach [1, pp. 59-61]) and the argument in Peyerimhoff [14, Section 8]) that

$$f \in (A,\lambda,p)_0^*$$

if and only if $\exists\, \psi \in BV[0,\infty)$ such that (20) holds. The result for $|A,\lambda,p|_0$ follows by Lemma 4(d) and a similar argument, using the representation for continuous linear functionals on $L^1[0,\infty)$ to obtain $m \in L^\infty[0,\infty)$ (the space of functions essentially bounded on $[0,\infty)$) such that (21) holds (c.f. Jakimovski and Tzimbalario [10, p. 380]).

3. PROOFS OF THE THEOREMS

Proof of Theorem 1. Sufficiency. We note first, integrating the definition of $F^p(e^i,\sigma)$ by parts p times, that

$$F^p(e^i,\sigma) = \frac{\sigma^p}{p!}\int_0^\infty e^{-t\sigma}\, dA^p(e^i,t) = \int_0^\infty e^{-t\sigma}\, dA^0(e^i,t) = \Delta e^{-\lambda_i \sigma} \quad (\sigma>0) . \tag{22}$$

Suppose now that $p\geq 1$ is fixed and that (4) holds. Then, by Corollary 1.1, (17) follows and if, in addition, $u \in \ell_\infty$, we obtain the conclusion of Corollary 3.1(b). In conjunction with Lemma 4(a) and (c), when (H_p) is satisfied, this gives us

$$\exists\, f \in (A,\lambda,p)_0^*, \quad \bar{u}_i = f(e^i) \qquad (i = 0,1,\ldots)\ .$$

It now follows by (22) and Lemma 5(a) that $\exists\, \varphi \in BV[0,\infty)$ such that

$$\bar{u}_i = \Delta u_i = \int_0^\infty F^p(e^i,\sigma)\,d\varphi(\sigma) = \int_0^\infty (\Delta e^{-\lambda_i\sigma})\,d\varphi(\sigma) \qquad (i = 0,1,\ldots)\ ,$$

so that

$$u_n = u_0 - \int_0^\infty d\varphi(\sigma) + \int_0^\infty e^{-\lambda_n\sigma}\,d\varphi(\sigma) \qquad (n = 0,1,\ldots)\ .$$

Now by altering the value of $\varphi(\sigma)$ at $\sigma = 0$, namely by taking

$$\psi(0) = \varphi(0) + \int_0^\infty d\varphi(\sigma) - u_0, \quad \psi(\sigma) = \varphi(\sigma) \quad (\sigma > 0)\ ,$$

we have

$$\psi \in BV[0,\infty) \quad \text{and} \quad u_n = \int_0^\infty e^{-\lambda_n\sigma}\,d\psi(\sigma) \qquad (n = 0,1,\ldots)\ .$$

For the case $p = 0$, (4) reduces to $\bar{u} \in \ell$; Corollary 1.1 is not required in the proof, and we use Corollary 3.2 instead of Corollary 3.1.

<u>Necessity</u>. We can readily show by partial summation that for any sequence $\{s_n\}$,

$$\text{(23)} \qquad \sum_{n=0}^{N-p-1} E_n^p\, \Delta D^p s_n = s_0 - \sum_{r=0}^{p} E_{N-r}^r\, D^r s_{N-r} \qquad (0 \le p \le N,\ \ N = 0,1,\ldots)\ .$$

Now if $\{u_n\}$ has the representation (3), we have

$$\text{(24)} \qquad \sum_{n=0}^{N-p-1} E_n^p |\Delta D^p u_n| \le \int_0^\infty |d\psi(\sigma)| \sum_{n=0}^{N-p-1} E_n^p |\Delta D^p e^{-\lambda_n\sigma}|\ .$$

By the mean-value theorem for divided differences, for any fixed $\sigma > 0$,

$$\text{(25)} \qquad D^p e^{-\lambda_n\sigma} = \frac{\sigma^p}{p!}\, e^{-\xi_n\sigma} \ge 0 \qquad (\text{where } \lambda_n \le \xi_n \le \lambda_{n+p})$$

and hence also

$$\text{(26)} \qquad \Delta D^p e^{-\lambda_n\sigma} \equiv (\lambda_{n+p+1} - \lambda_n) D^{p+1} e^{-\lambda_n\sigma} \ge 0\ .$$

Thus by (23), (25), (26), for $0 \le p \le N$, $N = 0,1,\ldots$, $\sigma \ge 0$.

$$\text{(27)} \qquad \sum_{n=0}^{N-p-1} E_n^p |\Delta D^p e^{-\lambda_n\sigma}| = 1 - \sum_{r=0}^{p} E_{N-r}^r D^r e^{-\lambda_{N-r}\sigma} \le 1\ .$$

Substituting (27) into (24) and letting $N \to \infty$, we obtain the required result (5)(i). Since, by (25) and (27),

$$0 \leq E_n^p D^p e^{-\lambda_n \sigma} \leq 1 ,$$

we obtain (5)(ii) from

$$|E_n^p D^p u_n| = \left| \int_0^\infty E_n^p D^p e^{-\lambda_n \sigma} \, d\psi(\sigma) \right| \leq \|\psi\|_{BV} .$$

Proof of Theorem 2. Sufficiency. Suppose first that $p \geq 1$ is fixed and that (7) holds. Define, say, $u_0 = 0$. Then, by Corollary 1.2, (18) follows with $\{\lambda_n u_n\}$ in place of $\{u_n\}$; if, in addition,

$$\{\lambda_n u_n\} \in \ell_\infty ,$$

we obtain the conclusion of Corollary 3.3(b) (with $\lambda_i u_i$ in place of u_i). In conjunction with Lemma 4(b) and (d), when (H_p) holds, this gives

$$\exists \; f \in |A,\lambda,p|_0^*, \quad \overline{\lambda_i u_i} = f(e^i) \qquad (i = 0,1,\ldots) .$$

It now follows by (22) and Lemma 5(b) that $\exists \; -m \in L^\infty[0,\infty)$ such that

$$\Delta(\lambda_i u_i) = \int_0^\infty -m(\sigma) \, dF^p(e^i,\sigma) = -\Delta \int_0^\infty m(\sigma) \, d(e^{-\lambda_i \sigma}) \qquad (i = 0,1,\ldots) ;$$

thus

$$\lambda_n u_n = \lambda_n \int_0^\infty m(\sigma) \, e^{-\lambda_n \sigma} \, d\sigma \qquad (n = 1,2,\ldots) ,$$

and (6) follows.

For the case $p = 0$, (7) reduces to

$$\lambda_n u_n = O(1) ;$$

Corollary 1.2 is not required in the proof, and we use Corollary 3.4 (with $\{\lambda_n u_n\}$ in place of $\{u_n\}$) instead of Corollary 3.3.

Necessity. We note first that, for any sequence $\{s_n\}$,

$$D^p(\lambda_n s_n) = \lambda_n D^p s_n - D^{p-1} s_{n+1} \qquad (n \geq 0, \quad p \geq 1) . \tag{28}$$

Since

$$\int_0^\infty e^{-\lambda_n \sigma} \, d\sigma = \lambda_n^{-1} \qquad (n \geq 1) ,$$

we have

$$\int_0^\infty D^p e^{-\lambda_n \sigma}\, d\sigma = D^p \lambda_n^{-1} = (\lambda_n \cdots \lambda_{n+p})^{-1} ;$$

and using also the relation in (26), we now see that

$$(29) \quad \begin{cases} \text{(i)} & \displaystyle\int_0^\infty E_n^p D^p e^{-\lambda_n \sigma}\, d\sigma = \lambda_n^{-1} , \\ \text{and} & \\ \text{(ii)} & \displaystyle\int_0^\infty E_n^p\, \Delta D^p e^{-\lambda_n \sigma}\, d\sigma = \lambda_n^{-1} - \lambda_{n+p+1}^{-1} . \end{cases} \qquad (n \geq 1, \quad p \geq 0)$$

Suppose that $\{u_n\}$ has the representation (6). Then, by (23) and (28), for $N \geq p$, $p = 0,1,\dots,$

$$\begin{aligned} \sum_{n=1}^{N-p-1} E_n^p\, \Delta D^p(\lambda_n u_n) &= \int_0^\infty m(\sigma) \sum_{n=1}^{N-p-1} E_n^p\, \Delta D^p(\lambda_n e^{-\lambda_n \sigma}) \cdot d\sigma \\ &= - \int_0^\infty m(\sigma) \sum_{r=0}^{p} E_{N-r}^r D^r(\lambda_{N-r} e^{-\lambda_{N-r}\sigma}) \cdot d\sigma \\ &= - \int_0^\infty m(\sigma) \Big\{ \sum_{r=0}^{p} E_{N-r}^r\, \lambda_{N-r} D^r(e^{-\lambda_{N-r}\sigma}) \\ &\qquad - \sum_{r=1}^{p} E_{N-r}^r D^{r-1}(e^{-\lambda_{N-r+1}\sigma}) \Big\}\, d\sigma \\ &= - \int_0^\infty m(\sigma)\, \lambda_{N-p} E_{N-p}^p D^p(e^{-\lambda_{N-p}\sigma})\, d\sigma . \end{aligned}$$

Hence, by (25) and (29)(i), we obtain (8)(i) from

$$\left| \sum_{n=1}^{N-p-1} E_n^p\, \Delta D^p(\lambda_n u_n) \right| \leq \|m\|_{L^\infty} \lambda_{N-p} \int_0^\infty E_{N-p}^p D^p(e^{-\lambda_{N-p}\sigma}) d\sigma = \|m\|_{L^\infty} .$$

Now (8)(ii) follows from (8)(i) and (10) (with $\{\lambda_n u_n\}$ in place of $\{u_n\}$).

<u>Proof of Theorem 3</u>. Here we need Hölder's inequality in the form

$$(30) \qquad \text{if} \int |g| \leq 1 \quad \text{then} \quad \left| \int fg \right|^q \leq \int |f|^q\, |g| \qquad (1 \leq q < \infty) .$$

Now if $\{u_n\}$ has the representation (11), where $1 \leq q < \infty$, we have

$$\sum_{n=1}^{N} (\lambda_n^{-1} - \lambda_{n+p+1}^{-1})^{-q+1}\, |E_n^p\, \Delta D^p u_n|^q =$$

$$= \sum_{n=1}^{N} (\lambda_n^{-1} - \lambda_{n+p+1}^{-1}) \left| \int_0^\infty m(\sigma) \cdot (\lambda_n^{-1} - \lambda_{n+p+1}^{-1})^{-1} E_n^p \Delta D^p e^{-\lambda_n \sigma} d\sigma \right|^q$$

$$\leq \int_0^\infty |m(\sigma)|^q \sum_{n=1}^{N} E_n^p \Delta D^p e^{-\lambda_n \sigma} d\sigma ,$$

by (26), (29)(ii), (30),

$$\leq \int_0^\infty |m(\sigma)|^q d\sigma ,$$

by (27).

The inequality (12)(ii) follows in a similar way.

REMARK. An observation made in a letter from Professor David Borwein in October, 1964, has some relevance to our theorems.

Consider the condition

$$(0) : \exists K \in (0,\infty) \text{ such that } \sum_{n=N}^{\infty} 1/\lambda_n \leq K/\lambda_N \quad (N = 1,2,\ldots) .$$

Borwein remarks (the proof is straightforward and takes only a few lines) that

$$(31) \qquad (0) \Longleftrightarrow (H_p) \text{ holds for some } p \geq 0 .$$

If we apply (31) fo Theorem I, for example (and use Remark (iii)), we get a solution of the moment problem in a form which is much more directly stated in terms of the convergence of $\Sigma\, 1/\lambda_n$, namely:

THEOREM I'. Let $0 = \lambda_0 < \lambda_1 < \cdots < \lambda_n \to \infty$. If, for a sequence $\{u_n\}_{n\geq 0}$,

$$(32) \quad \exists\, \psi(\cdot) \in BV[0,\infty) \text{ such that } u_n = \int_0^\infty e^{-\lambda_n \sigma} d\psi(\sigma) \quad (n = 0,1,\ldots) ,$$

then

$$(33) \qquad \sup_{p\geq 0} \sum_{n=0}^{\infty} E_n^p |\Delta D^p u_n| < \infty .$$

Conversely, if condition (0) holds, then (33) implies (32).

REFERENCES

1. S. Banach, Théorie des opérations linéaires. Warsaw: 1932 (reprinted New York (Chelsea).

2. K. Endl, On systems of linear inequalities in infinitely many variables and generalized Hausdorff means. Math. Z. 82 (1963), 1-7.

3. F. Hallenbach, Zur Theorie der Limitierungsverfahren von Doppelfolgen. Thesis: Bonn 1933.

4. F. Hausdorff, Summationmethoden und Momentfolgen. I. Math. Z. 9 (1921), 74-109.

5. F. Hausdorff, Summationmethoden und Momentfolgen. II. Math. Z. 9 (1921), 280-299.

6. A. Jakimovski and A. Livne, General Kojima-Toeplitz like theorems and consistency theorems. J. d'Anal. Math. 24 (1971), 323-368.

7. A. Jakimovski and D. C. Russell, Matrix mappings between BK-spaces. Bull. London Math. Soc. 4 (1972), 345-353.

8. A. Jakimovski and D. C. Russell, Best order conditions in linear spaces, with applications to limitation, inclusion, and high indices theorems for ordinary and absolute Riesz means. Studia Math. 56 (1976), 101-120.

9. A. Jakimovski and D. C. Russell, High indices theorems for Riesz and Abel typical means. Commentationes Math. (to appear).

10. A. Jakimovski and J. Tzimbalario, Inclusion relations for absolute Riesz means and on a conjecture of Maddox. Proc. London Math. Soc. (3) 30 (1975), 366-384.

11. B. I. Korenblyum, On a problem of interpolation. Doklady Akad. Nauk SSSR 81 (1951), 991-994 (in Russian).

12. H.-H. Körle, On absolute summability by Riesz and generalized Cesàro means. I. Canad. J. Math. 22, (1970), 202-208.

13. A. Meir, An inclusion theorem for generalized Cesàro and Riesz means. Canad. J. Math. 20 (1968), 735-738.

14. A. Peyerimhoff, Konvergenz- und Summierbarkeitsfaktoren. Math. Z. 55 (1951), 23-54.

15. D. C. Russell, On generalized Cesàro means of integral order. Tôhoku Math. J. (2) 17 (1965), 410-442; Corrigenda 18 (1966), 454.

16. D. C. Russell, Summability methods which include the Riesz typical means, II. Proc. Cambridge Phil. Soc. 69 (1971), 297-300.

17. J. A. Shohat and J. D. Tamarkin, The problem of moments. Amer. Math. Soc. Mathematical Surveys No. 1: 1943, 1950.

18. K. Zeller, Allegemeine Eigenschaften von Limitierungsverfahren. Math. Z. 53 (1951), 463-487.

Approximations and Probabilistic Inequalities

ON JACKSON-TYPE INEQUALITIES IN APPROXIMATION THEORY

P. L. Butzer
Lehrstuhl A für Mathematik
Rheinisch-Westfälische Technische Hochschule Aachen
5100 Aachen
WEST GERMANY

J. Junggeburth
Lehrstuhl A für Mathematik
Rheinisch-Westfälische Technische Hochschule Aachen
5100 Aachen
WEST GERMANY

ABSTRACT. The purpose of this paper is to present a brief survey of known results on Jackson inequalities for best approximation by trigonometric and algebraic polynomials in specific spaces, as well as for the approximation of functions by linear processes. It covers singular convolution integrals, semigroup and resolvent operators, and operators defined via a multiplier structure, with emphasis on the interconnections of Jackson-type inequalities with Voronovskaja-type relations and saturation.

1. INTRODUCTION

Concerning the large area of general inequalities, it seems only natural for approximation theorists to make contributions. Indeed, apart from the many standard inequalities, they in particular need those that are somehow characteristic to approximation and error estimates. Perhaps the most important group of such inequalities is that connected with the name of S.N. Bernstein. Of the standard works by Hardy-Littlewood-Pólya [27], Beckenbach-Bellman [3], Mitrinović [37], and Shisha [48], already [37] deals with Bernstein inequalities. A review of such inequalities by Görlich-Paulus [26], covering more than 250 papers on the subject, is in preparation.

It is somehow surprising that a second fundamental inequality of approximation, namely that used by D. Jackson [30] in 1911 in his proof of the now-called fundamental direct theorem of best (trigonometric or algebraic) approximation, does not yet seem to have caught the interest of those working in the field of inequalities. Nevertheless, we believe it to be appropriate to point out that Jackson inequalities, which involve the best approximation of a function in terms of its derivatives, be studied more closely. This is also confirmed by the fact that Jackson and Bernstein inequalities are in a certain sense dual to each other (see [15]), although this is not our goal here. The essential facts in the theory of best approximation are that once a Jackson inequality holds, then not only does the Jackson theorem follow but also the Steckin theorem on simultaneous approximation of a function and its derivatives. If a Bernstein inequality is also known to be valid, then the fundamental Bernstein inverse theorems and those of Zamansky can be deduced, as well as the converses of the Steckin and Zamansky theorems (due to Butzer-Pawelke-Sunouchi).

In the case of approximation of functions by linear approximation processes, the matter is analogous. Here a Jackson-type inequality, one involving specific approximation processes of a function in terms of its derivatives, is the basic hypothesis for the proofs of direct approximation theorems. Together with Bernstein-type inequalities, these are practically the only assumptions that determine the complete approximation behaviour, at least in the case of nonoptimal approximation by commutative processes that do not approximate the function too rapidly (e.g., with exponential order $\exp(\alpha x^{\tau})$, $\alpha,\tau > 0$ [21]). This is the situation not only for concrete processes in the specific spaces $C_{2\pi}$, $C[a,b]$, $L^p_{2\pi}$ or $L^p(\mathbb{R})$, $1 \leq p < \infty$, but also for general processes in the setting of arbitrary Banach spaces. It is actually the Banach-space setting that points out best the primary role played by Jackson- and Bernstein-type inequalities in approximation -- a theory built up in a series of papers ([13], [14], [16], [17]) since 1968 (but not to be dealt with here). Moreover, the distinction between a Jackson (-type) inequality and a Jackson (-type) direct approximation theorem comes out best in the Banach-space frame. Although a Jackson (-type) inequality could be regarded as a special (boundary) case of a Jackson (-type) theorem, once the inequality has been established, the Jackson theorem in its full generality follows rapidly.

It is therefore the purpose of this brief, first survey paper to consider Jackson inequalities both for best approximation and for the approximation by linear processes. Due to the amount of material available and to lack of space, only a small selection can be presented. For the same reasons, the reference list must be kept at a minimum. A detailed review is in preparation [9].

Concerning the actual contents, Section 2 deals with the Jackson inequality for best approximation, together with a proof, and also a statement and proof of the Jackson theorem. Section 3 treats Jackson-type inequalities for singular convolution integrals on $X_{2\pi}$ having positive, even kernels, for semigroup and resolvent operators as well as for general linear processes satisfying a Voronovskaja-type relation. The saturation concept, considered in Section 4, is important since it gives not only the best possible approximation order in Jackson-type inequalities but also the precise subspace of functions for which this best possible order is attained. Finally, Section 5 is devoted to Jackson-type inequalities and saturation for linear operators that are generated via multipliers in connection with Fourier expansions in Banach spaces. This essentially enables one to treat such inequalities for any of the classical orthogonal expansions, such as those of Laguerre, Hermite, Jacobi, Bessel, and Walsh.

Let us conclude with the hope that the matter is here presented in such a form that it catches the interest of those working in the area of general inequalities and not only in approximation theory.

The authors would like to thank Professor R.J. Nessel for his critical reading of the manuscript and for his many helpful suggestions.

2. JACKSON INEQUALITIES FOR BEST APPROXIMATION

2.1. THE PROBLEM FOR $C_{2\pi}$. Letting T_n be the set of all trigonometric polynomials

$$t_n(x) = \sum_{k=-n}^{n} c_k e^{ikx}$$

of degree at most n, $n \in \mathbb{N} = \{1,2,3,\ldots\}$, and $C_{2\pi}$ the space of all continuous 2π-periodic functions on the line $\mathbb{R}$ with norm

$$\|f\|_{C_{2\pi}} := \max_{-\pi \leq x \leq \pi} |f(x)| ,$$

we call

$$E_n(f;C_{2\pi}) := \inf_{t_n \in T_n} \|f - t_n\|_{C_{2\pi}} \qquad (n \in \mathbb{N}) \tag{2.1}$$

the n-th degree of approximation of $f \in C_{2\pi}$ by elements of T_n. Whereas the Weierstrass theorem states that $E_n(f;C_{2\pi})$ tends to zero as $n \to \infty$ for all $f \in C_{2\pi}$, Jackson's fundamental direct theorem shows that the smoother the function, the faster does $E_n(f;C_{2\pi})$ tend to zero. More specifically, one has the following result:

JACKSON'S THEOREM. For $f^{(r)} \in C_{2\pi}$, $r \in \mathbb{N}$, there holds

$$E_n(f;C_{2\pi}) \leq M_r n^{-r} \omega_2(f^{(r)};n^{-1};C_{2\pi}) \qquad (n \in \mathbb{N}), \tag{2.2}$$

where M_r is a constant. In particular, if $f^{(r)} \in \mathrm{Lip}_2(\alpha;C_{2\pi})$, $0 < \alpha \leq 2$, then $E_n(f;C_{2\pi}) = O(n^{-r-\alpha})$.

Here, $g \in \text{Lip}_s(\alpha;C_{2\pi})$, $0 < \alpha \leq s$, $s \in \mathbb{N}$, means that

$$(2.3) \qquad \omega_s(g;\delta;C_{2\pi}) := \sup_{|h|\leq\delta} \|\Delta_h^s g(\cdot)\|_{C_{2\pi}} = O(\delta^\alpha) \qquad (\delta \to 0+),$$

where $\Delta_h^s f(x) = \sum_{k=0}^{s} (-1)^{s-k}\binom{s}{k} f(x+kh)$, $h \in \mathbb{R}$.

To prove (2.2), Jackson [30],[29, p. 10] made use of JACKSON'S INEQUALITY:

$$(2.4) \qquad f^{(r)} \in C_{2\pi} \Rightarrow E_n(f;C_{2\pi}) \leq J_r n^{-r} \|f^{(r)}\|_{C_{2\pi}}$$

for all $n \in \mathbb{N}$, J_r being a constant independent of f and n.

Let us sketch a proof of (2.4), as it gives a number of insights. First, construct a sequence $t_n(x) \in T_n$ such that

$$(2.5) \qquad \|f - t_n\|_{C_{2\pi}} \leq J_1 n^{-1} \|f'\|_{C_{2\pi}} \qquad (f' \in C_{2\pi};\ n \in \mathbb{N})$$

with the aid of particular singular convolution integrals

$$(2.6) \qquad (X_n f)(x) := (1/2\pi)\int_{-\pi}^{\pi} f(x-u)\chi_n(u)du := (\chi_n * f)(x)$$

that are elements of T_n, $\{\chi_n(x); n \in \mathbb{N}\}$ being a kernel, i.e.,

$$\chi_n \in L^1_{2\pi} \text{ with } \int_{-\pi}^{\pi} \chi_n(u)du = 2\pi .$$

A good choice is the integral of Fejér-Korovkin, with kernel defined by

$$(2.7) \qquad k_n(x) = \frac{2}{n+2}\left[\frac{\sin(\pi/(n+2))\cdot\cos((n+2)x/2)}{\cos(\pi/(n+2)) - \cos x}\right]^2 .$$

This $k_n(x)$ is an even, positive element of T_n whose first Fourier coefficient is given (cf. [10, p. 79 ff.]) by

$$(2.8) \qquad k_n^\wedge(1) = \cos\left(\frac{\pi}{n+2}\right) ; \quad f^\wedge(k) := \frac{1}{2\pi}\int_{-\pi}^{\pi} f(u)e^{-iku}\,du \qquad (k \in \mathbb{Z}),$$

$\mathbb{Z} = \{0,\pm1,\pm2,\dots\}$. Now for any even, positive kernel there holds ([49])

$$(2.9) \qquad \frac{1}{2\pi}\int_{-\pi}^{\pi} |u|^\alpha \chi_n(u)du \leq \left(\frac{\pi}{\sqrt{2}}\sqrt{1-\chi_n^\wedge(1)}\right)^\alpha \qquad (0 < \alpha \leq 2) ;$$

thus, by (2.8), for $f' \in C_{2\pi}$,

$$(2.10) \qquad \begin{aligned} \|K_n f - f\|_{C_{2\pi}} &= \left\|\frac{1}{2\pi}\int_{-\pi}^{\pi} k_n(u)[f(x-u) - f(x)]\,du\right\|_{C_{2\pi}} \\ &\leq \frac{1}{2\pi}\int_{-\pi}^{\pi} |u| k_n(u)du\ \|f'\|_{C_{2\pi}} \leq \left(\frac{\pi}{\sqrt{2}}\right)^2 \frac{1}{n+2}\|f'\|_{C_{2\pi}} . \end{aligned}$$

To deduce (2.5) for higher derivatives, thus (2.4), replace $t_n(x) =$

$(k_n * f)(x)$ by a linear combination of iterates of K_n, namely [10, p. 97 ff.] by

$$(U_{r,n}f)(x) = \sum_{j=1}^{r} (-1)^{j+1}\binom{r}{j} K_n^j ,$$

the powers (of the operators K_n) being defined by

$$K_n^j = K_n^1 (K_n^{j-1}) , \quad j \in \mathbb{N} .$$

Again $U_{r,n}f$ is a convolution integral in T_n. Since

$$U_{r,n}f - f = (-1)^{r-1}(K_n - I)^r f ,$$

applying (2.5) iteratively to $(K_n - I)^{r-1} f$ then yields (2.4) by

$$\begin{aligned} E_n(f;C_{2\pi}) &\leq \|(K_n - I)[(K_n - I)^{r-1} f]\|_{C_{2\pi}} \\ &\leq \frac{\pi^2}{2} \frac{1}{n+2} \|(K_n - I)[(K_n - I)^{r-2} f']\|_{C_{2\pi}} \qquad (2.11) \\ &\leq \left(\frac{\pi^2}{2}\right)^r \frac{1}{(n+2)^r} \|f^{(r)}\|_{C_{2\pi}} . \end{aligned}$$

Note that Jackson's theorem, i.e. (2.2), can be deduced from (2.4) by using the integral means of $f \in C_{2\pi}$ of order 2,

$$(A_t^2 f)(x) := \frac{1}{t^2}\int_{-t/2}^{t/2}\int_{-t/2}^{t/2} f(x + u_1 + u_2)du_1\, du_2 , \qquad (2.12)$$

with the properties (cf. [10, p. 37 f.], [51, p. 163 ff.]) that

$$f^{(r)} \in C_{2\pi} \quad \text{implies} \quad (A_t^2 f)^{(r+2)}(x) \in C_{2\pi} ,$$

and

$$\|(A_t^2 f)^{(r+2)}\|_{C_{2\pi}} \leq t^{-2}\omega_2(f^{(r)};t;C_{2\pi}) ,$$

$$\|A_t^2 f - f\|_{C_{2\pi}} \leq (1/2)\omega_2(f;t;C_{2\pi}) .$$

Combining inequality (2.4) with these, setting $t = (n + 2)^{-1}$, we get

$$\begin{aligned} E_n(f;C_{2\pi}) &\leq E_n(A_t^2 f;C_{2\pi}) + E_n(f - A_t^2 f;C_{2\pi}) \\ &\leq \left(\frac{\pi^2}{2(n+2)}\right)^{r+2} \frac{1}{t^2}\, \omega_2(f^{(r)};t;C_{2\pi}) + \frac{1}{2}\left(\frac{\pi^2}{2(n+2)}\right)^r \omega_2(f^{(r)};t;C_{2\pi}) \\ &\leq \left[\left(\frac{\pi^2}{2}\right)^2 + \frac{1}{2}\right]\left(\frac{\pi^2}{2}\right)^r (n+2)^{-r}\, \omega_2(f^{(r)};(n+2)^{-1};C_{2\pi}) , \end{aligned}$$

which proves Jackson's theorem.

The constant $J_r = (\pi^2/2)^r$ in Jackson's inequality (2.4) is by no means the best possible. The latter is given (see [36, p. 116], [51, p. 78]) by

$$K_r' := \frac{4}{\pi} \sum_{k=1}^{\infty} \frac{(-1)^{k(r+1)}}{(2k+1)^{r+1}},$$

where

$$1 = K_0' < K_2' < \cdots < K_{2r}' < \cdots < 4/\pi < \cdots < K_{2r-1}' < \cdots < K_1' = \pi/2 .$$

2.2. VARIOUS EXTENSIONS. Jackson's inequality (J - INQ) can also be stated for the space $L^p_{2\pi}$ of functions that are 2π-periodic and Lebesgue integrable to the p-th power, $1 \le p < \infty$, with norm

$$\|f\|_{L^p_{2\pi}} := \left\{(2\pi)^{-1} \int_{-\pi}^{\pi} |f(x)|^p \, dx\right\}^{1/p} .$$

With $X_{2\pi}$ standing for the space $C_{2\pi}$ or $L^p_{2\pi}$, $1 \le p < \infty$, J - INQ reads ([1], [36]):

$$E_n(f;X_{2\pi}) \le J_r n^{-r} \|f^{(r)}\|_{X_{2\pi}} \qquad (f \in X^r_{2\pi}; n \in \mathbb{N}),$$

where

$$X^r_{2\pi} = \{f \in X_{2\pi};\ (d/dx)^j f \in X_{2\pi},\ 1 \le j \le r\} .$$

There also exists a somewhat sharper (since $E_n(f;X_{2\pi}) \le \|f\|_{X_{2\pi}}$) form of J - INQ:

$$E_n(f;X_{2\pi}) \le J_r(n+2)^{-r} E_n(f^{(r)};X_{2\pi}) \qquad (f \in X^r_{2\pi}; n \in \mathbb{N}) .$$

The classical J - INQ can be extended to <u>fractional-order derivatives</u>. With the fractional difference of order $\gamma > 0$ denoted by

$$\Delta_h^\gamma f(x) = \sum_{k=0}^{\infty} (-1)^k \binom{\gamma}{k} f(x - kh)$$

(which generates the same modulus of continuity as $\Delta_h^s f(x)$ for $\gamma = s$), the (strong) derivative of $f \in X_{2\pi}$ of fractional order $\gamma > 0$ is defined as the limit in $X_{2\pi}$-norm of

$$h^{-\gamma}(\Delta_h^\gamma f)(x) \quad \text{as} \quad h \to 0 .$$

If this limit exists, it is denoted (cf. [53], [18]) by $D^\gamma f$. In this setting, J - INQ reads (see [7]) thus:

$$E_n(f;X_{2\pi}) \le J_\gamma(n+2)^{-\gamma} \|D^\gamma f\|_{X_{2\pi}} \qquad (f \in X^\gamma_{2\pi}; n \in \mathbb{N}) ,$$

the proof following by modifying that for (2.4).

J - INQ can also be formulated in the case of approximation of non-periodic functions defined on a finite interval, $[-1,1]$ say, by <u>algebraic</u>

<u>polynomials</u> $p_n(x)$ of degree at most n. Namely, with $E_n(f;C[-1,1])$ defined in the obvious way (e.g., [19]), one has the following formulation:

If $f \in C^r[-1,1]$ and $n \geq r$, then

$$E_n(f;C[-1,1]) \leq (\pi^2/2)^r[(n+2)(n+1)\cdots(n+3-r)]^{-1}\|f^{(r)}\|_{C[-1,1]} .$$

This formulation may basically be obtained from the trigonometric one under the substitution $x = \cos\theta$.

There also exists a "pointwise" version, due to A.F. Timan [51] and S.A. Teljakovskii [50], giving a better quality of approximation at the end-points ± 1 of $[-1,1]$ (see also [23]):

If $f \in C^r[-1,1]$ and $n \geq 2r+1$, then there exists a sequence of p_n such that

$$|f(x) - p_n(x)| \leq L_r\left(\frac{\sqrt{1-x^2}}{n}\right)^r \|f^{(r)}\|_{C[-1,1]} \qquad (x \in [-1,1]) ,$$

the constant L_r depending only on r.

J-INQ can also be formulated in the framework of "homogeneous Banach spaces" (see [46, p. 206 ff.]). This point of view is useful when extending the inequality to functions of several variables, in particular to the space $L^p(T^m)$ of functions $f(x_1,\ldots,x_m)$ on the m-dimensional torus T^m (see [51, p. 273 ff.], [42], [47]).

As a final example of a large number of already classical results, let us point out that J-INQ may be established for the approximation of functions belonging to $C(\mathbb{R})$ (the space of functions that are bounded and uniformly continuous on $\mathbb{R}$) or $L^p(\mathbb{R})$, $1 \leq p < \infty$, with

$$\text{norm } \|f\|_p = \left(\int_{\mathbb{R}} |f(x)|^p \, dx\right)^{1/p} ,$$

this time T_n being replaced by the set of entire functions of exponential type whose restriction to $\mathbb{R}$ belongs again to $C(\mathbb{R})$ or $L^p(\mathbb{R})$, respectively (see [51, p. 259 ff., 301, 325], [1, p. 229], [43], [35]).

3. JACKSON-TYPE INEQUALITIES FOR LINEAR APPROXIMATION PROCESSES

3.1. GENERAL CONSIDERATIONS. In establishing the J-INQ for best approximation, namely (2.4), use was made of a (linear) convolution integral $(U_{r,n}f)(x)$ having the same order of approximation as $E_n(f;C_{2\pi})$ for $n \to \infty$ in the case that $f^{(r)} \in C_{2\pi}$. This suggests that one should study inequalities of the type (2.4) or (2.11) not only for $E_n(f;C_{2\pi})$ but also for large classes of linear convolution integrals on $X_{2\pi}$, or, more generally, for arbitrary linear approximation processes $\{T_\rho\}$ acting on a Banach space (B-space) X with norm $\|\cdot\|_X$.

Let ρ be a parameter over some set $\mathbb{A}$ (either an interval (a,b) with $0 \leq a < b \leq +\infty$ or the set $\mathbb{N}$), and let ρ_0 be one of the points a,b or $+\infty$. A family $\{T_\rho; \rho \in \mathbb{A}\}$ of bounded linear operators mapping X into itself, i.e., $T_\rho \in [X]$ for $\rho \in \mathbb{A}$, is called a linear approximation process on X if, for each $f \in X$,

$$\|T_\rho f\|_X \leq M\|f\|_X , \quad \lim_{\rho \to \rho_0} \|T_\rho f - f\|_X = 0 , \tag{3.1}$$

the constant M being independent of $\rho \in \mathbb{A}$ and f.

Replacing the differential operator $(d/dx)^r$ defined on $C^r_{2\pi}$ (cf. (2.4)) by an arbitrary closed linear operator B with domain $D(B) \subset X$ and range in X, and the order n^{-r} by some positive function $\varphi(\rho)$ tending to zero as $\rho \to \rho_0$, we call

$$\|T_\rho f - f\|_X \leq C\ \varphi(\rho)\ \|Bf\|_X \qquad (f \in D(B);\ \rho \in \mathbb{A}) \tag{3.2}$$

a Jackson-type inequality (J-T-INQ) or order $\varphi(\rho)$ on X for the process $\{T_\rho; \rho \in \mathbb{A}\}$.

The question now is whether it is possible to verify an inequality of type (3.2) for particular $\{T_\rho\}$ on specific spaces X. Different specializations follow below.

3.2. SINGULAR INTEGRALS ON $X_{2\pi}$ HAVING POSITIVE AND EVEN KERNELS. The J-T-INQ for such integrals of type (2.6) on $X_{2\pi}$ reads as follows:

PROPOSITION 3.1. Let the kernel of $(\chi_\rho f)(x)$ be even and positive. If $f \in X^j_{2\pi}$, $j = 1,2$, then

$$\|\chi_\rho f - f\|_{X_{2\pi}} \leq 2^{1-j}\Big(\frac{\pi}{\sqrt{2}}\ \sqrt{1 - \hat{\chi}_\rho(1)}\Big)^j\ \|f^{(j)}\|_{X_{2\pi}} \qquad (\rho \in \mathbb{A}) . \tag{3.3}$$

The proof is similar to that for (2.10), using

$$f(x + u) + f(x - u) - 2f(x) = \int_0^u \int_{-v}^{v} f''(x - y)dy\ dv$$

if $j = 2$ (see [10, pp. 69, 77]).

Let us apply Proposition 3.1 in case $j = 2$ to some well-known singular integrals that play an important role in approximation.

The integral of Fejér-Korovkin for $f \in X_{2\pi}$ (recall (2.7)) satisfies, in view of (2.8) and (3.3), the J-T-INQ with $\varphi(n) = (n + 2)^{-2}$:

$$(3.4)\qquad \|K_n f - f\|_{X_{2\pi}} \leq \frac{\pi^4}{4}(n+2)^2 \|f''\|_{X_{2\pi}} \qquad (f \in X^2_{2\pi}\ ;\ n \in \mathbb{N}).$$

For the singular integral of Jackson, defined for $f \in X_{2\pi}$ by $(J_n f)(x) := (j_n * f)(x)$, where

$$(3.5)\qquad j_n(x) = \frac{3}{n(2n^2+1)}\left[\frac{\sin(nx/2)}{\sin(x/2)}\right]^4 \in T_{2n-2}\,,$$

one has (see [10, pp. 60,70], [22, p. 79], [8])

$$j_n^{\wedge}(1) = 1 - 3(2n^2+1)^{-1}\,.$$

Then (3.3) gives the J-T-INQ

$$(3.6)\qquad \|J_n f - f\|_{X_{2\pi}} \leq \frac{3\pi^2}{8}(2n^2+1)^{-1}\|f''\|_{X_{2\pi}} \qquad (f \in X^2_{2\pi};\ n \in \mathbb{N}).$$

For the singular integral of de la Vallée Poussin, defined for $f \in X_{2\pi}$ by

$$(V_n f)(x) = (v_n * f)(x),$$

where

$$(3.7)\qquad v_n(x) = \frac{(n!)^2}{(2n)!}\left(2\cos\frac{x}{2}\right)^{2n} \in T_n$$

with (see [10, p. 112]) $v_n^{\wedge}(1) = n/(n+1)$, there holds the J-T-INQ

$$(3.8)\qquad \|V_n f - f\|_{X_{2\pi}} \leq \frac{\pi^2}{4}(n+1)^{-1}\|f''\|_{X_{2\pi}} \qquad (f \in X^2_{2\pi}\ ;\ n \in \mathbb{N}).$$

The integral means $(A_t^r f)(x)$ of $f \in X_{2\pi}$ of order $r \in \mathbb{N}$, defined analogously to (2.12) (see [10, pp. 37, 54, 77]), are often referred to by Riemann's method of summation of the Fourier series of f of order r when rewritten as (cf. [8, p. 373])

$$(A_t^r f)(x) = (a_t^r * f)(x)\,,$$

where

$$(3.9)\qquad a_t^r(x) = \sum_{k=-\infty}^{\infty}\left[\frac{\sin tk/2}{tk/2}\right]^r e^{ikx}$$

(with parameter $\rho = t \to 0+$). By (3.3), there holds [9], for $r \in \mathbb{N}$,

$$(3.10)\qquad \|A_t^r f - f\|_{X_{2\pi}} \leq \frac{\pi^2}{2}\left(1+\frac{1}{\pi^2}\right)^r t^2\|f''\|_{X_{2\pi}} \qquad (f \in X^2_{2\pi}\ ;\ 0 < t \leq 1)\,.$$

3.3. SEMIGROUP OPERATORS. An important class of approximation processes acting on a B-space X and satisfying a J-T-INQ are semigroups

$\{T_t\,;\ 0 \le t < \infty\}$ (see, e.g., [6]). This is a family of operators $T_t \in [X]$, $t \ge 0$, satisfying the semigroup property

$$T_{t_1+t_2} = T_{t_1} T_{t_2}$$

for $t_1, t_2 \ge 0$, with $T_0 = I$ (identity operator). The family $\{T_t\}$ is said to be uniformly bounded and of class (C_0) if

$$\|T_t f\|_X \le M\|f\|_X\ ,\ t \ge 0\ ,\ \text{and}\ \lim_{t\to 0+} \|T_t f - f\|_X = 0\ ,\quad \text{each}\quad f \in X\ .$$

The infinitesimal generator A of $\{T_t\}$ is defined by

$$\lim_{t\to 0+} \left\|\frac{T_t - f}{t} - Af\right\|_X = 0 \tag{3.11}$$

if this limit exists. The operator A is linear, closed (in general unbounded); its domain $D(A)$ is dense in X and becomes a normalized B-subspace (i.e., $\|f\|_X \le \|f\|_{D(A)}$) under the norm

$$\|f\|_{D(A)} := \|f\|_X + \|Af\|_X\ .$$

For semigroups, the basic formula

$$T_t f - f = \int_0^t T_u[Af]du \qquad (f \in D(A)\ ;\ t > 0)$$

immediately yields the J-T-INQ in the form

$$\|T_t f - f\|_X \le Mt\|Af\|_X \qquad (f \in D(A)\ ;\ t > 0)\ . \tag{3.12}$$

As an application, consider the general singular integral of Weierstrass, defined for $f \in L^p(\mathbb{R})$, $\kappa, t > 0$, by (see [10, pp. 257, 465, 506])

$$(W_t^\kappa f)(x) := \frac{1}{\sqrt{2\pi}} \int_{\mathbb{R}} f(x - u)\{t^{-1/\kappa} w^\kappa(t^{-1/\kappa} u)\}\, du\ , \tag{3.13}$$

its kernel $w^\kappa(x)$ being given via its Fourier transform,

$$[w^\kappa]^\wedge(v) := (1/\sqrt{2\pi}) \int_{\mathbb{R}} w^\kappa(x) e^{-ivx}\, dx = \exp(-|v|^\kappa) \qquad (v \in \mathbb{R})\ .$$

Now w^κ is positive for $0 < \kappa \le 2$, and w^κ belongs to $L^1(\mathbb{R})$ for all $\kappa > 0$, with

$$\int_{\mathbb{R}} w^\kappa(x)dx = \sqrt{2\pi}\ .$$

For $\kappa = 1,2$, w^κ has explicit representations ([10, p. 125 f.]), the integrals then being the well-known ones of Gauss-Weierstrass and Cauchy-Poisson. $\{W_t^\kappa\ ;\ 0 \le t < \infty\}$ forms a uniformly bounded semigroup of class (C_0)

with generator $A^\kappa f = f^{\{\kappa\}}$ (κ-th strong Riesz derivative of f), which is given for $1 \le p \le 2$ by

$$D(A^\kappa) = \{f \in L^p(\mathbb{R}); \; -|v|^\kappa f^\wedge(v) = g^\wedge(v), \;\; g \in L^p(\mathbb{R})\}, \quad g = f^{\{\kappa\}} .$$

So (3.12) yields the J-T-INQ

$$\|W_t^\kappa f - f\|_{L^p(\mathbb{R})} \le t\|f^{\{\kappa\}}\|_{L^p(\mathbb{R})} \qquad (f \in D(A^\kappa) \; ; \; t > 0) . \tag{3.14}$$

As another type of approximation process having a J-T-INQ, consider the family of <u>resolvent operators</u> $\{\lambda R(\lambda;A) \; ; \; \lambda > 0\}$ of the generator A of the uniformly bounded semigroup $\{T_t\}$ of class (C_0), namely

$$R(\lambda;A)f = \int_0^\infty e^{-\lambda u} \, T_u f \, du \qquad (f \in X \; ; \; \lambda > 0) .$$

There holds

$$\|\lambda R(\lambda;A)f\|_X \le M\|f\|_X , \quad \lambda > 0 ,$$

and

$$\|\lambda R(\lambda;A)f - f\|_X \to 0 \quad \text{as} \quad \lambda \to \infty ,$$

for each $f \in X$. In view of the relation

$$\lambda R(\lambda;A)f - f = R(\lambda;A)Af , \quad f \in D(A) ,$$

one has the J-T-INQ

$$\|\lambda R(\lambda;A)f - f\|_X \le M\lambda^{-1} \|Af\|_X \qquad (f \in D(A) \; ; \; \lambda > 0) . \tag{3.15}$$

For applications to particular singular integrals, see [6, p. 136 ff.].

3.4. VORONOVSKAJA-TYPE RELATIONS. A further sufficient condition such that a process $\{T_\rho \, ; \, \rho \in \mathbf{A}\}$ with $T_\rho \in [X]$ possesses a J-T-INQ is that it satisfies a relation of type (3.11), namely a <u>Voronovskaja-type relation</u>; i.e., there exists a positive function $\varphi(\rho)$ tending to zero for $\rho \to \rho_0$, and a closed linear operator B with domain $D(B)$ dense in X and range in X, such that, for each $f \in D(B)$,

$$\lim_{\rho \to \rho_0} \|\varphi^{-1}(\rho)\{T_\rho f - f\} - Bf\|_X = 0 . \tag{3.16}$$

Again $D(B)$ is a B-subspace of X under

$$\|f\|_{D(B)} := \|f\|_X + \|Bf\|_X ,$$

so, by (3.16), $\{\varphi^{-1}(\rho)\{T_\rho f - f\} \, ; \, \rho \in \mathbb{A}\}$ defines a family of bounded linear operators on $D(B)$ into X such that for each $f \in D(B)$ there is a constant M_f (depending at most on f) with

$$\varphi^{-1}(\rho)\|T_\rho f - f\|_X \leq M_f, \quad \text{all} \quad \rho \in \mathbb{A} .$$

The uniform-boundedness theorem then implies that

$$\|T_\rho f - I\|_{[D(B),X]} \leq M\,\varphi(\rho), \quad \rho \in \mathbb{A},$$

with a constant $M \neq M_f$. This is equivalent to the J-T-INQ (3.17), giving:

PROPOSITION 3.2. If $\{T_\rho\}$ satisfies relation (3.16) on X, then

$$\|T_\rho f - f\|_X \leq M\,\varphi(\rho)\,\|f\|_{D(B)} \qquad (f \in D(B)\,;\ \rho \in \mathbb{A}) . \tag{3.17}$$

Inequality (3.17) has, in comparison with (3.12), the same order, but the seminorm $\|Af\|_X$ is replaced by the norm $\|f\|_{D(B)}$, and the constant M is undetermined.

Since Voronovskaja-type relations are known for a large variety of approximation processes, they give a useful criterion for J-T-INQ. For example, in the case of the integral of de la Vallée Poussin $(V_nf)(x)$ of (3.7), (3.16) reads ([10, p. 449], [8]):

$$\lim_{n\to\infty} \|n\{V_nf - f\} - f''\|_{X_{2\pi}} = 0 \qquad (f \in X_{2\pi}^2) .$$

Let us finally mention that if a J-T-INQ is known for a given process $\{T_\rho\,;\ \rho \in \mathbb{A}\}$ on X, then one can establish a Jackson-type theorem for this process. Here we need the K-functional having properties similar to the modulus of continuity (2.3); it is its generalization to a B-space setting, defined by

$$K(\varphi(\rho),f;X,D(B)) := \inf_{g\in D(B)} \{\|f - g\|_X + \varphi(\rho)\|g\|_{D(B)}\} \qquad (f \in X) . \tag{3.18}$$

PROPOSITION 3.3. If $\{T_\rho\,;\ \rho \in A\}$ of (3.1) satisfies (3.17) on $D(B)$, then

$$\|T_\rho f - f\|_X \leq C'K(\varphi(\rho),f;X,D(B)) \qquad (f \in X;\ \rho \in \mathbb{A}) . \tag{3.19}$$

Indeed, let $f = (f - g) + g$, any $g \in D(B)$. Then, since T_ρ is linear,

$$\|[T_\rho - I]f\|_X \leq \|[T_\rho - I](f - g)\|_X + \|[T_\rho - I]g\|_X$$
$$\leq \max\{M + 1,C\} \cdot \{\|f - g\|_X + \varphi(\rho)\|g\|_{D(B)}\} .$$

Taking the infimum over all $g \in D(B)$ yields (3.19).

For particular spaces $X, D(B)$ such as $X_{2\pi}$, $X_{2\pi}^r$, $r \in \mathbb{N}$, and $\varphi(\rho) = \rho^{-r}$, $\rho \to \infty$, the K-functional is directly comparable to ω_r with constants C_1, C_2 by (see, e.g., [6], [17], [23]) ($f \in X_{2\pi}$, $\rho > 0$)

$$C_1\omega_r(f;\rho^{-1};X_{2\pi}) \leq K(\rho^{-r},f;X_{2\pi},X_{2\pi}^r) \leq C_2\omega_r(f;\rho^{-1};X_{2\pi}) .$$

Proposition 3.3 applied to $(J_n f)(x)$ of (3.5) gives, for $f \in X_{2\pi}$,

$$\|J_n f - f\|_{X_{2\pi}} \leq C_3\omega_2(f;n^{-1};X_{2\pi}) .$$

4. JACKSON-TYPE INEQUALITIES AND SATURATION

Having established J-T-INQ of the form (3.2) for $f \in D(B)$ in Section 3, one may first ask whether the class $D(B) \subset X$ can be enlarged for the same order $\varphi(\rho)$. Indeed, are any and, if so, what $f \notin D(B)$ are permitted in (3.2)? On the other hand, $E_n(f;C_{2\pi})$ tends the faster to zero the smoother $f \in C_{2\pi}$; thus the second question is whether the same occurs for $\|T_\rho f - f\|_X$.

An answer is supplied by so-called saturation theorems. Given a process $\{T_\rho;\ \rho \in \mathbb{A}\}$ on a B-space X, when improving the smoothness properties upon $f \in X$ it often happens that the approximation order reaches a critical index that cannot be surpassed, no matter how smooth f is. More precisely, if for a family $\{T_\rho;\ \rho \in \mathbb{A}\}$ on X there is a positive function $\varphi(\rho)$ on $\mathbb{A}$ tending to zero as $\rho \to \rho_0$, such that any $f \in X$ with

$$(4.1) \qquad \|T_\rho f - f\|_X = o(\varphi(\rho)) \qquad (\rho \to \rho_0)$$

is trivial in some sense, e.g., $T_\rho f = f$, all $\rho \in \mathbb{A}$, and if the set

$$(4.2) \qquad F[X;T_\rho] := \{f \in X;\ \|T_\rho f - f\|_X = O(\varphi(\rho)),\ \ \rho \to \rho_0\}$$

contains at least one nontrivial element $f \in X$, then the family $\{T_\rho;\ \rho \in \mathbb{A}\}$ is said to be <u>saturated</u> on X with order $O(\varphi(\rho))$, and $F[X;T_\rho]$ is called its <u>Favard</u> or <u>saturation</u> class. For this concept, see [10, p. 433 ff.], [6], [22], and the literature cited there. Note that $F[X;T_\rho]$ is a normalized B-subspace of X under the norm (see [10, p. 440 ff.])

$$(4.3) \qquad \|f\|_F := \|f\|_X \sup_{\rho \in \mathbb{A}} \{\varphi^{-1}(\rho)\|T_\rho f - f\|_X\} .$$

PROPOSITION 4.1. If $\{T_\rho;\ \rho \in \mathbb{A}\}$ is saturated on X with order $O(\varphi(\rho))$, then $\{T_\rho\}$ satisfies an optimal J-T-INQ of the form

$$(4.4) \qquad \|T_\rho f - f\|_X \leq M\,\varphi(\rho)\,\|f\|_F \qquad (f \in F;\ \rho \in \mathbb{A}) .$$

The proof proceeds as for Proposition 3.2, using the uniform-boundedness principle. Inequality (4.4) is optimal in the sense that

$$\|T_\rho f - f\|_X = o(\varphi(\rho)) , \quad \rho \to \rho_0 ,$$

yields $T_\rho f = f$, $\rho \in \mathbb{A}$.

The next theorem essentially states that the existence of a Voronovskaja-type relation (3.16) implies that the process is saturated with order $O(\varphi(\rho))$, and that the Favard class $F[X;T_\rho]$ can actually be characterized in terms of the relative completion $D(B)^{\sim X}$ of $D(B)$, which is independent of the process $\{T_\rho\}$ in question. If Y is a normalized B-subspace of X, the (relative) completion of Y relative to X, denoted by $Y^{\sim X}$, is the set of those $f \in X$ for which there exists a sequence $\{f_n; n \in \mathbb{N}\} \subset Y$ such that

$$\|f_n\|_Y \le M \quad \text{and} \quad \lim_{n\to\infty} \|f_n - f\|_X = 0 .$$

$Y^{\sim X}$ becomes a normalized B-subspace under the norm

$$\|f\|_{Y^{\sim X}} := \inf\left\{\sup_{n\in\mathbb{N}} \|f_n\|_Y ; \{f_n\} \subset Y, \lim_{n\to\infty} \|f_n - f\|_X = 0\right\} . \tag{4.5}$$

THEOREM 4.2. Let $f \in X$, $\{T_\rho; \rho \in \mathbb{A}\}$ be a commutative process on X, i.e.,

$$T_{\rho_1} \cdot T_{\rho_2} = T_{\rho_2} \cdot T_{\rho_1} , \quad \rho_1, \rho_2 \in \mathbb{A} ,$$

and B a closed linear operator satisfying (3.16). Suppose there exists a regularization process $\{J_n; n \in \mathbb{N}\}$, i.e., $J_n \in [X]$ with

$$J_n(X) \subset D(B) , \quad n \in \mathbb{N} , \quad \lim_{n\to\infty} \|J_n f - f\|_X = 0 , \quad \text{each } f \in X ,$$

and the operators J_n and T_ρ commute for each $n \in \mathbb{N}$, $\rho \in \mathbb{A}$.

(a) If $\|T_\rho f - f\|_X = o(\varphi(\rho))$ for $f \in X$, then $f \in D(B)$ and $Bf = 0$.

(b) The following assertions are equivalent:

(i) $\|T_\rho f - f\| = O(\varphi(\rho))$, $\rho \to \rho_0$, i.e., $f \in F[X;T_\rho]$;

(ii) $F \in D(B)^{\sim X}$;

(iii) $f \in D(B)$, provided X is reflexive.

For a proof, see [10, p. 502 f.], [4]. In comparison with (4.4), under the hypothesis of Theorem 4.2 there holds

$$\|T_\rho f - f\|_X \le M\, \varphi(\rho)\, \|f\|_{D(B)^{\sim X}} \quad (f \in D(B)^{\sim X};\ \rho \in \mathbb{A}) , \tag{4.6}$$

so that the norm $\|f\|_{\mathbb{F}}$ of (4.4) (defined in terms of the process $\{T_\rho\}$) is replaced by the concrete (equivalent) $\|f\|_{D(B)^{\sim X}}$. Moreover, one now knows that the order $O(\varphi(\rho))$ is best possible.

Before illustrating our results by several examples, let us summarize the chain of J-T-INQ for $\{T_\rho\}$ deduced so far for all $\rho \in \mathbb{A}$:

$$\varphi^{-1}(\rho)\|T_\rho f - f\|_X \le M \begin{cases} \|Bf\|_X & (f \in D(B)) , & (3.2)\\ \|f\|_{D(B)} & (f \in D(B)) , & (3.17)\\ \|f\|_F & (f \in F[X;T_\rho]) , & (4.4)\\ \|f\|_{D(B)^{\sim X}} & (f \in D(B)^{\sim X}) , & (4.6)\end{cases}$$

the constant M (which may differ at each occurrence) being independent of f and ρ. In comparing (3.17) with (4.6), notice that the former follows directly by (3.16), the latter by the saturation theorem, this time for all f in $D(B)^{\sim X}$, the maximal extension of $D(B)$.

Concerning the integral $(K_n f)(x)$ (cf. (2.7)) for $f \in C_{2\pi}$, it is known that (3.2) is satisfied with

$$Bf := (\pi^2/2)f'' \quad \text{and} \quad D(B) = C^2_{2\pi}$$

(see (3.4)). By (finite) Fourier-transform methods, one can show (see [10, p. 374 ff., 370]) that

$$D(B)^{\sim C_{2\pi}} = V[2;C_{2\pi}] := \{f \in C_{2\pi}\,;\, f \in AC^1_{2\pi},\, f'' \in L^\infty_{2\pi}\} ,$$

where $V[2;C_{2\pi}]$ is a normalized B-subspace of $C_{2\pi}$ under the norm

$$\|f\|_V := \|f\|_{C_{2\pi}} + \|f''\|_{L^\infty_{2\pi}} .$$

Moreover, it is known that

$$V[2;C_{2\pi}] = \mathrm{Lip}_2(2;C_{2\pi}),$$

the latter having the equivalent norm with $\alpha = 2$:

$$\|f\|_{\mathrm{Lip}_2(\alpha;C_{2\pi})} := \|f\|_{C_{2\pi}} + \sup_{u>0} \|u^{-\alpha}\Delta^2_u f(x)\|_{C_{2\pi}} \qquad (0 < \alpha \le 2)$$

(see [10, pp. 370, 376]). Hence Theorem 4.2 implies that

$$F[C_{2\pi};K_n] = D(B)^{\sim C_{2\pi}} = \mathrm{Lip}_2(2;C_{2\pi}) ,$$

so that, by (4.6), $(K_n f)(x)$ satisfies, for $f \in F[C_{2\pi};K_n]$ and $n \in \mathbb{N}$, the optimal J-T-INQ:

$$\|K_n f - f\|_{C_{2\pi}} \le \frac{M}{(n+2)^2} \begin{cases} \|f\|_{C_{2\pi}} + \|f''\|_{L^\infty_{2\pi}} \\ \|f\|_{C_{2\pi}} + \sup_{u>0} \|u^{-2}\Delta_u^2 f(x)\|_{C_{2\pi}} \end{cases} .$$

The first of the two equivalent norms on the right-hand side is to be compared (except for the additive term $\|f\|_{C_{2\pi}}$) with (3.3) in case $j = 2$, having seminorm $\|f''\|_{C_{2\pi}}$; the second norm gives an estimate that avoids the derivatives of f. However, both could already be obtained by slightly modifying the proof of (3.3). Indeed, for $f \in Lip_2(2;X_{2\pi})$,

$$\|X_\rho f - f\|_{X_{2\pi}} \le \frac{\pi^2}{4} (1 - X_\rho^\wedge(1)) \sup_{u>0} \|u^{-2}\Delta_u^2 f(x)\|_{X_{2\pi}} . \tag{4.7}$$

But the singular integral of Fejér, for, e.g.,

$$X_{2\pi} = L^1_{2\pi} ,$$

gives an example of a process whose Favard class is an actual extension of $Lip_2(2;L^1_{2\pi})$. It is defined for $f \in X_{2\pi}$ by

$$(\sigma_n f)(x) := (F_n * f)(x) ,$$

where

$$F_n(x) = \frac{1}{n+1} \left[\frac{\sin((n+1)x/2)}{\sin(x/2)}\right]^2, \quad F_n^\wedge(1) = 1 - \frac{1}{n+1} \qquad (n \in \mathbb{N}) . \tag{4.8}$$

In particular, (4.7) holds for $(\sigma_n f)(x)$ if $f \in Lip_2(2;L^1_{2\pi})$. But, for the saturation class, it is known that $F[L^1_{2\pi};\sigma_n]$ is equal to the spaces (see [10, pp. 446, 451, 349, 376, 372])

$$\{f \in L^1_{2\pi}; f^{\sim\prime} \in BV_{2\pi}\} = \{f \in L^1_{2\pi}; f^\sim \in Lip_1(1;L^1_{2\pi})\}$$

with equivalent norms (see (4.9)). Here $BV_{2\pi}$ denotes the space of bounded Borel measures on the circle group, and $f^\sim$ the conjugate function of $f \in L^1_{2\pi}$ (see [10, p. 334]):

$$f^\sim(x) = \lim_{\varepsilon \to 0+} \frac{1}{\pi} \int_\varepsilon^\pi [f(x+u) - f(x-u)] \cot \frac{u}{2}\, du .$$

Therefore (4.6) gives, for $f \in F[L^1_{2\pi};\sigma_n]$ and $n \in \mathbb{N}$,

$$\|\sigma_n f - f\|_{L^1_{2\pi}} \le \frac{M}{n+1} \begin{cases} \|f\|_{L^1_{2\pi}} + \|f^{\sim\prime}\|_{BV_{2\pi}} \\ \|f\|_{L^1_{2\pi}} + \sup_{u>0} \|u^{-1}\Delta_u^1 f^\sim(x)\|_{L^1_{2\pi}} \end{cases} . \tag{4.9}$$

This is a definite improvement upon the above remark since

$$Lip_2(2;L^1_{2\pi}) \subset \{f \in L^1_{2\pi}; f^\sim \in Lip_1(1;L^1_{2\pi})\} ,$$

the inclusion being proper (see [54], [10]); and it justifies the concept of saturation theory in connection with J-T-INQ.

The same matter may also be considered for the integral $(W_t^{\kappa}f)(x)$ of (3.13) for $f \in L^p(\mathbb{R})$, $1 \le p < \infty$.

Let us finally consider one of the best-known processes in algebraic approximation, namely the Bernstein polynomials, defined for $f \in C[0,1]$ by

$$(4.10) \qquad (B_nf)(x) := \sum_{k=0}^{n} f\left(\frac{k}{n}\right)\binom{n}{k}x^k(1-x)^{n-k} \qquad (x \in [0,1];\ n \in \mathbb{P}) .$$

It is easily seen that they satisfy a "pointwise" J-T-INQ of the form (see [36, p. 102])

$$(4.11) \qquad |(B_nf)(x) - f(x)| \le \frac{x(1-x)}{2n} \begin{cases} \|f''\|_{L^\infty[0,1]} \\ \|f'\|_{Lip_1(1;C[0,1])} \end{cases} \qquad (n \in \mathbb{N}) .$$

This may be used to deduce a direct approximation theorem for $(B_nf)(x)$, namely,

$f \in Lip_2(\alpha;C[0,1]) \Rightarrow |(B_nf)(x) - f(x)| \le M\left[\frac{x(1-x)}{n}\right]^{\alpha/2}$, $0 < \alpha \le 2$; $x \in [0,1]$; $n \in \mathbb{N}$. It is also known (cf. [36], [5], [23]) that here the converse holds for any $0<\alpha\le 2$, the case $\alpha = 2$ giving the saturation theorem.

5. JACKSON-TYPE INEQUALITIES FOR MULTIPLIER OPERATORS

A further approach to J-T-INQ is delivered by the theory of multipliers for systems of orthogonal projections as developed in B-spaces by P.L. Butzer, E. Görlich, R.J. Nessel, and W. Trebels (cf. [11], [12], [25], [52], and the literature cited there). By this method, the results to be established are valid for summation processes of arbitrary orthogonal systems under relatively general assumptions. Although the matter can be generalized to locally convex spaces (cf. [32], [33]), let us set up a B-space version.

5.1. GENERAL THEORY. Given a Banach space X, let $\{P_k;\ k \in \mathbb{P}\} \subset [X]$ be a total sequence of mutually orthogonal projections on X, in short a system $\{P_k\}$, i.e.,

(i) $P_kf = 0$ for all $k \in \mathbb{P}$ implies $f = 0$ (total),

(ii) $P_jP_k = \delta_{jk}P_k$, δ_{jk} being the Kronecker symbol (mutually orthogonal).

Then to each $f \in X$ one may associate its unique Fourier-series expansion

$$(5.1) \qquad f \sim \sum_{k=0}^{\infty} P_kf \qquad (f \in X) .$$

The sequence $\{P_k\}$ is said to be fundamental if the set Π of all polynomials, i.e., the set of all finite linear combinations $\sum_{k=0}^{n} f_k$ with $f_k \in P_k(X)$, is dense in X.

With s the set of all sequences $\tau = \{\tau_k;\ k \in \mathbb{P}\}$ of scalars, $\tau \in s$ is called a <u>multiplier</u> for $f \in X$ (with respect to $\{P_k\}$), if for each $f \in X$ there exists an element $f^\tau \in X$ such that $P_k f^\tau = \tau_k P_k f$ for all $k \in \mathbb{P}$. Since $\{P_k\}$ is total on X, f^τ is uniquely determined by f. Denote the set of all multipliers τ for X by $M = M(X;\{P_k\})$. To each $\tau \in M$ there corresponds a closed, and by the closed-graph theorem, a bounded multiplier operator $T^\tau \in [X]$, defined by $T^\tau f = f^\tau$. (In general we do not distinguish between multipliers and the corresponding multiplier operators.) The set M with the natural vector operations, coordinate-wise multiplication, and norm

$$(5.2) \qquad \|\tau\|_M := \sup\{\|f^\tau\|_X;\ f \in X;\ \|f\|_X \leq 1\} = \|T^\tau\|_{[X]}$$

is a commutative Banach algebra, isometrically isomorphic to the subspace $[X]_M \subset [X]$ of multiplier operators on X. For any $\psi \in s$, let

$$(5.3) \quad X^\psi := \{f \in X;\ \text{there exists an } f^\psi \in X \text{ with } \psi_k P_k f = P_k f^\psi,\ \text{all } k \in \mathbb{P}\}.$$

Evidently $X^\psi \subset X$, and the linear operator $B^\psi : X^\psi \to X$, defined by $B^\psi f = f^\psi$ for $f \in X^\psi$, is closed for each $\psi \in s$. Furthermore, $P_k(X) \subset X^\psi$ for each $k \in \mathbb{P}$, so that B^ψ is densely defined if $\{P_k\}$ is fundamental on X. The space X^ψ becomes a normalized B-subspace of X under the norm

$$\|f\|_\psi := \|f\|_X + \|B^\psi f\|_X .$$

PROPOSITION 5.1. Let $\varphi(\rho)$ be a positive function on $\mathbb{A}$ with

$$\lim_{\rho \to \rho_0} \varphi(\rho) = 0 .$$

Let

$$\{T_\rho; \rho \in \mathbb{A}\} \subset [X]_M$$

be a linear approximation process of multiplier operators with associated multipliers

$$\{\tau(\rho);\ \rho \in \mathbb{A}\} \subset M,$$

and let $\psi \in s$. Furthermore, let $\{\lambda(\rho);\ \rho \in \mathbb{A}\} \subset s$ be given via the condition

$$(5.4) \qquad \varphi^{-1}(\rho)\{\tau_k(\rho) - 1\} = \psi_k \lambda_k(\rho) \qquad (k \in \mathbb{P}\,;\ \rho \in \mathbb{A}) .$$

If

$$\{\lambda(\rho)\} \subset M \quad \text{and} \quad \sup_{\rho \in \mathbb{A}} \|\lambda(\rho)\|_M \leq C ,$$

then there holds the J-T-INQ

$$(5.5) \qquad \varphi^{-1}(\rho)\|T_\rho f - f\|_X \leq \sup_{\rho \in \mathbb{A}} \|\lambda(\rho)\|_M \, \|B^\psi f\|_X \qquad (f \in X^\psi;\ \rho \in \mathbb{A}) .$$

Concerning the proof, the assumptions immediately imply

$$\varphi^{-1}(\rho)\{T_\rho f - f\} = L_\rho B^\psi f \qquad (f \in X^\psi;\ \rho \in \mathbb{A}) ,$$

where the family $\{L_\rho;\ \rho \in \mathbb{A}\} \subset [X]_M$ corresponding to $\{\lambda(\rho)\} \subset M$ is uniformly bounded in $\rho \in \mathbb{A}$.

Note that one may treat Bernstein, Zamansky, Bohr, and Nikolskii-type inequalities in a similar way (cf. [12], [24], [25], [41]).

In Section 3 it was seen that a Voronovskaja-type relation (3.16) on $D(B)$ implies the J-T-INQ (3.17). In case $\{T_\rho;\ \rho \in \mathbb{A}\}$ is a family of multiplier operators, (3.16) implies that moreover

$$(5.6) \qquad \lim_{\rho \to \rho_0} \varphi^{-1}(\rho)\{\tau_k(\rho) - 1\} = \psi_k \qquad (k \in \mathbb{P}) .$$

But now the converse is also true; i.e., a J-T-INQ (3.17) (for multiplier operators) with (5.6) implies (3.16), provided Π is dense in X^ψ. Indeed, by (5.6),

$$(5.7) \qquad \lim_{\rho \to \rho_0} \|\varphi^{-1}(\rho)\{T_\rho f_k - f_k\} - B^\psi f_k\|_X = 0$$

holds first for $f_k \in P_k(X) \subset X^\psi$ and each $k \in \mathbb{P}$, and so on Π. Since Π is dense in X^ψ by assumption, (5.7) together with (3.17) gives (3.16) by the Banach-Steinhaus theorem. Hence, a J-T-INQ (3.17) and (5.6) are necessary and sufficient conditions for a Voronovskaja-type relation to hold.

Let us now consider the saturation problem of Section 4 in the present multiplier frame. For this purpose, we begin with (5.4) and set

$$\mathbb{K} := \{k \in \mathbb{P};\ \tau_k(\rho) = 1 \ \text{ for all } \ \rho \in \mathbb{A}\} ,$$

supposing $\mathbb{K} \neq \mathbb{P}$. Then $\psi_k \neq 0$ for $k \notin \mathbb{K}$. If furthermore

$$\lim_{\rho \to \rho_0} \lambda_k(\rho) = 1 ,$$

this gives (5.6). Now Theorem 4.2 takes on the following simple form:

PROPOSITION 5.2. (a) Let $f \in X$, and

$$\{T_\rho;\ \rho \in \mathbb{A}\} \subset [X]_{\mathbf{M}}$$

be a process such that the associated sequence

$$\{\tau(\rho);\ \rho \in \mathbb{A}\} \subset \mathbf{M}$$

satisfies (5.6) for $k \in \mathbb{P}\backslash\mathbb{K}$. If

$$\|T_\rho f - f\|_X = o(\varphi(\rho)) \ ,\quad \rho \to \rho_0 \ ,$$

then

$$f \in \cup_{k\in\mathbb{K}} P_k(X) \quad \text{and} \quad T_\rho f = f$$

for all $\rho \in \mathbb{A}$.

(b) Let $\{P_k\}$ be fundamental. Let $\{\lambda(\rho);\ \rho \in \mathbb{A}\}$ be given via (5.4) with

$$\sup_{\rho \in \mathbb{A}} \|\lambda(\rho)\|_{\mathbf{M}} \leq C \ ,$$

and suppose in addition (5.6) to be satisfied for $k \in \mathbb{P}\backslash\mathbb{K}$. Then

$$F[X;T_\rho] = (X^\psi)^{\sim X}$$

with equivalent norms.

5.2. MULTIPLIER CRITERIA FOR J-T-INQ. With Proposition 5.1 and Proposition 5.2 on J-T-INQ thus established, the actual problem for the applications is to check whether a given sequence $\eta = \{\eta_k\} \in s$ belongs to $\mathbf{M}$, i.e., whether the assumptions given on $\{\tau(\rho)\}$ and $\{\lambda(\rho)\}$ in (5.4) are satisfied with respect to the orthogonal system $\{P_k;\ k \in \mathbb{P}\} \subset [X]$ in question. In this section, some criteria for subclasses of $\mathbf{M}(X;\{P_k\})$ are presented in terms of the uniform boundedness of the (C,κ)-means; these are just the classes $bv_{\kappa+1}$, well known in connection with the theory of divergent series.

Let the (C,κ)-means of (5.1) be defined for $f \in X$ and $\kappa \geq 0$ by

$$(5.8)\quad (C,\kappa)_n f := (A_n^\kappa)^{-1} \sum_{k=0}^{n} A_{n-k}^\kappa P_k f \ ;\ A_n^\kappa := \frac{\Gamma(n+\kappa+1)}{\Gamma(n+1)\Gamma(\kappa+1)} \qquad (n \in \mathbb{P}) .$$

$(C,\kappa)_n f$ coincides for $\kappa = 0$ with the partial sums $\sum_{k=0}^{n} P_k f$, for $\kappa = 1$ with their first arithmetic means.

C^κ-CONDITION. The pair $X,\{P_k\}$ is said to satisfy the C^κ-condition, $\kappa \geq 0$, if there exists a constant $C_\kappa > 0$ such that

$$(5.9)\qquad \|(C,\kappa)_n f\|_X \leq C_\kappa \|f\|_X \qquad (f \in X;\ n \in \mathbb{P}) .$$

If (5.9) is satisfied for a fixed $\kappa \geq 0$, then it holds for all $\beta > \kappa$; in particular, $\|(C,\beta)_n f\|_X \leq C_\kappa \|f\|_X$ for $f \in X$, $n \in \mathbb{P}$.

To derive the appropriate multiplier criterion, we introduce the (scalar-) sequence spaces $bv_{\kappa+1}$ as subspaces of ℓ^∞ (the set of bounded sequences) by

$$(5.10) \quad bv_{\kappa+1} := \{\eta \in \ell^\infty;\ \|\eta\|_{bv_{\kappa+1}} := \sum_{k=0}^{\infty} A_k^\kappa |\Delta^{\kappa+1}\eta_k| + \lim_{k\to\infty} |\eta_k| < \infty\} ,$$

where the (fractional) difference operator Δ^β is defined by

$$(5.11) \qquad \Delta^\beta \eta_k = \sum_{\ell=0}^{\infty} A_\ell^{-\beta-1}\, \eta_{k+\ell} .$$

The series (5.11) converges absolutely if $\eta \in \ell^\infty$ and $\beta \geq 0$. It follows that $\lim_{k\to\infty} \eta_k = \eta_\infty$ exists for $\eta \in bv_{\kappa+1}$, and $bv_{\kappa+1} \subset bv_{\gamma+1}$, $0 \leq \gamma < \kappa$, in the sense of continuous embedding. Furthermore, for each $\eta \in bv_{\kappa+1}$, $\kappa \geq 0$,

$$(5.12) \qquad \eta_n - \eta_\infty = \sum_{k=0}^{\infty} A_k^\kappa\, \Delta^{\kappa+1}\, \eta_{k+n} \qquad (n \in \mathbb{P}) .$$

For these foregoing fundamentals, see [52] and the literature cited there.

LEMMA 5.3. Let $X,\{P_k\}$ satisfy the C^κ-condition (5.9) for some $\kappa \geq 0$. Then every $\eta \in bv_{\kappa+1}$ is a multiplier; i.e., $bv_{\kappa+1} \subset M$ and

$$\|\eta\|_M \leq C_\kappa \|\eta\|_{bv_{\kappa+1}} \qquad (\eta \in bv_{\kappa+1}) .$$

As in [11], for $f \in X$ and $\eta \in bv_{\kappa+1}$, set

$$f^\eta := \sum_{k=0}^{\infty} A_k^\kappa \Delta^{\kappa+1}\eta_k (C,\kappa)_k f + \eta_\infty f .$$

Then, by (5.9) and (5.10), $f^\eta \in X$, and so, by (5.12),

$$P_n f^\eta = P_n f \left\{ \sum_{k=n}^{\infty} A_{k-n}^\kappa \Delta^{\kappa+1}\eta_k + \eta_\infty \right\} = \eta_n P_n f ,$$

proving the lemma. For this result, compare also [44].

To give some sufficient conditions for $\eta \in s$ to belong to $bv_{\kappa+1}$, we proceed as in [11; II] and [52]. For this purpose, we define the spaces BV_{j+1}, $j \in \mathbb{P}$. Let C_0 be the set of functions $e(x)$ uniformly continuous on $[0,\infty)$ with $\lim_{x\to\infty} e(x) = 0$, and $AC_{loc}(0,\infty)$ or $BV_{loc}(0,\infty)$ the set of functions that are locally absolutely continuous or locally of bounded variation on $(0,\infty)$. Then

$$(5.13) \quad BV_{j+1} := \{e \in C_0; e,\ldots,e^{(j-1)} \in AC_{loc}(0,\infty), e^{(j)} \in BV_{loc}(0,\infty)\} ,$$

and

$$\|e\|_{BV_{j+1}} := \sup_{x\geq 0} |e(x)| + \frac{1}{\Gamma(j+1)}\int_0^\infty x^j |de^{(j)}(x)| < \infty .$$

The spaces $BV_{\kappa+1}$ for arbitrary $\kappa \geq 0$ can be defined analogously using the concept of fractional derivatives (cf. [12]). Then $BV_{\kappa+1} \subset BV_{\mu+1}$ for $0 \leq \mu < \kappa$. As a consequence, one has the following criterion concerning uniformly bounded multipliers depending on the parameter $\rho \in \mathbb{A}$. For simplicity, take $\mathbb{A} = (0,\infty)$ with $\rho_0 = \infty$.

LEMMA 5.4. Let the pair $X, \{P_k\}$ satisfy the C^κ-condition (5.9) for some $\kappa \geq 0$. Let $\{\eta(\rho)\} \subset \mathbf{s}$ be such that there exists $\{e(\cdot;\rho)\} \subset BV_{\kappa+1}$ and $\eta_k(\rho) = e(k;\rho)$ for each $k \in \mathbb{P}$ and $\rho \in \mathbb{A}$. Then

$$\|\eta(\rho)\|_M \leq C_\kappa \sum_{k=0}^\infty A_k^\kappa |\Delta^{\kappa+1}\eta_k(\rho)| \leq C_\kappa^* \int_0^\infty x^\kappa |de^{(\kappa)}(x;\rho)| .$$

In particular, if $e(x) \in BV_{\kappa+1}$ and $e(x;\rho)$ is of Fejér's type, i.e., $e(x;\rho) = e(x/\rho)$, then $\{\eta(\rho)\}$ is a family of multipliers uniformly bounded in $\rho \in \mathbb{A}$.

Concerning the applications to trigonometric Fourier expansions, the criterion delivers only radial multipliers. But this handicap can be avoided by proceeding as in [12] or [40].

5.3. APPLICATIONS TO VARIOUS APPROXIMATION PROCESSES. To illustrate the actual efficiency of the multiplier theory sketched, let us verify J-T-INQ for some families of multiplier operators $\{T_\rho;\ \rho \in \mathbb{A}\}$. At first, approximation with saturation order $O(\varphi(\rho))$ is considered for particular choices of $\varphi(\rho)$ and $\{\psi_k;\ k \in \mathbb{P}\}$. On the other hand, Proposition 5.1 also yields J-T-INQ with nonoptimal ("weaker") approximation order for other (larger) subspaces than the saturation class. For this, an example is given in Corollary 5.6. Let us take the following means:

<u>Picard means</u> of order $\nu > 0$ of the Fourier series (5.1) of $f \in X$:

$$C_\rho^\nu f \sim \sum_{k=0}^\infty c_\nu(k/\rho)P_k f \ ; \quad c_\nu(x) = (1 + x^\nu)^{-1} \quad (x \geq 0,\ \ \nu > 0) . \tag{5.14}$$

Using (5.4) with $\tau_k(\rho) = c_\nu(k/\rho)$, $\varphi^{-1}(\rho) = \rho^\nu$, $\psi_k = k^\nu$, and $\lambda_k(\rho) = c_\nu(k/\rho)$ given via (5.4), one has $c_\nu(x) \in BV_{j+1}$, $j \in \mathbb{P}$.

<u>Bessel potentials</u> of order $\mu > 0$:

$$B_\rho^\mu f \sim \sum_{k=0}^\infty b_\mu(k/\rho)P_k f \ ; \quad b_\mu(x) = (1 + x^2)^{-\mu} \qquad (x \geq 0) . \tag{5.15}$$

Here $b_\mu(x) \in BV_{j+1}$, $j \in \mathbb{P}$, $\ell(x) := x^{-\gamma}(b_\mu(x) - 1) \in BV_{j+1}$, $j \in \mathbb{P}$ and $0 < \gamma \leq 2$, and

$$\lambda_k(\rho) = \ell(k/\rho), \quad \psi_k = k^\gamma, \quad \text{and} \quad \varphi(\rho) = \rho^{-\gamma} .$$

Abel-Cartwright means of order $\sigma > 0$:

$$(5.16) \qquad W_t^\sigma f \sim \sum_{k=0}^{\infty} w_\sigma(kt)P_k f; \; w_\sigma(x) = \exp(-x^\sigma) \qquad (x \geq 0; \; t \to 0+) ,$$

where $w_\sigma(x)$, $\ell(x) := x^{-\gamma}(w_\sigma(x) - 1) \in BV_{j+1}$, $j \in \mathbb{P}$, $j \geq \kappa$ and $0<\gamma\leq\sigma$, and $\lambda_k(t) = \ell(kt)$, $\psi_k = k^\gamma$, and $\varphi(t) = t^\gamma$.

Riesz means of order $\alpha, \lambda > 0$:

$$(5.17) \qquad R_\rho^{\alpha,\lambda} f \sim \sum_{k<\rho} r_{\alpha,\lambda}(k/\rho)P_k f; \; r_{\alpha,\lambda}(x) = \begin{cases} (1 - x^\alpha)^\lambda & 0 \leq x \leq 1 \\ 0 & x \geq 1 , \end{cases}$$

where $r_{\alpha,\lambda}(x)$ belongs to $BV_{\kappa+1}$ for $\kappa < \lambda < \infty$ provided $\kappa \notin \mathbb{N}$, and for $\kappa \leq \lambda$ if $\kappa \in \mathbb{N}$ and $\alpha > 0$ arbitrary;

$$\ell(x) := x^{-\gamma}\{r_{\alpha,\lambda}(x) - 1) \in BV_{\kappa+1} \quad \text{if } \alpha, \kappa, \lambda \text{ as above and } 0 < \gamma \leq \alpha ,$$

and

$$\lambda_k(\rho) = \ell(k/\rho), \quad \psi_k = k^\gamma, \quad \text{and} \quad \varphi(\rho) = \rho^{-\gamma} .$$

As a typical result, let us state a J-T-INQ for the Riesz means:

COROLLARY 5.5. Let $X, \{P_k\}$ satisfy (5.9) for some $\kappa \geq 0$. Under the above parameter restrictions, there holds the J-T-INQ, for $0 < \gamma \leq \alpha$,

$$\|R_\rho^{\alpha,\lambda} f - f\|_X \leq C_\kappa \|\ell\|_{BV_{\kappa+1}} \rho^{-\gamma} \|B^\psi f\|_X \qquad (f \in X^\psi) ,$$

$\rho^{-\alpha}$ being the saturation order on $(X^\psi)^{\sim X}$ if $\gamma = \alpha$.

For a further example of a nonoptimal J-T-INQ, consider the Abel-Cartwright means in (5.16) for $\sigma = 2$, and choose

$$\varphi(t) = [\log(1 + t^{-1})]^{-1}, \quad \psi_k = 1 + \log(1 + k^2)^{1/2};$$

this gives:

COROLLARY 5.6. Let $X, \{P_k\}$ satisfy the C^κ-condition for $\kappa = 1$. Then the Abel-Cartwright means $\{W_t^2 f; \; t \geq 0\}$ satisfy the (nonoptimal) J-T-INQ

$$\|W_t^2 f - f\|_X \le C_1[\log(1 + t^{-1})]^{-1} \|B^\psi f\|_X \qquad (f \in X^\psi) .$$

Note that the optimal J-T-INQ holds with $\varphi(t) = t^2$ on $(X^\psi)^{\sim X}$, where $\psi_k = k^2$. The above result follows in view of the fact that

$$\lambda_t(x) = \log(1 + t^{-1})\{\exp(-x^2t^2) - 1\}\{1 + \log(1 + x^2)^{1/2}\}^{-1} \qquad (t > 0)$$

belongs to BV_2.

5.4. APPLICATIONS TO ORTHOGONAL EXPANSIONS. Now we present some examples of projections $\{P_k;\ k \in \mathbb{P}\}$ that are defined via classical orthogonal expansions in B-spaces, and that satisfy a C^κ-condition (5.9) for $\kappa > 0$, but generally not for $\kappa = 0$.

Hermite series. Given the B-space $L^p(\mathbb{R})$, $1 \le p < \infty$, and the Hermite polynomials

$$H_k(x) := (-1)^k \exp(x^2)(d/dx)^k \exp(-x^2) , \quad k \in \mathbb{P},$$

the functions

$$\mathcal{H}_k(x) := \exp(-x^2/2)(2^k k!\, \sqrt{\pi})^{-1/2}\, H_k(x)$$

form an orthonormal family on $\mathbb{R}$. Thus the projections

$$P_k f(x) := \left[\int_{-\infty}^{\infty} f(y)\, \mathcal{H}_k(y)dy\right] \mathcal{H}_k(x) \qquad (k \in \mathbb{P}) \tag{5.18}$$

are mutually orthogonal, furthermore total and fundamental on $L^p(\mathbb{R})$, $1 \le p < \infty$. The C^κ-condition (5.9) for these $\{P_k;\ k \in \mathbb{P}\}$ is satisfied with $\kappa = 1$ if $1 \le p < \infty$ and even with $\kappa = 0$ if $4/3 < p < 4$ (cf. [38], [45]).

Note that $\mathcal{H}_k(x)$, $k \in \mathbb{P}$, is an eigenfunction of the operator $\mathfrak{R} := (d/dx)^2 + 1 - x^2$ with corresponding eigenvalue $-2k$.

COROLLARY 5.7. For the Abel-Cartwright means $\{W_t^\sigma f\}$ in (5.16) with $\sigma > 0$ and $\gamma = 1$, there holds the J-T-INQ

$$\left\| \sum_{k=0}^{\infty} \exp(-(kt)^\sigma)P_k f - f\right\|_p \le C_1\|\ell\|_{BV_2}\, \rho^{-\sigma} \left\|\sum_{k=0}^{\infty} k^\sigma P_k f\right\|_p \qquad (f \in X^\psi) ,$$

the second norm on the right-hand side being $\|B^\psi g\|_p = \|\mathfrak{R}^r f\|_p$ for $\sigma = r \in \mathbb{N}$.

In particular, for $\sigma = 1$ on the left-hand side (with $e^{-t} = v \to 1-$) (cf. [28; p. 572]), one has

$$\sum_{k=0}^{\infty} v^k P_k f(x) = \int_{-\infty}^{\infty} f(y)\{\pi(1 - v^2)\}^{-1/2} \exp\left(- \frac{v^2x^2 - 2xvy + y^2}{1 - v^2}\right) dy ,$$

hence, with a constant $M > 0$, the J-T-INQ

$$\left\|[\pi(1-v^2)]^{-1/2}\int_{-\infty}^{\infty} f(y)\exp\left(-\frac{v^2x^2-2xvy+y^2}{1-v^2}\right)dy - f(x)\right\|_p \leq M(1-v)\left\|\left|\left(\frac{d}{dx}\right)^2+1-x^2\right|f(x)\right\|_p .$$

Jacobi polynomials. Let X be either the B-space $C[-1,1]$ or $L^p_{(\alpha,\beta)}(-1,1)$ of p-th power Lebesgue integrable functions on $(-1,1)$ with respect to the weight

$$w_{\alpha,\beta}(x) := (1-x)^{\alpha}(1+x)^{\beta}, \quad \alpha,\beta > -1,$$

and norm

$$\|f\|_p := \left\{\int_{-1}^{1} |f(x)|^p \, w_{\alpha,\beta}(x)dx\right\}^{1/p}; \quad 1 \leq p < \infty .$$

The Jacobi polynomials $\varphi_k^{(\alpha,\beta)}(x)$ of degree k and order (α,β) are defined by

$$\varphi_k^{(\alpha,\beta)}(x) := (1-x)^{-\alpha}(1+x)^{-\beta}(2^k k!)^{-1}(-d/dx)^k\{(1-x)^{k+\alpha}(1+x)^{k+\beta}\} .$$

These are orthogonal on $(-1,1)$ with respect to $w_{\alpha,\beta}(x)$, and

$$\int_{-1}^{1}[\varphi_k^{(\alpha,\beta)}(x)]^2 w_{\alpha,\beta}(x)dx = \frac{2^{\alpha+\beta+1}\Gamma(k+\alpha+1)\Gamma(k+\beta+1)}{(2k+\alpha+\beta+1)\Gamma(k+\alpha+\beta+1)\Gamma(k+1)} := [h_k^{\alpha,\beta}]^{-1}$$

for $k \in \mathbb{P}$. It follows that the projections $(k \in \mathbb{P})$

$$(5.19) \quad P_k^{(\alpha,\beta)}f(x) := \left[\int_{-1}^{1} f(y)\varphi_k^{(\alpha,\beta)}(y)w_{\alpha,\beta}(y)\,dy\right] h_k^{\alpha,\beta}\,\varphi_k^{(\alpha,\beta)}(x)$$

are mutually orthogonal, total, and fundamental in X. The C^{κ}-condition (5.9) is satisfied for $\kappa = 0$ if (cf. [39])

$$(\alpha+1)\left|\frac{1}{p}-\frac{1}{2}\right| < \min\left(\frac{1}{4}, \frac{1+\alpha}{2}\right) \quad \text{and} \quad (\beta+1)\left|\frac{1}{p}-\frac{1}{2}\right| < \min\left(\frac{1}{4}, \frac{1+\beta}{2}\right) .$$

Furthermore, (5.9) is satisfied if (cf. [2], [52])

$$\kappa > (\alpha+\tfrac{1}{2})\left|1-\tfrac{2}{p}\right|, \quad \alpha \geq \beta > -1 \quad \text{and} \quad \alpha+\beta \geq 1, \quad 1 \leq p < \infty .$$

Thus in these cases the multiplier criteria can be used. Note that $\varphi^{(\alpha,\beta)}(x)$ is an eigenfunction of the operator

$$R^{\alpha,\beta} := (1-x^2)(d/dx)^2 + \{\beta-\alpha-x(\alpha+\beta+2)\}(d/dx)$$

with eigenvalue $-k(k+\alpha+\beta+1)$.

Again Corollary 5.5 may be applied. In order to concretize the operator B^{ψ}, let us consider approximation by the Bessel potentials with an appropriate modified kernel, i.e.,

$$B_\rho^{\mu,\alpha,\beta} f \sim \sum_{k=0}^{\infty} \left(1 + \frac{k(k+\alpha+\beta+1)}{\rho(\rho+\alpha+\beta+1)}\right)^{-\mu} P_k^{(\alpha,\beta)} f ,$$

which gives (5.15) in case $\alpha + \beta = -1$. Then one may show that this kernel is an element of bv_{j+1}, $j \in \mathbb{P}$, and so is $\lambda_k(\rho)$, given via (5.4) with $\varphi^{-1}(\rho) = \rho(\rho + \alpha + \beta + 1)$ and $\psi_k = k(k + \alpha + \beta + 1)$, $k \in \mathbb{P}$.

COROLLARY 5.8. Given (5.9) for some $\kappa \geq 0$; then under the above choices of $\varphi(\rho)$ and $\{\psi_k;\ k \in \mathbb{P}\}$ there holds on X the J-T-INQ

$$\|B_\rho^{\mu,\alpha,\beta} f - f\|_X \leq C_\kappa \|\lambda(\rho)\|_{bv_{\kappa+1}} [\rho(\rho + \alpha + \beta + 1)]^{-1} \|\mathfrak{R}^{\alpha,\beta} f\|_X \quad (f \in X^\psi;\ \rho > 0).$$

Trigonometric Fourier series in $\mathbb{R}^m$. Let $\mathbb{R}^m$ be the Euclidean m-space with elements r, $x = (x_1, \ldots, x_m)$, inner product

$$\langle r,x\rangle = \sum_{j=1}^{m} r_j x_j , \quad \langle r,r\rangle := |r|^2 ,$$

$\mathbb{Z}^m$ the set of all integer lattice points $r \in \mathbb{R}^m$, and

$$\mathbb{T}^m := \{x \in \mathbb{R}^m ;\ |x_j| \leq \pi,\ 1 \leq j \leq m\}$$

the m-dimensional torus. On $X(\mathbb{T}^m)$, equal to $C(\mathbb{T}^m)$ or $L^p(\mathbb{T}^m)$, $1 \leq p < \infty$, with usual norms, one defines projections $\{P_k;\ k \in \mathbb{P}\}$ by

$$\text{(5.20)} \qquad \begin{aligned} P_k f(x) &:= \sum_{|r|^2=k} f^\wedge(r) e^{i\langle r,x\rangle} && (k \in \mathbb{P}), \\ f^\wedge(r) &:= (2\pi)^{-m} \int_{\mathbb{T}^m} f(x) e^{-i\langle r,x\rangle}\, dx && (r \in \mathbb{Z}^m) \end{aligned}$$

being the r-th Fourier coefficient. Then the system $\{P_k\}$ is mutually orthogonal, total, and fundamental on $X(\mathbb{T}^m)$. The C^κ-condition (5.9) is satisfied in $C(\mathbb{T}^m)$ for $\kappa > (m-1)/2$ and in $L^p(\mathbb{T}^m)$, $1 \leq p < \infty$, for $\kappa > (m-1)|1/p - 1/2|$.

COROLLARY 5.9. Given (5.9) for some $\kappa \geq 0$; then there holds on $X(\mathbb{T}^m)$ the J-T-INQ, with constant $M > 0$,

$$\left\| \sum_{r\in\mathbb{Z}^m} e^{-t|r|^2} f^\wedge(r) e^{i\langle r,x\rangle} - f(x) \right\|_X \leq Mt \|\Delta f(x)\|_X ,$$

where $\Delta = \sum_{j=1}^m (\partial^2/\partial^2 x_j)$ is the standard Laplacian.

Note that the theory sketched so far works particularly in case $p = 1$; this is an especially interesting case, as it is not covered by the theory of multipliers in terms of Markinciewicz-type criteria (cf. [20]).

Further examples of systems of orthogonal projections which satisfy a C^{κ}-condition (5.9) are listed for B-spaces in [25], [52], and for some locally convex spaces in [31], [32], [33]. They include summation processes of Laguerre expansions, Bessel series, and Haar series, hence giving J-T-INQ for processes of multiplier operators with respect to these systems. For another criterion concerning systems $\{P_k;\ k \in \mathbb{P}\}$ of Abel-bounded expansions, see [11; III] and [34].

REFERENCES

1. N. I. Achieser, Vorlesungen über Approximationstheorie, Akademie Verlag, Berlin, 1967.

2. R. Askey and S. Wainger, A convolution structure for Jacobi series, Amer. J. Math. 91 (1969), 463-485.

3. E. F. Beckenbach and R. Bellman, Inequalities, 3rd revised printing, Springer, New York, 1971.

4. H. Berens, Interpolationsmethoden zur Behandlung von Approximations-prozessen auf Banachräumen, Lecture notes in mathematics 64, Springer, Berlin, 1968.

5. H. Berens and G. G. Lorentz, [illegible]erse theorems for Bernstein polynomials, Indiana Univ. Math. J. 21 (19[illegible]), 693-708.

6. P. L. Butzer and H. Berens, Semi-groups of operators and approximation, Springer, New York, 1967.

7. P. L. Butzer, H. Dyckhoff, E. Görlich, and R. L. Stens, Best trigonometric approximation, fractional order derivatives and Lipschitz classes (to appear).

8. P. L. Butzer and E. Görlich, Saturationsklassen und asymptotische Eigenschaften trigonometrischer singulärer Integrale, in: Festschrift zur Gedächnisfeier Karl Weierstrass 1815-1965, H. Behnke and K. Kopfermann (eds.), 339-392, Wiss. Abh. Arbeitsgemeinschaft für Forschung des Landes Nordrhein Westfalen 33, Westdeutscher Verlag, Köln-Opladen, 1966.

9. P. L. Butzer and J. Junggeburth, A review of Jackson-type inequalities for best approximation and for linear approximation processes (in preparation).

10. P. L. Butzer and R. J. Nessel, Fourier analysis and approximation, Vol. I., Birkhäuser and Academic Press, Basel and New York, 1971.

11. P. L. Butzer, R. J. Nessel, and W. Trebels, On summation processes of Fourier expansions in Banach spaces, I: Comparison theorems; II: Saturation theorems; III: Jackson- and Zamansky-type inequalities for Abel-bounded expansions, Tôhoku Math. J. 24 (1972), 127-140; 551-569; 27 (1975), 213-223.

12. P. L. Butzer, R. J. Nessel, and W. Trebels, Multipliers with respect to spectral measures in Banach spaces and approximation, I: Radial multipliers in connection with Riesz-bounded spectral measures; II: One

dimensional Fourier multipliers, J. Approximation Theory 8 (1973), 335-356; 14 (1973), 23-29.

13. P. L. Butzer and K. Scherer, Über die Fundamentalsätze der klassischen Approximationstheorie in abstrakten Räumen, in: Abstract spaces and approximation, P. L. Butzer and B. Sz.-Nagy (eds.), 113-125, ISNM 10, Birkhäuser, Basel, 1969.

14. P. L. Butzer and K. Scherer, On the fundamental approximation theorems of D. Jackson, S. N. Bernstein and theorems of M. Zamansky and S. B. Steckin, Aequationes Math. 3 (1969), 170-185.

15. P. L. Butzer and K. Scherer, On fundamental theorems of approximation theory and their dual versions, J. Approximation Theory 3 (1970), 87-100.

16. P. L. Butzer and K. Scherer, Approximation theorems for sequences of commutative operators in Banach spaces, in: Proc. Conf. Constructive Function Theory, Varna (Bulgaria), 137-146 (1970).

17. P. L. Butzer and K. Scherer, Jackson- and Bernstein-type inequalities for families of commutative operators in Banach spaces, J. Approximation Theory 5 (1972), 308-342.

18. P. L. Butzer and U. Westphal, An access to fractional differentiation via fractional difference quotients, in: Fractional calculus and its applications (Proc. Conf., New Haven, 1974), Lecture notes in mathematics 457, 116-145, Springer Verlag, Berlin, 1975.

19. E. W. Cheney, Introduction to approximation theory, McGraw-Hill, New York, 1966.

20. W. C. Connett and A. L. Schwartz, A multiplier theorem for ultraspherical series, Studia Math. 51 (1974), 51-70.

21. W. Dahmen, Asymptotisch optimale Approximation für klassische und exponentielle Konvergenzgeschwindigkeiten, Dissertation, RWTH Aachen,1976.

22. R. A. DeVore, The approximation of continuous functions by positive, linear operators, Lecture notes in mathematics 293, Springer, Berlin,1972.

23. R. A. DeVore, Degree of approximation, in: Approximation theory II, G. G. Lorentz, C. K. Chui, and L. L. Schumacher (eds.), 117-161, Academic Press, New York, 1976.

24. E. Görlich, Bohr-type inequalities for Fourier expansions in Banach spaces, in: Approximation theory, G. G. Lorentz (ed.), 359-363, Academic Press, New York, 1973.

25. E. Görlich, R. J. Nessel, and W. Trebels, Bernstein-type inequalities for families of multiplier operators in Banach spaces with Cesàro decompositions, I: General theory; II: Applications, Acta Sci. Math. (Szeged) 34 (1973), 121-130; 36 (1974), 39-48.

26. E. Görlich and G. Paulus, A review of Bernstein and Nikolskii-type inequalities (in preparation).

27. G. H. Hardy, J. E. Littlewood, and G. Pólya, Inequalities, Cambridge Univ. Press, London and New York, 1952.

28. E. Hille and R. S. Phillips, Functional analysis and semi-groups (revised edition), Amer. Math. Soc. Colloq. Publ. Vol. 31, Providence, Rhode Island, 1957.

29. D. Jackson, The theory of approximation, Amer. Math. Soc. Colloq. Publ. Vol. 11, New York, 1930.

30. D. Jackson, Über die Genauigkeit der Annäherung stetiger Funktionen durch ganze rationale Funktionen gegebenen Grades und trigonometrische Summen gegebener Ordnung, Dissertation, Universität Göttingen, 1911.

31. J. Junggeburth, Multiplikatorkriterien in lokalkonvexen Räumen mit Anwendungen auf Orthogonalentwicklungen in Gewichtsräumen, Dissertation, RWTH Aachen, 1975.

32. J. Junggeburth, Multipliers for (C,κ)-bounded Fourier expansions in weighted locally convex spaces and approximation, Rev. Un. Mat. Argentina 27 (1975), 127-146.

33. J. Junggeburth and R. J. Nessel, Approximation by families of multipliers for (C,α)-bounded Fourier expansions in locally convex spaces, I: Order-preserving operators, J. Approximation Theory 13 (1975), 167-177.

34. J. Junggeburth and R. J. Nessel, Multipliers with respect to Abel-bounded spectral measures in locally convex spaces, Rev. Un. Mat. Argentina (in print).

35. J. Junggeburth, K. Scherer, and W. Trebels, Zur besten Approximation auf Banachräumen mit Anwendungen auf ganze Funktionen, Forschungsberichte des Landes Nordrhein Westfalen, Nr. 2311, 51-75, Westdeutscher Verlag, Opladen, 1973.

36. G. G. Lorentz, Approximation of functions, Holt, Rinehart and Winston, New York, 1966.

37. D. S. Mitrinović, Analytic inequalities, Springer, New York, 1970.

38. B. Muckenhoupt, Mean convergence of Hermite and Laguerre series, Trans. Amer. Math. Soc. 147 (1970), I: 419-431; II: 433-460.

39. B. Muckenhoupt, Mean convergence of Jacobi series, Proc. Amer. Math. Soc. 23 (1969), 306-310.

40. R. J. Nessel and G. Wilmes, A multiplier criterion in Euclidean n-space with applications to Bernstein inequalities, Abh. Math. Sem. Univ. Hamburg. 44 (1975), 143-151.

41. R. J. Nessel and G. Wilmes, On Nikolskii-type inequalities for orthogonal expansions, in: Approximation theory II, G. G. Lorentz, C. K. Chui, and L. L. Schumacher (eds.), 479-484, Academic Press, New York, 1976.

42. D. J. Newman and H. S. Shapiro, Jackson's theorem in higher dimensions, in: On approximation theory, P. L. Butzer and J. Korevaar (eds.), 208-217, ISNM 5, Birkhäuser, Basel, 1964.

43. S. M. Nikolskii, Inequalities for entire functions of finite degree and their application to the theory of differentiable functions of several variables, Amer. Math. Soc. Transl. Ser. 2, 80 (1969), 1-38, (Trudy Mat. Inst. Steklov 38 (1951), 244-278).

44. W. Orlicz, Über k-fach monotone Folgen, Studia Math. 6 (1936), 149-159.

45. E. L. Poiani, Mean Cesàro summability of Laguerre and Hermite series, Trans. Amer. Math. Soc. 173 (1972), 1-31.

46. H. S. Shapiro, Topics in approximation theory, Lecture notes in mathematics 187, Springer, Berlin, 1971.

47. H. S. Shapiro, Approximation by trigonometric polynomials to periodic functions of several variables, in: Abstract spaces and approximation, P. O. Butzer and B. Sz.-Nagy (eds.), 203-217, ISNM 10, Birkhäuser, Basel, 1969.

48. O. Shisha, Inequalities, Vol. I., II., III., Academic Press, New York and London, 1967, 1970, 1972 (Proc. of the Symposiums held at Wright-Patterson Air Force Base, Ohio, in 1965, 1967, and at the University of California, Los Angeles, in 1969).

49. E. L. Stark, Inequalities for trigonometric moments and for Fourier coefficients of positive cosine polynomials in approximation (in print).

50. S. A. Teljakowskii, Two theorems on the approximation of functions by algebraic polynomials, Amer. Math. Soc. Transl. 77 (1968), 162-178.

51. A. F. Timan, Theory of approximation of functions of a real variable, Pergamon Press, Oxford, 1963.

52. W. Trebels, Multipliers for (C,α)-bounded Fourier expansions in Banach spaces and approximation theory, Lecture notes in mathematics 329, Springer, Berlin, 1973.

53. U. Westphal, An approach to fractional powers of operators via fractional differences, Proc. London Math. Soc. 29 (1974), 557-576.

54. A. Zygmund, Trigonometric series, Vol. I, Cambridge Univ. Press, Cambridge, 1968.

ON LIPSCHITZ CONDITION AND ZYGMUND'S PROPERTY FOR FUNCTIONS OF SEVERAL VARIABLES

O. Shisha
Department of Mathematics
University of Rhode Island
Kingston, Rhode Island 02881
U.S.A.

G. R. Verma
Department of Mathematics
University of Rhode Island
Kingston, Rhode Island 02881
U.S.A.

ABSTRACT. The literature contains results on the relation between the smoothness of a real function f of several real variables and the degree of approximation to f by splines. The purpose of the present note is to study two elementary cases and to derive the corresponding results by simple methods, allowing the range of f to lie in a general space.

1. DEFINITIONS AND NOTATION

Let d be a positive integer; let C denote the (open) unit cube in (real) Euclidean d-space $\mathbb{R}^d$:

$$C = \{(x_1,x_2,\ldots,x_d) : 0 < x_j < 1, \quad j = 1,2,\ldots,d\} ;$$

and let $\overline{C}$ denote the closure of C. Given positive integers $n,k_1,k_2,\ldots,k_d$, with each $k_j \leq n$, consider

$$C^{(n)}_{k_1,k_2,\ldots,k_d} = \{(x_1,x_2,\ldots,x_d) : \frac{k_j - 1}{n} < x_j < \frac{k_j}{n}, \quad j = 1,2,\ldots,d\} ,$$

a cube in $\mathbb{R}^d$.

Let f be a mapping of $S \subseteq \mathbb{R}^d$ into a metric space with metric ρ. Then f is said to satisfy in S a <u>Lipschitz condition of order</u> α, where $\alpha > 0$, if and only if there exists a constant L such that

$$\rho(f(x),f(y)) \leq L\|x - y\|^{\alpha} \tag{1}$$

whenever $x,y \in S$. If $s \in S$, if the line segment sx lies in S whenever $x \in S$, and if f satisfies in S a Lipschitz condition of order $\alpha > 1$, then f is constant in S.

A mapping f of a convex $S \subseteq \mathbb{R}^d$ into a (real or complex) normed linear space L is said to have <u>Zygmund's property in</u> S if and only if f is continuous there and there exists a constant A such that

(2) $$\|f(x) - 2f(x + h) + f(x + 2h)\|_L \leq A\|h\|_{\mathbb{R}^d}$$

whenever $x, x + 2h \in S$.

2. LIPSCHITZ CONDITION

We start with the following result.

THEOREM 1. Let f be a mapping of C into a metric space M with metric ρ, and let $\alpha > 0$. A necessary and sufficient condition for f to satisfy in C a Lipschitz condition of order α is:

(*) existence, for $n = 1,2,\ldots,$ of a mapping f_n of

$$\bigcup_{\substack{1 \leq k_j \leq n \\ j=1,2,\ldots,d}} c^{(n)}_{k_1,k_2,\ldots,k_d}$$

into M, which throughout each $c^{(n)}_{k_1,k_2,\ldots,k_d}$ is constant and satisfies

$$\rho(f(x), f_n(x)) \leq \frac{B}{n^\alpha},$$

B being a constant.

REMARKS. Assume the first sentence of Theorem 1.

(i) Suppose that M is complete, and that f satisfies in C a Lipschitz condition of order $\alpha > 0$. Then f is uniformly continuous in C and hence can be extended to $\overline{C}$ in such a way as to be continuous there [1, Corollary 8.12, p. 134]; it will clearly satisfy in $\overline{C}$ a Lipschitz condition of order α.

(ii) Let $\alpha > 1$ and assume (*). Then f satisfies in C a Lipschitz condition of order α, and hence is constant in C.

Proof of Theorem 1. To prove necessity, suppose that (1) holds throughout C, L being a constant. For $n = 1,2,\ldots,$ define f_n on each

$$c = c^{(n)}_{k_1,k_2,\ldots,k_d}$$

to be the value of f at the center of c; then for each $x \in c$,

$$\rho(f(x), f_n(x)) \leq L(\sqrt{d}/2)^\alpha/n^\alpha .$$

To prove sufficiency, let x and y $(\neq x)$ be points of C, and let n_0 be the greatest positive integer n for which x and y lie in some (same) $c^{(n)}_{k_1,k_2,\dots,k_d}$.

For $j = 1,2,\dots,$ let p_j be the jth prime.

We shall show that

$$\|x - y\| \geq (p_d p_{d+1} n_0)^{-1} . \tag{3}$$

From (3) we then obtain:

$$\rho(f(x),f(y)) \leq \rho(f(x),f_{n_0}(x)) + \rho(f_{n_0}(y),f(y)) \leq 2Bn_0^{-\alpha}$$
$$\leq 2B(p_d p_{d+1})^{\alpha} \|x - y\|^{\alpha} .$$

Suppose (3) is false. Let $x = (x_1,x_2,\dots,x_d)$ and $y = (y_1,y_2,\dots,y_d)$ lie in

$$c^{(n_0)}_{\tilde{k}_1,\tilde{k}_2,\dots,\tilde{k}_d} ,$$

and let s_j denote the convex hull of $\{x_j,y_j\}$, $j = 1,2,\dots,d$. If there were an integer m, $1 \leq m \leq d$, such that each s_j is contained in one of the intervals

$$((\tilde{k}_j - 1)n_0^{-1} + (r-1)(n_0 p_m)^{-1},\ (\tilde{k}_j - 1)n_0^{-1} + r(n_0 p_m)^{-1}), \quad r = 1,2,\dots,p_m ,$$

then we would reach a contradiction to the maximality of n_0. Thus for each integer m, $1 \leq m \leq d$, there is an integer j_m, $1 \leq j_m \leq d$, for which s_{j_m} contains a point

$$z_m = (\tilde{k}_{j_m} - 1)n_0^{-1} + r_m(n_0 p_m)^{-1}, \quad r_m \text{ a positive integer} < p_m.$$

Observe that if $1 \leq m_1 < m_2 \leq d$, then

$$j_{m_1} \neq j_{m_2} .$$

Let j be an integer, $1 \leq j \leq d$. Then for some m, $1 \leq m \leq d$, we have $j = j_m$, and so $z_m \in s_j$. The nearest point to z_m of the form

$$(\tilde{k}_j - 1)n_0^{-1} + r(n_0 p_{d+1})^{-1} , \quad r \text{ a positive integer } < p_{d+1} ,$$

is at a

$$\text{distance} \geq (n_0 p_m p_{d+1})^{-1} \geq (n_0 p_d p_{d+1})^{-1} > \text{the length of } s_j ,$$

from z_m. Hence s_j lies in one of the intervals

$$((\tilde{k}_j - 1)n_0^{-1} + (r-1)(n_0 p_{d+1})^{-1}, (\tilde{k}_j - 1)n_0^{-1} + r(n_0 p_{d+1})^{-1}), \ r = 1,2,\ldots,p_{d+1}.$$

Since this holds for $j = 1,2,\ldots,d,$ it contradicts the maximality of n_0.

3. ZYGMUND'S PROPERTY

Our second result is the following.

THEOREM 2. **Let** f, with **domain** C, **map** C **into** a (real or complex) **complete normed linear space** L. **A necessary and sufficient condition for** f to have Zygmund's property in C is existence, for $n = 1,2,\ldots,$ of a **mapping** f_n of

$$\bigcup_{\substack{1 \le k_j \le n \\ j=1,2,\ldots,d}} C^{(n)}_{k_1,k_2,\ldots,k_d}$$

into L, which throughout each $C^{(n)}_{k_1,k_2,\ldots,k_d}$ is of the form

$$f_n(x_1,x_2,\ldots,x_d) \equiv \alpha_0 + \sum_{j=1}^{d} x_j \alpha_j ,$$

with each $\alpha_j \in L$, and satisfies

$$\|f(x) - f_n(x)\| \le D/n ,$$

D being a constant.

Proof (sufficiency). Given an

$$(x_1,x_2,\ldots,x_d) \in C$$

and a positive integer n^*, there is clearly an integer $n > n^*$ such that no x_j is of the form k/n, k an integer. [For $j = 1,2,\ldots,d,$ let

$n_j = n^*$ if x_j is irrational; if x_j is rational, let $x_j = k_j/n_j$,

where $n_j \ge n^*$ and k_j are integers. Then

$$n = 1 + n_1 n_2 \cdots n_d \ge 1 + n^{*d} > n^* .$$

If some x_j were k/n, k an integer, then we would have

$$n_j(k - Nk_j) = k_j, \quad N \text{ an integer} ,$$

implying that k_j/n_j is an integer.]

To establish the continuity of f in C, let

$$x^* = (x_1, x_2, \ldots, x_d) \in C ,$$

and let $\varepsilon > 0$. Let n be an integer $> 3D/\varepsilon$ such that no x_j is of the form k/n, k an integer. Then x^* lies in some

$$c = c^{(n)}_{k_1, k_2, \ldots, k_d} .$$

Let $\delta > 0$ be such that if $\|x - x^*\| < \delta$, then $x \in c$, and

$$\|f_n(x) - f_n(x^*)\| < \varepsilon/3 .$$

Then $\|x - x^*\| < \delta$ implies

$$\|f(x) - f(x^*)\| \le \|f(x) - f_n(x)\| + \|f_n(x) - f_n(x^*)\| + \|f_n(x^*) - f(x^*)\|$$
$$< 2Dn^{-1} + 3^{-1}\varepsilon < \varepsilon .$$

Let $x, x + 2h$ ($\neq x$) be points of C. Let n_0 be the greatest positive integer n for which x and $x + 2h$ lie in some (same)

$$c^{(n)}_{k_1, k_2, \ldots, k_d} .$$

By (3),

$$2\|h\| \ge (p_d p_{d+1} n_0)^{-1} .$$

Hence, using the linear form of f_{n_0}, we have

$$\|f(x) - 2f(x + h) + f(x + 2h)\| = \|\{f(x) - f_{n_0}(x)\} - 2\{f(x + h) - f_{n_0}(x + h)\}$$
$$+ \{f(x + 2h) - f_{n_0}(x + 2h)\}\| \le \|f(x) - f_{n_0}(x)\| + 2\|f(x + h) - f_{n_0}(x + h)\|$$
$$+ \|f(x + 2h) - f_{n_0}(x + 2h)\| \le 4D/n_0 \le 8Dp_d p_{d+1}\|h\| .$$

4. PROOF OF NECESSITY

To prove necessity in Theorem 2, we need two lemmas.

LEMMA 1. Assume the first sentence of Theorem 2, and suppose that f has Zygmund's property in C. Then f can be extended to $\overline{C}$ in such a way as to have Zygmund's property there.

LEMMA 2. Let f be a continuous mapping of $\overline{C}$ into a (real or complex) normed linear space L, and suppose that, for some constant A, (2) holds whenever $x, x + 2h \in \overline{C}$. Then there exists a linear function

$$(4) \qquad \ell(x_1,x_2,\ldots,x_d) \equiv a_0 + \sum_{j=1}^{d} x_j a_j \qquad (a_0,a_1,\ldots,a_d \in L)$$

such that for every $x = (x_1,x_2,\ldots,x_d) \in \overline{C}$,

$$\|f(x) - \ell(x)\| \leq A[c_d + 2^{-1}\sqrt{d}] ,$$

where $c_1 = 0$, and $c_k = \Sigma_{j=2}^{k} \sqrt{j}$ if $k > 1$.

Proof of Theorem 2 (necessity). By Lemma 1, we may assume that f is continuous in $\overline{C}$, and that, for some constant A, (2) holds whenever $x, x + 2h \in \overline{C}$. Let $n, k_1, k_2, \ldots, k_d$ be integers ≥ 1, with $k_j \leq n$, $j = 1,2,\ldots,d$. Define on $\overline{C}$:

$$f^{(n)}(x_1,x_2,\ldots,x_d) \equiv f((k_1 - 1 + x_1)/n,\ldots,(k_d - 1 + x_d)/n) .$$

Then $f^{(n)}$ is continuous in $\overline{C}$, and if $x, x + 2h \in \overline{C}$, we have

$$\|f^{(n)}(x) - 2f^{(n)}(x + h) + f^{(n)}(x + 2h)\|_L \leq (A/n)\|h\|_{\mathbb{R}^d} .$$

By Lemma 2, there is a linear function (4) satisfying

$$\|f^{(n)}(x) - \ell(x)\| \leq D/n$$

for each

$$x = (x_1,x_2,\ldots,x_d) \in \overline{C} ,$$

where $D = A[c_d + 2^{-1}\sqrt{d}]$. For each

$$x = (x_1,x_2,\ldots,x_d) \in C^{(n)}_{k_1,k_2,\ldots,k_d} ,$$

set

$$f_n(x) = \ell(nx_1 - k_1 + 1 ,\ldots, nx_d - k_d + 1) .$$

Then for every such x,

$$\|f(x) - f_n(x)\| = \|f^{(n)}(nx_1 - k_1 + 1 ,\ldots, nx_d - k_d + 1) - \ell(nx_1 - k_1 + 1 ,\ldots, nx_d - k_d + 1)\| \leq \frac{D}{n} .$$

5. PROOF OF LEMMA 1

To prove Lemma 1, it is enough to show that f is uniformly continuous in C, for then it can be extended to $\overline{C}$ in such a way as to be continuous there; and if $x, x + 2h \in \overline{C}$, then, with c_0 the center of C, A some constant, and $0 < \varepsilon < 1$, we have

$$\|f(x + \varepsilon(c_0 - x)) - 2f(x + \varepsilon(c_0 - x) + h(1 - \varepsilon)) + f(x + \varepsilon(c_0 - x) + 2h(1 - \varepsilon))\|$$

$$\leq A(1 - \varepsilon) \|h\|_{\mathbb{R}^d} ,$$

and therefore (2).

Suppose f is not uniformly continuous in C. Then there are $\varepsilon > 0$, $x \in \overline{C}$, and $x_n, y_n \in C$, $n = 1,2,\ldots$, such that $x_n \to x$, $y_n \to x$, and

$$\|f(y_n) - f(x_n)\| \geq \varepsilon , \quad n = 1,2,\ldots . \tag{5}$$

Let

$$h_0 = 2^{-1}(c_0 - x) ,$$

so that $x + \mu h_0 \in C$ whenever $0 < \mu \leq 2$. Set

$$\lambda = \min\{1, \varepsilon/[4(A + 1)(\|h_0\| + 1)]\} .$$

Let $N \geq 0$ be such that if $n > N$, then $x_n + 2\lambda h_0$, $y_n + 2\lambda h_0 \in C$, and

$$\|f(y_n + \lambda h_0) - f(x_n + \lambda h_0)\| < \varepsilon/8 ,$$

$$\|f(y_n + 2\lambda h_0) - f(x_n + 2\lambda h_0)\| < \varepsilon/4 .$$

If $n > N$, then

$$\begin{aligned} \|f(y_n) - f(x_n)\| = \|&\{f(y_n) - 2f(y_n + \lambda h_0) + f(y_n + 2\lambda h_0)\} \\ &- \{f(x_n) - 2f(x_n + \lambda h_0) + f(x_n + 2\lambda h_0)\} \\ &+ 2\{f(y_n + \lambda h_0) - f(x_n + \lambda h_0)\} \\ &- \{f(y_n + 2\lambda h_0) - f(x_n + 2\lambda h_0)\}\| \\ < 2A\lambda\|h_0\| &+ 2^{-1}\varepsilon < \varepsilon , \end{aligned}$$

contradicting (5).

6. PROOF OF LEMMA 2

Let

$$\varepsilon_0 = \underbrace{(0,0,0,\ldots,0)}_{d} ,$$

and, for $j = 1,2,\ldots,d$,

$$\varepsilon_j = (\delta_{1j}, \delta_{2j}, \ldots, \delta_{dj}) \quad \text{(Kronecker's delta)} .$$

We can assume that

$$f(\varepsilon_0) = f(\varepsilon_1) = \cdots = f(\varepsilon_d) = 0 ,$$

since otherwise we may replace f by

$$f(x_1,x_2,\ldots,x_d) - f(\varepsilon_0) + \sum_{j=1}^{d} x_j[f(\varepsilon_0) - f(\varepsilon_j)] .$$

We shall prove Lemma 2 with $\ell(x_1,x_2,\ldots,x_d) \equiv 0$.

Let $1 \le k \le d$, and let η_k be a point of $\mathbb{R}^d$ having k coordinates 1, and the rest (if $k < d$) zero. Then

$$\|f(\eta_k)\| \le Ac_k .$$

This is true for $k = 1$. Suppose it is true for some k, $1 \le k < d$. We shall prove it for $k + 1$.

Let the first coordinate of η_{k+1} that is 1 be the jth, and set

$$\eta_k = \eta_{k+1} - \varepsilon_j .$$

Then

$$\begin{aligned} 2\|f(2^{-1}\eta_{k+1})\|_L &= \|f(\varepsilon_j) - 2f(2^{-1}\eta_{k+1}) + f(\eta_k) - f(\eta_k)\|_L \\ &\le A\|2^{-1}(\eta_k - \varepsilon_j)\|_{\mathbb{R}^d} + \|f(\eta_k)\|_L \\ &\le A(2^{-1}\sqrt{k+1} + c_k) . \end{aligned}$$

We have

$$\begin{aligned} \|f(\eta_{k+1})\| - 2\|f(2^{-1}\eta_{k+1})\| &\le \|f(\varepsilon_0) - 2f(2^{-1}\eta_{k+1}) + f(\eta_{k+1})\| \\ &\le 2^{-1} A\|\eta_{k+1}\|_{\mathbb{R}^d} = 2^{-1} A\sqrt{k+1} , \end{aligned}$$

and so

$$\|f(\eta_{k+1})\| \le A(\sqrt{k+1} + c_k) = Ac_{k+1} .$$

To prove that

$$\|f(x)\| \le A[c_d + 2^{-1}\sqrt{d}]$$

throughout $\overline{C}$, we first show, for $n = 0,1,2,\ldots,$ that:

(**) if $x = (x_1,x_2,\ldots,x_d) \in \overline{C}$, and each x_j is of the form

$$\sum_{k=0}^{n} a_{j,k}\, 2^{-k} ,$$

where every $a_{j,k}$ is 0 or 1, then

$$(6) \qquad \|f(x_1,x_2,\ldots,x_d)\| \le A[c_d + 2^{-1}\sqrt{d}\,(1 - 2^{-n})] \,.$$

Then if $(x_1,x_2,\ldots,x_d)$ is an arbitrary point of $\overline{C}$, say,

$$x_j = \sum_{k=0}^{\infty} a_{j,k}\, 2^{-k} \,, \quad j = 1,2,\ldots,d \,,$$

where each $a_{j,k}$ is 0 or 1, the inequality (6) implies

$$\|f(x_1,x_2,\ldots,x_d)\| = \lim_{n\to\infty} \|f(\Sigma_{k=0}^{n} a_{1,k}\, 2^{-k},\ldots,\Sigma_{k=0}^{n} a_{d,k}\, 2^{-k})\| \le A[c_d + 2^{-1}\sqrt{d}] \,.$$

Now (**) is true for $n = 0$. Suppose it holds for some $n \ge 0$. For $j = 1,2,\ldots,d$, let

$$x_j = \sum_{k=0}^{n+1} a_{j,k}\, 2^{-k} \in [0,1], \quad u_j = x_j - a_{j,n+1}\, 2^{-n-1}, \quad v_j = x_j + a_{j,n+1}\, 2^{-n-1},$$

where each $a_{j,k}$ is 0 or 1. Then each u_j and each v_j belongs to $[0,1]$ and is of the form

$$\sum_{k=0}^{n} a'_k\, 2^{-k} \,, \quad \text{each } a_k \text{ being } 0 \text{ or } 1 \,.$$

Hence

$$\|f(u_1,u_2,\ldots,u_d)\| \le A[c_d + 2^{-1}\sqrt{d}\,(1 - 2^{-n})] \,,$$

$$\|f(v_1,v_2,\ldots,v_d)\| \le A[c_d + 2^{-1}\sqrt{d}\,(1 - 2^{-n})] \,.$$

Since

$$\begin{aligned} 2f(x_1,x_2,\ldots,x_d) = f(u_1,\ldots,u_d) + f(v_1,\ldots,v_d) - [&f(u_1,\ldots,u_d) \\ &- 2f(u_1 + a_{1,n+1}2^{-n-1},\ldots,u_d + a_{d,n+1}2^{-n-1}) \\ &+ f(u_1 + 2a_{1,n+1}2^{-n-1},\ldots,u_d + 2a_{d,n+1}2^{-n-1})] \,, \end{aligned}$$

we have

$$\begin{aligned} 2\|f(x_1,x_2,\ldots,x_d)\| &\le 2A[c_d + 2^{-1}\sqrt{d}\,(1 - 2^{-n})] + A2^{-n-1}\sqrt{d} \\ &= 2A[c_d + 2^{-1}\sqrt{d}\,(1 - 2^{-n-1})] \,, \end{aligned}$$

as desired.

REFERENCE

1. D. W. Hall and G. L. Spencer, II, Elementary Topology, John Wiley and Sons, New York, 1955.

INEQUALITIES RELATED TO THE NORMAL LAW

Alexander M. Ostrowski
CH-6926 Certenago
Montagnola, Ti.
SWITZERLAND

Raymond M. Redheffer
University of California
Los Angeles, California 90024
U.S.A.

ABSTRACT. In the important case of a symmetric distribution, it is shown that the familiar approximation leading to the normal law is actually an estimate from above. A more elementary inequality is presented first; this is much easier to prove than the final result, but it leads, nevertheless, to the solution of a nontrivial maximizing problem.

1. AN ELEMENTARY INEQUALITY

Throughout this note, n and ν are integers with $n \geq 2$, $0 \leq \nu \leq n$, and x is real, $0 < x < 1$. We shall establish the following:

THEOREM 1. If $q = \nu/n$, then

$$\binom{n}{\nu} x^{\nu}(1 - x)^{n-\nu} < e^{-2n(x-q)^2} .$$

To see this, let us locate the maximum of

$$f(x) = x^{\nu}(1 - x)^{n-\nu} e^{2n(x-q)^2} ,$$

or, equivalently, of $F(x) = n^{-1} \log f(x)$. From

$$F(x) = q \log x + (1 - q)\log(1 - x) + 2(x - q)^2 ,$$

it follows that

$$F'(x) = \frac{q}{x} - \frac{1 - q}{1 - x} + 4(x - q) = (q - x) \frac{1 - 4x(1 - x)}{x(1 - x)} .$$

Since the second factor on the right is positive for $x \neq 1/2$, the function $F(x)$ is increasing on $(0,q)$ and decreasing on $(q,1)$. This shows that $F(x)$, and hence $f(x)$, has its maximum at $x = q$.

The inequality $f(x) \leq f(q)$ can be written

$$\binom{n}{\nu} x^{\nu}(1 - x)^{n-\nu} \leq \binom{n}{\nu} q^{\nu}(1 - q)^{n-\nu} e^{-2n(x-q)^2} . \tag{1}$$

For any fixed q on $0 \leq q \leq 1$, the expression

$$C(n,q) = \binom{n}{\nu} q^{\nu}(1 - q)^{n-\nu} \tag{2}$$

is a term in the expansion of $[q + (1 - q)]^n$. Hence $C(n,q) < 1$, and this gives Theorem 1.

The result is sharp in the following sense. If

$$\binom{n}{\nu} x^{\nu}(1 - x)^{n-\nu} \leq \alpha e^{\beta n(x-q)^2} , \tag{3}$$

where α and β are absolute constants, then $\alpha \geq 1$ and $\beta \geq -2$. The inequality $\alpha \geq 1$ follows from $\nu = q = 0$, $x \to 0+$. The inequality $\beta \geq -2$ follows from a familiar asymptotic formula which forms the basis for the normal law [2], [3]. Instead of using this rather sophisticated formula, however, we shall give an elementary proof based on the relation

$$\binom{n}{\nu}^{1/n} = \frac{1 + o(1)}{q^q(1 - q)^{1-q}} , \qquad q = \frac{\nu}{n} , \tag{4}$$

where $o(1)$ denotes a function of (ν,n) which tends uniformly to 0 as $n \to \infty$.

Although (4) is well known, the easy derivation will be given here. The formula

$$\log n! = \sum_{j=1}^{n} \log j = n \log n - n + o(n)$$

follows by comparing the sum with an integral. Applying this to n, ν, and $n-\nu$ gives

$$\log \binom{n}{\nu} = n \log n - \nu \log \nu - (n - \nu)\log(n - \nu) + o(n) . \tag{5}$$

If we divide (5) by n and replace ν throughout by qn, the result is equivalent to (4).

By forming the nth root of both sides in (3), and using (4), it is found that (3) can hold for $n \to \infty$ only if

$$\left(\frac{x}{q}\right)^q \left(\frac{1 - x}{1 - q}\right)^{1-q} \leq e^{\beta(x-q)^2} [1 + o(1)] . \tag{6}$$

Given any rational number $q = \nu_0/n_0$ on $(0,1)$, let $n = jn_0$, $\nu = j\nu_0$, $j \to \infty$. This shows that (6) holds without the term $o(1)$ for rational q, and then by continuity for all q, $0 < q < 1$.

Taking logarithms when $x \neq q$, we get

$$\beta \geq \frac{q \log(x/q) + (1 - q) \log(1-x)/(1-q)}{(x - q)^2} . \tag{7}$$

The choice $q = 1/2$, $x = (1+t)/2$ gives

$$\beta \geq \frac{2}{t^2} \log(1 - t^2) = -2 + O(t^2),$$

and $\beta \geq -2$ follows when $t \to 0$.

On the other hand, Theorem 1 shows that $\beta = -2$ must satisfy (7). Writing $q = y$, we summarize as follows:

THEOREM 2. Let $S = \{(x,y) \mid 0 < x < 1,\ 0 < y < 1,\ y \neq x\}$. Then

$$\sup_{(x,y)\in S} \frac{y \log(x/y) + (1 - y) \log(1-x)/(1-y)}{(y - x)^2} = -2 .$$

It is left for the reader to explore the problem of proving Theorem 2 without using Theorem 1.

2. SHARPER FORMS OF THE INEQUALITY

Although the constant $\alpha = 1$ in (3) cannot be improved when $\nu = 0$ or $\nu = n$, an improvement is possible when $1 \leq m \leq \nu \leq n-m$. For example, if $1 \leq \nu \leq n-1$, then the inequality

$$\binom{n}{\nu} x^{\nu}(1 - x)^{n-\nu} \leq \alpha e^{-2n(x-q)^2}$$

holds for all n when $\alpha = 1/2$, it holds for all but finitely many n if $1/e < \alpha < 1/2$, and it fails for all n if $\alpha = 1/e$.

For proof, let us write (2) in the form

$$C(n,q) = \frac{f(\nu)f(n - \nu)}{f(n)}, \quad \text{where} \quad f(n) = \frac{n^n}{n!} . \tag{8}$$

We shall investigate the beharior of $C(n,q)$ when ν is increased to $\nu + 1$. To this end, observe that the inequality

$$f(\nu + 1)f(n - \nu - 1) \leq f(\nu)f(n - \nu) \tag{9}$$

is equivalent to

$$\frac{f(\nu + 1)}{f(\nu)} \leq \frac{f(n - \nu)}{f(n - \nu - 1)} ,$$

and hence to

$$\left(1 + \frac{1}{\nu}\right)^{\nu} \leq \left(1 + \frac{1}{n - \nu - 1}\right)^{n-\nu-1} .$$

On the left, we recognize the function which occurs in the limit definition of e. Since that function is monotone, (9) holds if, and only if,

$\nu \leq n - \nu - 1$. Hence, $C(n,q)$ decreases as ν progresses from the ends of its interval toward the center, and the best value (independent of ν) is obtained when $\nu = m$ or $n - m$. The special case $m = 1$ gives the following theorem, which implies the results for $1 \leq \nu \leq n - 1$ noted above:

THEOREM 3. If $1 \leq \nu \leq n - 1$ and $q = \nu/n$, then

$$\binom{n}{\nu} x^{\nu}(1 - x)^{n-\nu} \leq \left(1 - \frac{1}{n}\right)^{n-1} e^{-2n(x-q)^2} .$$

Equality holds if, and only if, $x = q$ and $\nu = 1$ or $\nu = n - 1$.

According to familiar results in the theory of probability, the constant α ought to have the order of magnitude $O(n^{-1/2})$ for a broad range of values of x and q. We shall establish the following:

THEOREM 4. If $1 \leq \nu \leq n - 1$ and $q = \nu/n$, then

$$\binom{n}{\nu} x^{\nu}(1 - x)^{n-\nu} < \left(\frac{1}{2\pi nq(1 - q)}\right)^{1/2} e^{-2n(x-q)^2} .$$

The constant 2π is sharp as $n \to \infty$ for every value of q.

To prove this, define $\theta(n)$ by

$$n! = \left(\frac{n}{e}\right)^n \sqrt{2\pi n}\ \theta(n). \tag{10}$$

Substituting (10) into (8) gives

$$C(n,q) = \left(\frac{n}{2\pi\nu(n - \nu)}\right)^{1/2} \frac{\theta(n)}{\theta(\nu)\theta(n - \nu)} . \tag{11}$$

Given $q = \nu_0/n_0$, we set $n = jn_0$, $\nu = j\nu_0$, $x = q$, and we let $j \to \infty$. Since $\lim \theta(n) = 1$ as $n \to \infty$ by the Stirling - de Moivre formula, it follows that the constant 2π in Theorem 4 is sharp.

The main assertion in Theorem 4 can be deduced from an interesting formula of Binet [1], namely,

$$\log n! = \left(n + \frac{1}{2}\right)\log n - n + \frac{1}{2}\log(2\pi) + \int_0^{\infty} \frac{2 \tan^{-1}(t/n)}{e^{2\pi t} - 1}\, dt$$

Comparison with (10) shows that $\log \theta(n)$ equals the integral on the right; hence $\theta(j) > 1$ for $j \geq 1$, and $\theta(j)$ is decreasing. These two properties imply

$$\theta(n) < \theta(\nu)\theta(n - \nu) , \quad 1 \leq \nu \leq n - 1 .$$

Theorem 4 now follows from (11).

The greatest value of the constant α in Theorem 4 is $\pi^{-1/2}$, obtained when $\nu = 1$, $n = 2$. Hence Theorem 4 sharpens Theorem 1. When $\nu = 1$ or $n - 1$, the constant exceeds the optimum value (given in Theorem 3) by about 10%. If $2 \leq \nu \leq n - 2$, however, Theorem 4 is sharper than Theorem 3 for every value of n.

3. CONCLUDING REMARKS

In conclusion, we mention that the constant $\alpha = \alpha(n,q)$ in Theorem 4 has the general form that might be expected on the basis of the normal law. In a like manner, if β is allowed to depend on q, a natural choice is

$$\beta = -\frac{1}{2q(1 - q)} .$$

Unfortunately, the function corresponding to $f(x)$ in the proof of Theorem 1 now satisfies $f(x) \leq f(1 - q)$, and hence the exponential factor does not drop out as it did before. This is the reason why preference has been given to the case $\beta = -2$ here.

Theorems 1 and 2 were presented at Oberwolfach in May, 1976. Research of the respective authors has been done in Basel under auspices of the Swiss National Science Foundation, and in Karlsruhe under auspices of the United States Special Program, Alexander von Humbolt Stifung.

REFERENCES

1. A. Erdélyi, W. Magnus, F. Oberhettinger, and F. G. Tricomi, Higher transcendental functions, McGraw-Hill, N.Y., 1953, Vol. 1, p. 22.

2. G. G. Lorentz, Bernstein polynomials, University of Toronto Press, 1953, pp. 15, 18.

3. Ivan Sokolnikoff and R. M. Redheffer, Mathematics of physics and modern engineering, McGraw-Hill, N.Y., 1966, p. 624.

BOUND ON THE MEASURE OF A SET IN A PRODUCT SPACE

L. L. Campbell
Department of Mathematics
Queen's University
Kingston, Ontario
CANADA

ABSTRACT. A simple bound is obtained for the measures of certain sets in product spaces. The bound is a kind of Chernoff bound.

1. MAIN RESULT

Let (X,S,μ) be a measure space, and let $f(\cdot,\cdot)$ be a real-valued measurable function on X^2. Denote by μ_N the product measure on X^N, and denote by f_N the function defined on X^{2N} by

$$f_N(x,y) = N^{-1} \sum_{i=1}^{N} f(x_i,y_i). \tag{1}$$

For each $y \in X^N$ and each real r, define the set

$$B_N(r;y) = \{x \in X^N : f_N(x,y) \leq r\}.$$

Finally, for $s \geq 0$, define

$$Z(s) = \sup_{y \in X} \int_X \exp[-sf(x,y)]d\mu(x). \tag{2}$$

THEOREM. With the notation above, we have

$$\mu_N[B_N(r;y)] \leq [e^{sr}Z(s)]^N \tag{3}$$

for all $s \geq 0$.

Proof. Writing B for $B_N(r;y)$, we have

$$\mu_N(B) = \int_B d\mu_N(x)$$

$$\leq \int_B \exp Ns[r - f_N(x,y)]d\mu_N(x),$$

since the exponent is nonnegative on B. Because the integrand is nonnegative everywhere, we can expand the integration set to X^N, getting

$$\mu_N(B) \le e^{Nsr} \int_{X^N} \exp[-Nsf_N(x,y)]d\mu_N(x).$$

Thus, from (1),

$$\mu_N(B) \le e^{Nsr} \prod_{i=1}^{N} \int_X \exp[-sf(x_i,y_i)]d\mu(x_i).$$

The theorem now follows from (2).

2. REMARKS

The basic idea of the proof is the same as that used by Chernoff [1]. For applications of two special cases of this result to information theory, see [2]. When f_N is a metric, or a monotone function of a metric, on X^N, the theorem provides an exponentially growing bound on the sizes of balls in X^N. This could be useful in estimating the metric entropy [3] of sets in a product space.

REFERENCES

1. H. Chernoff, A measure of asymptotic efficiency for tests of a hypothesis based on a sum of observations, Ann. Math. Statist. 23 (1952), 493-507.

2. L. L. Campbell, Kraft inequality for decoding with respect to a fidelity criterion, IEEE Trans. Information Theory IT-19 (1973), 68-73.

3. G. G. Lorentz, Approximation of functions, Holt, Rinehart and Winston, New York, 1966.

INEQUALITIES AMONG OPERATIONS ON PROBABILITY DISTRIBUTION FUNCTIONS

R. Moynihan
Dept. of Math.
Bowdoin College
Brunswick, ME 04011
U.S.A.

B. Schweizer
Dept. of Math. and Stat.
Univ. of Massachusetts
Amherst, MA 01002
U.S.A.

A. Sklar
Dept. of Math.
Illinois Inst. of Tech.
Chicago, IL 60616
U.S.A.

ABSTRACT. Inequalities are established among certain binary operations on a space of probability distribution functions. These operations arise naturally in the theory of probabilistic metric spaces, in the generalized theory of information, and in other contexts of the theory of probability. It is further shown that, in most cases, equality holds in the inequalities if and only if at least one of the arguments is a unit step function, i.e., that the associated functional equations have in general only essentially trivial solutions.

1. INTRODUCTION

In this paper we establish some inequalities among certain natural binary operations on a space of probability distribution functions. These inequalities, which have proved useful in the theory of probabilistic metric spaces and in the generalized theory of information (see [3], [5], and [7]), are to be interpreted as follows: If β_1 and β_2 are two such binary operations, then $\beta_1 \leq \beta_2$ means that for all distribution functions F,G in the space $\mathcal{D}^+$ (defined in Section 2), the distribution functions $\beta_1(F,G)$ and $\beta_2(F,G)$ satisfy $\beta_1(F,G) \leq \beta_2(F,G)$, i.e., $(\beta_1(F,G))(t) \leq (\beta_2(F,G))(t)$ for all real numbers t.

To simplify the exposition, we have deliberately avoided striving for the greatest possible generality. In particular, we note that: (a) the operations defined in Section 3 can be extended to spaces of distribution functions more inclusive than $\mathcal{D}^+$, and some, though not all, of our principal results remain valid in these more inclusive spaces; (b) the class $\mathcal{L}$ of functions introduced in Section 2 can be enlarged; (c) additional operations could be considered. Some of these points are treated elsewhere: see [4] and [6].

2. PRELIMINARIES

A distribution function is a nondecreasing function mapping the extended real line $[-\infty,\infty]$ into the closed unit interval $[0,1]$. A random variable X gives rise to a distribution function F_X via: For all real t, $F_X(t)$ is the probability that $X < t$. But it is possible, and often desirable, to consider distribution functions and operations on them independently of any association with random variables (cf. [6]).

We denote by $\mathcal{D}^+$ the set of distribution functions F such that $F(0) = 0$ and $\sup F = F(\infty) = 1$. It will prove convenient here to normalize the functions in $\mathcal{D}^+$ by requiring them to be left-continuous. Among the functions in $\mathcal{D}^+$ are the unit step-functions ε_a defined for $0 \leq a < \infty$ by

$$\varepsilon_a(t) = \begin{cases} 0, & t \leq a , \\ 1, & t > a . \end{cases}$$

Closely connected to the distribution functions are the <u>copulas</u> (cf. [6] and [7]). These are the functions C that map the closed unit square $[0,1] \times [0,1]$ onto the closed unit interval $[0,1]$ and satisfy the conditions:

$$C(0,x) = C(x,0) = 0 \ , \ C(1,x) = C(x,1) = x \tag{2.1}$$

for all x in $[0,1]$;

$$C(x_1,y_1) - C(x_1,y_2) - C(x_2,y_1) + C(x_2,y_2) \geq 0 \tag{2.2}$$

whenever $x_1 \leq x_2$ and $y_1 \leq y_2$.

If we set $y_1 = 0$, $y_2 = y$ in (2.2), then (2.1) yields:

$$C(x_2,y) - C(x_1,y) \geq 0 \quad \text{whenever} \quad x_1 \leq x_2 \ . \tag{2.3}$$

Alternatively, setting $y_1 = y$, $y_2 = 1$ in (2.2) and applying (2.1) gives:

$$C(x_1,y) - C(x_2,y) - x_1 + x_2 \geq 0 \quad \text{whenever} \quad x_1 \leq x_2 \ . \tag{2.4}$$

Upon combining (2.3) and (2.4), we obtain:

$$0 \leq C(x_2,y) - C(x_1,y) \leq x_2 - x_1 \quad \text{whenever} \quad x_1 \leq x_2 \ , \tag{2.5}$$

and similarly,

$$0 \leq C(x,y_2) - C(x,y_1) \leq y_2 - y_1 \quad \text{whenever} \quad y_1 \leq y_2 \ . \tag{2.6}$$

It follows from the left-hand inequalities in (2.5) and (2.6) that C is nondecreasing in each place. Furthermore, applying (2.5) and (2.6) together yields:

$$|C(x_1,y_1) - C(x_2,y_2)| \leq |x_1 - x_2| + |y_1 - y_2|$$

for arbitrary (x_1,y_1), (x_2,y_2), whence it follows that C is continuous.

It also follows from (2.1) and (2.2) that for every (x,y) in $[0,1] \times [0,1]$, we have:

$$T_m(x,y) \leq C(x,y) \leq \text{Min}(x,y) \ , \tag{2.7}$$

where $T_m(x,y) = \text{Max}(x + y - 1,0)$. Both T_m and Min are copulas themselves, whence the set of copulas has a minimal and a maximal element. Another important copula is the function Prod, defined by: $\text{Prod}(x,y) = x \cdot y$ for all (x,y) in $[0,1] \times [0,1]$.

The connection between copulas and distribution functions arises as follows: Let X and Y be random variables with respective distribution functions F_X, F_Y and joint distribution function F_{XY}, so that for all (u,v) in $[-\infty,\infty] \times [-\infty,\infty]$, $F_{XY}(u,v)$ is the probability that $X < u$ and $Y < v$. Then there is a copula C_{XY} such that

$$F_{XY}(u,v) = C_{XY}(F_X(u),F_Y(v)) \quad \text{for all} \quad (u,v) \ . \tag{2.8}$$

If F_X and F_Y are continuous, then C_{XY} is unique. Another way of putting this, independent of any reference to random variables, is that a copula is a function that links a joint distribution function to its one-dimensional margins.

With every copula C, we associate its <u>dual copula</u> $\overline{C}$, defined by:

$$\overline{C}(x,y) = x + y - C(x,y) \quad \text{for all} \quad (x,y) \quad \text{in} \quad [0,1] \times [0,1] \ . \tag{2.9}$$

In particular, the dual of Min is Max. Since copulas are continuous, so are their duals; and it follows from (2.9), (2.5), and (2.6) that a dual copula is nondecreasing in each place. Furthermore, from (2.7) and (2.9), we see that for any dual copula $\overline{C}$,

$$\text{Max}(x,y) \leq \overline{C}(x,y) \leq \overline{T}_m(x,y) = \text{Min}(x + y,1) \tag{2.10}$$

for all (x,y) in $[0,1] \times [0,1]$.

Finally, we introduce one other family $\mathcal{L}$ of functions. An element of $\mathcal{L}$ is a function L satisfying the following conditions:

(2.11)
- (i) The domain of L is the closed first quadrant $[0,\infty] \times [0,\infty]$, and the range of L is the closed half-line $[0,\infty]$.
- (ii) L is nondecreasing in each place, whence it follows that $L(0,0) = 0$ and $L(\infty,\infty] = \infty$.
- (iii) L is continuous on $[0,\infty) \times [0,\infty)$, and $\lim_{x\to\infty} L(x,x) = \infty$.

Among the functions in $\mathcal{L}$ are the restrictions to the first quadrant of the ordinary operations of maximum, minimum, addition, and multiplication (with $0 \cdot \infty$ and $\infty \cdot 0$ defined arbitrarily). We denote these restrictions, respectively, by Sup, Inf, Sum, and Mult.

Given a function L in $\mathcal{L}$, and a positive number x, we define $L[x]$ to be the set of points (u,v) in the closed first quadrant such that (a) $L(u,v) = x$, and (b) every neighborhood of (u,v) contains a point (u',v') such that $L(u',v') < x$. It follows that $L[x]$ is nonempty and is a continuous curve. At every point (u,v) "southwest" of a point of $L[x]$, we have $L(u,v) < x$; at every point (u,v) "northeast" of a point of $L[x]$ we have $L(u,v) \geq x$. Loosely speaking, $L[x]$ is the left or lower boundary of the level set $\{(u,v) \mid L(u,v) = x\}$, or equivalently, the upper boundary in $[0,\infty] \times [0,\infty)$ of the set

$$L\{x\} = \{(u,v) \mid L(u,v) < x\} . \tag{2.12}$$

Note also that for any y such that $0 < y < x$, the curve $L[y]$ lies in the region $L\{x\}$.

3. BINARY OPERATIONS ON $\mathcal{D}^+$

Let C be a fixed copula and L a fixed element of $\mathcal{L}$. Then, given a pair F,G of distribution functions in $\mathcal{D}^+$, we define the function $\sigma_{C,L}(F,G)$ by:

$$(\sigma_{C,L}(F,G))(x) = \begin{cases} 0, & x \leq 0 , \\ \iint_{L\{x\}} dC(F(u),G(v)), & x > 0 , \end{cases} \tag{3.1}$$

where the integral in (3.1) is simply the Lebesgue-Stieltjes measure of the set $L\{x\}$ with respect to the measure induced by the function $C(F(u),G(v))$ [briefly, <u>the</u> $C(F,G)$-<u>measure of</u> $L\{x\}$]. The $C(F,G)$-measure of a Borel subset of the first quadrant is defined in the standard manner, starting with the observations that, given a point (x,y) in the first quadrant, the

$C(F,G)$-measure of the region $R(x,y)$ defined by

(3.2) $$R(x,y) = \{(u,v) \mid 0 \le u < x, \quad 0 \le v < y\}$$

is $C(F(x),G(y))$; the $C(F,G)$-measure of the set $\{(u,v) \mid 0 \le u \le x, 0 \le v < y\}$ is $C(F(x^+),G(y))$; the $C(F,G)$-measure of the set $\{(u,v) \mid 0 \le u < x, \ 0 \le v \le y\}$ is $C(F(x),G(y^+))$; and the $C(F,G)$-measure of the singleton set $\{(x,y)\}$ is given by

$$[C(F(x^+)) - C(F(x^+),G(y)) - C(F(x),g(y^+)) + C(F(x),G(y))] .$$

From (2.1), (2.2), and (2.11) it is easy to show that $\sigma_{C,L}(F,G)$ is in $\mathcal{D}^+$ (for the special case L = Sum, see [4]), so that $\sigma_{C,L}$ is in fact a binary operation on $\mathcal{D}^+$.

The significance of the operations $\sigma_{C,L}$ stems from the following fact: Let X and Y be nonnegative random variables with respective distribution functions F_X, F_Y and joint distribution function F_{XY}. Then F_X and F_Y are in $\mathcal{D}^+$; and for any L in $\mathcal{L}$, $L(X,Y)$ is a nonnegative random variable whose distribution function is precisely $\sigma_{C_{XY},L}(F_X,F_Y)$, with C_{XY} as in (2.8). (For a proof of this, see [6].) In particular, $\sigma_{C,\mathrm{Sum}}$ (which we abbreviate to σ_C) yields the distribution function of the sum of two random variables; if the random variables are independent, then C = Prod, and it is immediate from (3.1) that σ_{Prod} is convolution. For L's other than Sum, the operations $\sigma_{\mathrm{Prod},L}$ correspond to certain of the generalized convolutions of Urbanik [8].

The parentheses around the expression $\sigma_{C,L}(F,G)$ on the left-hand side of (3.1) are there to emphasize the fact that the expression denotes a single function; such parentheses will generally be omitted in the sequel.

Two other families of operations on $\mathcal{D}^+$ are defined as follows: As before, let C be a fixed copula, and L a fixed element of $\mathcal{L}$. Then for any F,G in $\mathcal{D}^+$, $\tau_{C,L}(F,G)$ is the function given by:

(3.3) $$\tau_{C,L}(F,G)(x) = \begin{cases} 0 , & x \le 0 , \\ \sup\{C(F(u),G(v)) \mid (u,v) \text{ in } L[x]\} , & x > 0 ; \end{cases}$$

and $\rho_{C,L}(F,G)$ is the function given by

(3.4) $$\rho_{C,L}(F,G)(x) = \begin{cases} 0 , & x \le 0 , \\ \inf\{\overline{C}(F(u),G(v)) \mid (u,v) \text{ in } L[x]\} , & x > 0 . \end{cases}$$

It is not difficult to show that both $\tau_{C,L}(F,G)$ and $\rho_{C,L}(F,G)$ are in $\mathcal{D}^+$. In particular, the left-continuity of $\tau_{C,L}(F,G)$ follows from a slight modification of the proof of Theorem 4.1 of [4], and that of $\rho_{C,L}(F,G)$ from an extension of Theorem 5.1 of [4]. As in the case of the operations $\sigma_{C,L}$, we abbreviate the symbols $\tau_{C,Sum}$ and $\rho_{C,Sum}$ to τ_C and ρ_C, respectively.

The operations τ_C are of fundamental importance in the theory of probabilistic metric spaces (see [4]). The operations ρ_C and $\tau_{C,L}$ (for some L's other than Sum) also play a role in this theory, while the operations $\rho_{C,L}$, for those L's which are "composition laws," are of importance in the generalized theory of information of Kampé de Fériet and Forte (cf. [2], [5]).

4. THE GENERAL INEQUALITIES

Let L_1 and L_2 be elements of $\mathfrak{L}$ such that $L_1 \leq L_2$, i.e., $L_1(u,v) \leq L_2(u,v)$ for all (u,v) in the closed first quadrant. It follows immediately that, for any $x > 0$, the set $L_2\{x\}$ is a subset of the set $L_1\{x\}$ and the curve $L_2[x]$ lies "southwest" of the curve $L_1[x]$. Now any integral of the form $\iint dC(F(u),G(v))$, as the $C(F,G)$-measure of its region of integration, is a nondecreasing functional of that region. We therefore have

$$\sigma_{C,L_2}(F,G)(x) \leq \sigma_{C,L_1}(F,G)(x) \tag{4.1}$$

for any copula C, every F,G in $\mathcal{D}^+$, and every $x \geq 0$. Similarly, since for any point (u_2,v_2) on $L_2[x]$ there is a point (u_1,v_1) on $L_1[x]$ such that $C(F(u_2)G(v_2)) \leq C(F(u_1),G(v_1))$, we have

$$\tau_{C,L_2}(F,G)(x) \leq \tau_{C,L_1}(F,G)(x) \tag{4.2}$$

and

$$\rho_{C,L_2}(F,G)(x) \leq \rho_{C,L_1}(F,G)(x) \tag{4.3}$$

for any copula C, every F,G in $\mathcal{D}^+$, and every $x \geq 0$.

We combine (4.1), (4.2), and (4.3) into:

THEOREM 1. Let C be a fixed copula, and let L_1,L_2 be functions in $\mathfrak{L}$ such that $L_1 \leq L_2$. Then we have:

$$\rho_{C,L_2} \leq \rho_{C,L_1}\ ,\ \sigma_{C,L_2} \leq \sigma_{C,L_1}\ ,\ \tau_{C,L_2} \leq \tau_{C,L_1}\ . \tag{4.4}$$

Our second set of inequalities is also easy to derive. If C_1 and C_2 are copulas and $C_1 \leq C_2$ (whence $\overline{C}_2 \leq \overline{C}_1$), then an immediate consequence (3.3) and (3.4) is:

THEOREM 2. For any fixed L in $\mathfrak{L}$, and any copulas C_1, C_2 such that $C_1 \leq C_2$, we have:

$$\rho_{C_2,L} \leq \rho_{C_1,L}, \quad \tau_{C_1,L} \leq \tau_{C_2,L}. \tag{4.5}$$

No such result holds for $\sigma_{C,L}$, it being in general impossible to pass from an inequality between copulas to a corresponding inequality between integrals of the type appearing in (3.1). A simple counterexample is the following:

Let F be the function in $\mathcal{D}^+$ defined by:

$$F(x) = \begin{cases} 0, & x \leq 0, \\ x, & 0 \leq x \leq 1, \\ 1, & 1 \leq x. \end{cases}$$

Taking L to be Sum, we find after some straightforward computations that

$$\sigma_{T_m}(F,F) = \varepsilon_1,$$

while $\sigma_{Prod}(F,F)$ and $\sigma_{Min}(F,F)$ are given, respectively, by:

$$\sigma_{Prod}(F,F)(x) = \begin{cases} 0, & x \leq 0, \\ \frac{1}{2}x^2, & 0 \leq x \leq 1, \\ 1 - \frac{1}{2}(2-x)^2, & 1 \leq x \leq 2, \\ 1, & 2 \leq x; \end{cases}$$

$$\sigma_{Min}(F,F)(x) = \begin{cases} 0, & x \leq 0, \\ \frac{1}{2}x, & 0 \leq x \leq 2, \\ 1, & 2 \leq x. \end{cases}$$

Thus none of the three functions $\sigma_{T_m}(F,F)$, $\sigma_{Prod}(F,F)$, $\sigma_{Min}(F,F)$ is comparable with either of the others. In particular, we have:

$$\sigma_{T_m}(F,F)(\tfrac{1}{2}) = 0 < \sigma_{Prod}(F,F)(\tfrac{1}{2}) = \tfrac{1}{8} < \sigma_{Min}(F,F)(\tfrac{1}{2}) = \tfrac{1}{4},$$

but

$$\sigma_{T_m}(F,F)(\tfrac{3}{2}) = 1 > \sigma_{Prod}(F,F)(\tfrac{3}{2}) = \tfrac{7}{8} > \sigma_{Min}(F,F)(\tfrac{3}{2}) = \tfrac{3}{4}.$$

We can, however, make the following observarion: For any copula C, any F,G in $\mathcal{D}^+$, and any point (u,v) in the first quadrant, we have

$$C(f(u),G(v)) = \iint_{R(u,v)} dC(F(s),G(t)) \ , \tag{4.6}$$

where $R(u,v)$ is given by (3.2). Now if for a given L in $\mathcal{L}$ and some $x \geq 0$, (u,v) is on the curve $L[x]$, then for every point (s,t) in $R(u,v)$ we have $L(s,t) < x$. The same argument that led to (4.1) therefore yields

$$C(F(u),G(v)) \leq \sigma_{C,L}(F,G)(x) \quad \text{for any} \quad (u,v) \quad \text{on} \quad L[x] \ ,$$

whence we have:

$$\tau_{C,L}(F,G)(x) \leq \sigma_{C,L}(F,G)(x) \quad \text{for any} \quad F,G \quad \text{in} \quad \mathcal{D}^+, \quad x \geq 0 \ . \tag{4.7}$$

Similarly, we note that

$$\overline{C}(F(u),G(v)) = \iint_{\Gamma(u,v)} dC(F(s),G(t)) \ , \tag{4.8}$$

where $\Gamma(u,v)$ is the L-shaped region that is the union of the two strips $0 \leq s < u$ and $0 \leq t < v$. Now if (u,v) is on $L[x]$, then the region of integration $L\{x\}$ in (3.1) is a subset of $\Gamma(u,v)$, and so we have

$$\sigma_{C,L}(F,G)(x) \leq \overline{C}(F(u),G(v)) \quad \text{for any} \quad (u,v) \quad \text{on} \quad L[x] \ .$$

It follows that

$$\sigma_{C,L}(F,G)(x) \leq \rho_{C,L}(F,G)(x) \quad \text{for any} \quad F,G \quad \text{in} \quad \mathcal{D}^+, \quad x \geq 0 \ . \tag{4.9}$$

We combine (4.7) and (4.9) into:

THEOREM 3. Let C be a copula, and L a function in $\mathcal{L}$. Then we have:

$$\tau_{C,L} \leq \sigma_{C,L} \leq \rho_{C,L} \ . \tag{4.10}$$

By using (2.7) and (4.5), we extend (4.10) to the following:

COROLLARY. With the notation of Theorem 3, we have:

$$\tau_{T_m,L} \leq \tau_{C,L} \leq \sigma_{C,L} \leq \rho_{C,L} \leq \rho_{T_m,L} \ . \tag{4.11}$$

5. SPECIAL CASES OF EQUALITY

For some particular choices of L, it is possible to assert equality on one or the other side of (4.10). For this purpose, it is convenient to introduce a new and very simple set of binary operations on $\mathcal{D}^+$, as follows: Let H be a function mapping the closed unit square onto the closed unit interval that is nondecreasing in each place (whence $H(0,0) = 0$ and

$H(1,1) = 1)$ and continuous. Define π_H by:

(5.1) $\pi_H(F,G)(x) = H(F(x),G(x))$ for F,G in $\mathcal{D}^+$, and any x.

Now, from the fact that $\mathrm{Sup}\{x\} = R(x,x)$ and $\mathrm{Inf}\{x\} = \Gamma(x,x)$, an argument similar to the one in Theorem 3 yields:

THEOREM 4. Let C be an arbitrary copula. Then we have:

(5.2) $$\tau_{C,\mathrm{Sup}} = \sigma_{C,\mathrm{Sup}} = \pi_C\ ,\ \sigma_{C,\mathrm{Inf}} = \rho_{C,\mathrm{Inf}} = \pi_{\overline{C}}\ .$$

Combining Theorems 1, 2, 3, and 4 yields the following very useful result:

COROLLARY. Let M be a function in $\mathcal{L}$ such that $M \geq \mathrm{Sup}$ (e.g., $M = \mathrm{Sum}$). Then for any copula C, we have:

(5.3) $$\tau_{T_m,M} \leq \tau_{C,M} \leq \sigma_{C,M} \leq \pi_C \leq \pi_{\mathrm{Min}}\ .$$

On the other hand, if L in $\mathcal{L}$ is such that $L \leq \mathrm{Inf}$, then for any copula C, we have:

(5.4) $$\pi_{\mathrm{Max}} \leq \pi_{\overline{C}} \leq \sigma_{C,L} \leq \rho_{C,L} \leq \rho_{T_m,L}\ .$$

Note that since $\pi_{\mathrm{Min}} \leq \pi_{\mathrm{Max}}$, (5.3) and (5.4) combine into a single string of nine inequalities.

When $C = \mathrm{Min}$, equality in (4.10) holds on both sides; more precisely, we have:

THEOREM 5. For any L in $\mathcal{L}$,

(5.5) $$\tau_{\mathrm{Min},L} = \sigma_{\mathrm{Min},L} = \rho_{\mathrm{Min},L}\ .$$

Proof. If (5.5) fails, then, in view of (4.10), for some F,G in $\mathcal{D}^+$ and some $t > 0$,

(5.6) $$\tau_{\mathrm{Min},L}(F,G)(t) < \rho_{\mathrm{Min},L}(F,G)(t)\ .$$

Let $w = \tau_{\mathrm{Min},L}(F,G)(t)$, and let

(5.7) $$a = \sup\{x \mid F(x) \leq w\} \text{ and } b = \sup\{x \mid G(x) \leq w\}\ .$$

Since $w < 1$ by (5.6), we have $0 \leq a < \infty$ and $0 \leq b < \infty$; and by left-continuity, $F(a) \leq w$ and $G(b) \leq w$.

Suppose $L(a,b) < t$. Then for some $c > 0$ the point $(a + c, b + c)$ lies on $L[t]$. It then follows from (5.7) that

$$\tau_{Min,L}(F,G)(t) \geq Min(F(a + c), G(b + c)) > w ,$$

which cannot be. Thus $L(a,b) \geq t > 0$, whence there exists a point (a',b') on $L[L(a,b)]$ for which $0 \leq a' \leq a$ and $0 \leq b' \leq b$. Hence

$$w < \rho_{Min,L}(F,G)(t) \leq \rho_{Min,L}(F,G)(L(a,b))$$
$$\leq Max(F(a'),G(b')) \leq Max(F(a),G(b)) \leq w ,$$

which is a contradiction and completes the proof.

A proof of the equality $\tau_{Min} = \sigma_{Min}$ for the space of all distribution functions appears in [1].

An immediate consequence of Theorem 5 is this:

THEOREM 6. Let F,G in $\mathcal{D}^+$ and the copula C be such that

$$C(f(u),G(v)) = Min(F(u),G(v)) \tag{5.8}$$

for all (u,v) in the first quadrant (note that this need not imply $C = Min$). Then for any L in $\mathcal{L}$ we have:

$$\tau_{C,L}(F,G) = \tau_{Min,L}(F,G) , \tag{5.9}$$

$$\sigma_{C,L}(F,G) = \sigma_{Min,L}(F,G) , \tag{5.10}$$

$$\rho_{C,L}(F,G) = \rho_{Min,L}(F,G) , \tag{5.11}$$

and

$$\tau_{C,L}(F,G) = \sigma_{C,L}(F,G) = \rho_{C,L}(F,G) . \tag{5.12}$$

<u>Proof</u>. Clearly, (5.8) implies (5.9), (5.10), and (5.11); and an appeal to Theorem 5 yields (5.12).

From (2.1) and the definition of ε_a, it follows that the hypothesis of Theorem 6 is satisfied for any copula C whenever one of F,G is a unit step function in $\mathcal{D}^+$. Hence we have the following:

COROLLARY. If F,G in $\mathcal{D}^+$ are such that either $F = \varepsilon_a$ for some $a \geq 0$ or $G = \varepsilon_b$ for some $b \geq 0$, then (5.9) - (5.12) hold for any copula C and any L in $\mathcal{L}$.

6. STRICT INEQUALITIES

In this section we prove a number of theorems which, taken together, are the statement that when (5.8) fails for a single pair (u,v) in $(0,\infty) \times (0,\infty)$, and L is strictly increasing (on either $(0,\infty) \times (0,\infty)$ or $[0,\infty) \times [0,\infty)$), then, apart from the essentially trivial cases in which at least one of F,G is a unit step function, the equalities in (5.9), (5.11), (5.12) all fail. As a consequence, it follows that when

$$C(s,t) < \mathrm{Min}(s,t) \quad \text{for all} \quad (s,t) \quad \text{in} \quad (0,1) \times (0,1) , \tag{6.1}$$

then (5.9), (5.11), (5.12), viewed as functional equations for the unknown distribution functions F and G in Δ^+, have only essentially trivial solutions.

Given functions F,G in Δ^+, we write $F < G$ if $F \leq G$ and there is a number x such that $F(x) < G(x)$.

THEOREM 7. Let L in $\mathcal{L}$ be strictly increasing in each place on the open first quadrant $(0,\infty) \times (0,\infty)$. Let C be a copula and F,G be in Δ^+. If there is a point (u,v) in the first quadrant such that

$$C(F(u),G(v)) < \mathrm{Min}(F(u),G(v)) , \tag{6.2}$$

then

$$\tau_{C,L}(F,G) < \sigma_{C,L}(F,G) . \tag{6.3}$$

<u>Proof</u>. By virtue of (4.7), it suffices to show that (6.2) implies the existence of a positive number z such that

$$\tau_{C,L}(F,G)(z) < \sigma_{C,L}(F,G)(z) . \tag{6.4}$$

We begin by observing that (6.2) implies that u and v are positive and finite, that $F(u) > C(F(u),G(v))$, and that $G(v) > C(F(u),G(v))$. Hence, by the left-continuity of F and G and the continuity of C, there are positive numbers u_0,v_0, with $u_0 < u$ and $v_0 < v$, such that

$$F(u_0) > C(F(u),G(v)) \geq C(F(u_0),G(v))$$

$$\text{and} \quad G(v_0) > C(F(u),G(v)) \geq C(F(u),G(v_0)) .$$

Since

$$G(v_0) = \lim_{x\to\infty} C(F(x),G(v_0)) ,$$

it follows that there is a number $u_3 > u$ such that

$$C(F(u_3),G(v_0)) > C(F(u),G(v_0)) .$$

Now define u_2 by

$$u_2 = \inf\{x \mid C(F(x),G(v_0)) = C(F(u_3),G(v_0))\} .$$

It follows that $u_2 \geq u > u_0$, and that

$$C(F(u_2^+),G(v_0)) > C(F(x),G(v_0)) \quad \text{for all} \quad x < u_2 . \tag{6.5}$$

Similarly, there is a number v_2 such that $v_2 \geq v > v_0$, and

$$C(F(u_0),G(v_2^+)) > C(F(u_0),G(y)) \quad \text{for all} \quad y < v_2 . \tag{6.6}$$

Now consider the numbers $L(u_0,v_2)$, $L(u_2,v_0)$. Suppose $L(u_0,v_2) \geq L(u_2,v_0)$, and set $z = L(u_0,v_2)$. Then the point (u_0,v_2) is on the curve $L[z]$, and the point (u_2,v_0) is either on $L[z]$ or in the region $L\{z\}$. Since $u_0 < u_2$, for every number u_1 satisfying $u_0 < u_1 < u_2$, there is a unique number v_1 such that (u_1,v_1) is on $L[z]$; moreover, v_1 satisfies $v_0 < v_1 < v_2$. Choose a particular point (u_1,v_1), and define the rectangular regions R_1 and R_2 as follows:

$$R_1 = \{(x,y) \mid 0 \leq x < u_0,\ \ v_1 \leq y \leq v_2\} ,$$

$$R_2 = \{(x,y) \mid u_1 \leq x \leq u_2,\ \ 0 \leq y < v_0\} .$$

Then R_1 and R_2 are disjoint subsets of $L\{z\}$; and letting m_1 denote the $C(F,G)$-measure of R_1, we have

$$m_1 = C(F(u_0),G(v_2^+)) - C(F(u_0),G(v_1)) ,$$

which, in view of (6.6), is positive. Similarly, m_2, the $C(F,G)$-measure of R_2, is given by

$$m_2 = C(F(u_2^+),G(v_0)) - C(F(u_1),G(v_0)) ,$$

which, in view of (6.5), is also positive.

Now let (x,y) be any point on $L[z]$. If $x \geq u_1$, then $R(x,y)$ is disjoint from R_1 and we have:

$$\sigma_{C,L}(F,G)(z) \geq \iint_{R(x,y)} dC(F(u'),G(v')) + \iint_{R_1} dC(F(u'),G(v'))$$

$$= C(F(x),G(y)) + m_1 .$$

Similarly, if $x \le u_1$, then $R(x,y)$ is disjoint from R_2, and we have:

$$\sigma_{C,L}(F,G)(z) \ge C(F(x),G(y)) + m_2 .$$

Since m_1 and m_2 are positive and independent of x and y, we therefore obtain

$$\sigma_{C,L}(F,G)(z) \ge \tau_{C,L}(F,G)(z) + \mathrm{Min}(m_1,m_2) > \tau_{C,L}(F,G)(z) ,$$

which is (6.4).

Finally, if $L(u_0,v_2) \le L(u_2,v_0)$, we set $z = L(u_2,v_0)$ and proceed as before. This completes the proof.

If F in $\mathcal{D}^+$ is not equal to ε_a for any $a \ge 0$, nor G in $\mathcal{D}^+$ to ε_b for any $b \ge 0$, then there is a point (u,v) in the first quadrant such that the point $(F(u),G(v))$ is in the interior of the unit square. Upon combining this observation with Theorem 7 and the corollary to Theorem 6, we obtain the following:

COROLLARY. If L is as in Theorem 7, and C satisfies (6.1), then

$$\tau_{C,L}(F,G) = \sigma_{C,L}(F,G)$$

for F,G in $\mathcal{D}^+$ if and only if either $F = \varepsilon_a$ for some $a \ge 0$, or $G = \varepsilon_b$ for some $b \ge 0$.

THEOREM 8. Let L in $\mathcal{L}$ be strictly increasing in each place on the half-closed first quadrant $[0,\infty) \times [0,\infty)$. Let C be a copula and F,G be in $\mathcal{D}^+$. If (6.2) holds for some point (u,v), then

$$\sigma_{C,L}(F,G) < \rho_{C,L}(F,G) . \tag{6.7}$$

(Note that the hypothesis on L here is stronger than the corresponding hypothesis in Theorem 7.)

<u>Proof</u>. Assume (6.2) and consider the function k_1 defined by

$$k_1(x) = F(x) - C(F(x),G(v)) .$$

If $x_1 < x_2$, then $k_1(x_2) - k_1(x_1)$ is equal to

$$C(F(x_2),G(\infty)) - C(F(x_2),G(v)) - C(F(x_1),G(\infty)) + C(F(x_1),G(v)) ,$$

which is nonnegative by virtue of (2.2). Since $k_1(0) = 0$, and (6.2) implies that $k_1(u) > 0$, it follows that the number u_0 defined by

$$u_0 = \sup\{x \mid k_1(x) = 0\}$$

has the properties: $0 \le u_0 < u$, and

(6.8) $$k_1(x) - k_1(u_0) = k_1(x) > 0 \quad \text{for all} \quad x > u_0 .$$

Similarly, the function k_2 defined by

$$k_2(y) = G(y) - C(F(u),G(y))$$

is nondecreasing, and there is a number v_0 such that $0 \le v_0 < v$, and

(6.9) $$k_2(y) - k_2(v_0) = k_2(y) > 0 \quad \text{for all} \quad y > v_0 .$$

Now consider the numbers $L(u_0,v)$ and $L(u,v_0)$. If $L(u_0,v) \le L(u,v_0)$, set $z = L(u_0,v)$. Then (u_0,v) is on $L[z]$ and (u,v_0) is in the complement of the set $L\{z\}$. We can therefore find a point (u_1,v_1) on $L[z]$ such that $u_0 < u_1 \le u$ and $v_0 < v_1 < v$. We now define rectangular regions R_1 and R_2 via:

$$R_1 = \{(x,y) \mid u_0 \le x < u_1, \ v \le y\} ,$$

$$R_2 = \{(x,y) \mid u \le x, \ v_0 \le y < v_1\} .$$

It follows that R_1,R_2, and $L\{z\}$ are pairwise disjoint. Moreover, the $C(F,G)$-measure of R_1 is $k_1(u_1) - k_1(u_0)$, which by (6.8) is a fixed positive number; and the $C(F,G)$-measure of R_2 is $k_2(v_1) - k_2(v_0)$, which by (6.9) is again a fixed positive number. Now for any point (x,y) on $L[z]$, it is readily seen that the set $\Gamma(x,y)$ must include one or the other of the sets R_1,R_2, in addition to the set $L\{z\}$. Hence it follows that the numbers $\rho_{C,L}(F,G)(z)$ and $\sigma_{C,L}(F,G)(z)$ must differ by at least a fixed positive quantity, which yields $\sigma_{C,L}(F,G) < \rho_{C,L}(F,G)$. The same conclusion follows in an analogous manner if $L(u_0,v) \ge L(u,v_0)$, and so the theorem is proved.

COROLLARY. If L is as in Theorem 8, and C satisfies (6.1), then

$$\sigma_{C,L}(F,G) = \rho_{C,L}(F,G)$$

for F,G in $\mathcal{D}^+$ if and only if either $F = \varepsilon_a$ for some $a \ge 0$, or $G = \varepsilon_b$ for some $b \ge 0$.

The following example shows that the hypothesis on L in Theorem 8 cannot be weakened to the corresponding hypothesis in Theorem 7.

Let $C = \text{Proc}$ and $L = \text{Mult}$. Then C satisfies (6.1), and L is strictly increasing on $(0,\infty) \times (0,\infty)$ but not on $[0,\infty) \times [0,\infty)$ (being constant and equal to 0 on the coordinate axes). Hence the hypotheses of Theorem 7, but not those of Theorem 8, are satisfied. Let F in $\mathcal{D}^+$ be given by:

$$F(x) = \begin{cases} 0\,, & x \leq 0\,, \\ 1/2\,, & 0 < x \leq 1\,, \\ 1\,, & 1 < x\,. \end{cases} \tag{6.10}$$

Then straightforward calculations yield the following:

$$\sigma_{\text{Prod,Mult}}(F,F)(x) = \begin{cases} 0\,, & x \leq 0\,, \\ 3/4\,, & 0 < x \leq 1\,, \\ 1\,, & 1 < x\,; \end{cases}$$

$\tau_{\text{Prod,Mult}}(F,F) = F < \sigma_{\text{Prod,Mult}}(F,F)$, in accordance with Theorem 7; but $\rho_{\text{Prod,Mult}}(F,F) = \sigma_{\text{Prod,Mult}}(F,F)$, contrary to the conclusion of Theorem 8.

THEOREM 9. Under the hypotheses of Theorem 8, we have

$$\tau_{C,L}(F,G) < \tau_{\text{Min},L}(F,G)\,. \tag{6.11}$$

Proof. As in the proof of Theorem 7, there exist u_0, v_0, with $0 < u_0 < u$ and $0 < v_0 < v$, such that

$$\text{Min}(F(u_0),G(v_0)) > C(F(u),G(v))\,. \tag{6.12}$$

Assume, without loss of generality, that $F(u_0) \geq G(v_0)$. Let

$$c = \sup\{x \mid F(x) < G(v_0)\}\,,\ d = \sup\{y \mid G(y) < G(v_0)\}\,,$$

and note that $0 \leq c \leq u_0 < u$ and $0 \leq d \leq v_0 < v$. Let $z = L(c,d)$. Since $L(c,v) > z$ and $L(u,d) > z$, there is a number w such that $z < w < \text{Min}(L(c,v),L(u,d))$. Let x_0 be the unique solution (if it exists) of the equation $L(x_0,v) = w$; and let y_0 be the unique solution (if it exists) of $L(u,y_0) = w$.

Now let (x,y) be a point on $L[w]$. If $y \geq v$, then $L(x,v) \leq w < L(c,v)$, whence x_0 exists and $x \leq x_0 < c$. Consequently,

$$C(F(x),G(y)) \leq F(x) \leq F(x_0) < G(v_0)\,. \tag{6.13}$$

Similarly, if $x \geq u$ then y_0 exists, $y \leq y_0 < d$, and

$$C(F(x),G(y)) \leq G(y) \leq G(y_0) < G(v_0)\,. \tag{6.14}$$

Otherwise, $x < u$ and $y < v$, so that by (6.12) we have

$$C(F(x),G(y)) \leq C(F(u),G(v)) < G(v_0) . \tag{6.15}$$

It follows from (6.13), (6.14), (6.15), and the definition of w that

$$\tau_{C,L}(F,G)(z+) \leq \tau_{C,L}(F,G)(w) < G(v_0) .$$

On the other hand, by considering the limit of the sequence of points $\{c + 2^{-n}, d + 2^{-n})\}$, we easily find that

$$G(v_0) \leq \tau_{\text{Min},L}(F,G)(z+) \leq \tau_{\text{Min},L}(F,G)(w) ,$$

which, combined with the preceding inequality, completes the proof.

COROLLARY 1. If L is as in Theorem 9, and C satisfies (6.1), then

$$\tau_{C,L}(F,G) = \tau_{\text{Min},L}(F,G)$$

for F,G in $\mathcal{D}^+$ if and only if either $F = \varepsilon_a$ for some $a \geq 0$ or $G = \varepsilon_b$ for some $b \geq 0$.

If C is any copula, then (4.11) yields

$$\tau_{T_m,L}(F,G) \leq \tau_{C,L}(F,G) \leq \sigma_{C,L}(F,G) \tag{6.16}$$

for all L in $\mathcal{L}$, and F,G in $\mathcal{D}^+$. If either F or G is a unit step function, then the Corollary to Theorem 6 implies that we have equality in both places in (6.16). On the other hand, if L is strictly increasing on $[0,\infty) \times [0,\infty)$, and neither F nor G is a unit step function, then there are two possibilities: either (5.8), or its negation, (6.2), is satisfied. If (5.8) is satisfied, then Theorem 6 yields $\tau_{C,L}(F,G) = \tau_{\text{Min},L}(F,G)$, but the preceding corollary to Theorem 9 yields $\tau_{T_m,L}(F,G) < \tau_{\text{Min},L}(F,G)$, whence we have strict inequality on the left of (6.16). If (6.2) is satisfied, then Theorem 7 gives us strict inequality on the right of (6.16). Hence we have the following result, which plays a crucial role in the characterization of Wald-betweenness in pseudometrically generated probabilistic metric spaces (see [3]):

COROLLARY 2. If L is as in Theorem 8 (in particular, if $L = \text{Sum}$), and C is any copula, then

$$\tau_{T_m,L}(F,G) = \sigma_{C,L}(F,G)$$

for F,G in $\mathcal{D}^+$ if and only if either $F = \varepsilon_a$ for some $a \geq 0$ or $G = \varepsilon_b$ for some $b \geq 0$.

An argument similar to, but simpler than, the one given in the proof of Theorem 9 yields:

THEOREM 10. Under the hypotheses of Theorem 7, we have

$$\rho_{Min,L}(F,G) < \rho_{C,L}(F,G) . \tag{6.17}$$

COROLLARY. If L is as in Theorem 7, and C satisfies (6.1), then

$$\rho_{Min,L}(F,G) = \rho_{C,L}(F,G) ,$$

for F,G in $\mathcal{D}^+$ if and only if either $F = \varepsilon_a$ for some $a \geq 0$ or $G = \varepsilon_b$ for some $b \geq 0$.

Finally, a simple calculation shows that if F is the distribution function defined by (6.10), then

$$\tau_{T_m,Mult}(F,F) = F = \tau_{Min,Mult}(F,F) ,$$

whence, again, the hypothesis on L of Theorem 9 cannot be weakened to that of Theorem 10.

ACKNOWLEDGMENT. The second author gratefully acknowledges the support of the Research Council of the University of Massachusetts.

REFERENCES

1. M. J. Frank, Associativity in a class of operations on distribution functions, Aequat. Math. 12 (1975), 121-144.

2. J. Kampé de Fériet et B. Forte, Information et probabilité, C. R. Acad. Sci. Paris, Sér. A 265 (1967), 110-114, 142-146, 350-353.

3. R. Moynihan and B. Schweizer, Betweenness relations in probabilistic metric spaces, to appear.

4. B. Schweizer, Multiplications on the space of probability distribution functions, Aequat. Math. 12 (1975), 156-183.

5. B. Schweizer and A. Sklar, Mesure aléatoire de l'information et mesure de l'information par un ensemble d'observateurs, C. R. Acad. Sci. Paris Sér. A 272 (1971), 149-152.

6. B. Schweizer and A. Sklar, Operations on distribution functions not derivable from operations on random variables, Studia Math. 52 (1974), 43-52.

7. A. Sklar, Random variables, joint distribution functions, and copulas, Kybernetika, 9 (1973), 449-460.

8. K. Urbanik, Generalized convolutions, Studia Math. 23 (1964), 217-245.

Functional Inequalities

FUNCTIONAL INEQUALITIES

T. Howroyd
Department of Mathematics
University of New Brunswick
Frederiction, N. B.
CANADA

ABSTRACT. It is shown that if b_1 and b_2 are real functions which map the half-open interval $[\alpha,\beta)$ into the closed interval $[0,1]$, and the functions f_1 and f_2 are suitably defined, then the system of linear homogeneous functional inequalities

$$\varphi(x) \leq b_1(x)\varphi(f_1(x)) \quad ,$$

$$\varphi(f_2(x)) \leq b_2(x)\varphi(x) \quad , \qquad \alpha \leq x < \beta \quad ,$$

requires each continuous solution φ to be bounded above by the constant $\max\{0,\varphi(\alpha)\}$. If b_1 and b_2 are both identically unity, then φ is constant. Relationships are obtained between solutions of a system of linear functional inequalities and the corresponding system of functional equations or the corresponding system obtained by reversing the inequalities. The nonexistence of continuous periodic solutions of some linear functional inequalities is demonstrated.

1. INTRODUCTION

If w and δ are positive real numbers and their ratio is irrational, then any continuous real function φ with periods w and δ is necessarily constant. The same result is valid if the biperiodicity of φ is replaced by the weaker requirement that φ satisfy the system of functional inequalities (Montel [5], Popoviciu [6])

$$\varphi(x + w) \leq \varphi(x) \leq \varphi(x + \delta) \quad .$$

On the other hand, Kuczma and Szymiczek [4] have shown that a continuous periodic solution φ of a functional equation of the form

$$\varphi(x) = g(x,\varphi(x))$$

is uniquely determined by its value at a single point.

The results in this paper generalize results in [5] and [6], and are related to results in [4], in that periodic solutions of some linear functional inequalities are shown not to exist.

2. SYSTEMS OF FUNCTIONAL INEQUALITIES

The systems of functional inequalities take the general form

$$\begin{aligned} &\varphi(x) \leq F_1(x,\varphi(f_1(x))) \quad , \\ &\varphi(f_2(x)) \leq F_2(x,\varphi(x)) \quad , \qquad \alpha \leq x < \beta \quad , \end{aligned} \tag{1}$$

where φ is an unknown function which is required to be continuous and real. The functions f_1 and f_2 are continuous maps of the half-open interval $[\alpha,\beta)$ into itself; they enjoy one or both of the following properties (H_{12}), (H_{21}):

(H_{ij}) If I is a nonempty open subinterval of the interval $[\alpha,\beta)$, then there exist positive integers p and q such that the qth iterate $f_j^q(\alpha)$ lies in the interval $f_i^p(I)$.

If, for example, $\beta = \infty$, $f_1(x) = x + w$, and $f_2(x) = x + \delta$, where w and δ are positive and w/δ is irrational, then it is well known that both (H_{12}) and (H_{21}) hold. Alternatively, the pairs of functions appearing in [4], one of which is a periodic function, satisfy one of these properties; a trivial example of such a pair of functions which satisfies (H_{12}) is given by $f_2(x) = x + w$ and $f_1(x) = cx$, where $w > 0$, $c > 1$, $\alpha = 0$, and $\beta = \infty$.

THEOREM 1. Let f_1 and f_2 satisfy the condition (H_{12}). Let F_1 and F_2 be real functions on $[\alpha,\beta) \times R$ which satisfy the conditions

$$y \leq 0 \Rightarrow F_j(x,y) \leq 0 \quad , \tag{2}$$

$$y > 0 \Rightarrow F_j(x,y) \leq y \quad , \tag{3}$$

$$F_j(x,y + z) \leq F_j(x,y) + F_j(x,z) \quad . \tag{4}$$

If φ is a continuous solution of the system of functional inequalities (1), then

$$\varphi(x) \leq \max\{0,\varphi(\alpha)\} \quad , \qquad \alpha \leq x < \beta \quad . \tag{5}$$

Proof. (i) Suppose $\varphi(\alpha) \leq 0$. Then repeated application of (1) and

(2) yields the inequalities

$$\varphi(f_2(\alpha)) \leq F_2(\alpha,\varphi(\alpha)) \leq 0 \quad ,$$

$$\varphi(f_2^2(\alpha)) \leq F_2(f_2(\alpha),\varphi(f_2(\alpha))) \leq 0 \quad ,$$

and it follows by induction that $\varphi(f_2^n(\alpha)) \leq 0$ $(n = 1,2,\ldots)$. However, if φ is continuous and positive at some point, then φ is positive in some open interval I, whence (1) and (2) imply the inequalities

$$F_1(x,\varphi(f_1(x))) > 0 \quad , \quad x \in I \quad ,$$

$$\varphi(f_1(x)) > 0 \quad , \quad x \in I \quad .$$

Hence φ is positive in $f_1(I)$, and by induction is positive in $f_1^n(I)$ $(n = 1,2,\ldots)$. But (H_{12}) implies that at least one of the iterates $f_2^n(\alpha)$ is in at least one of the intervals $f_1^n(I)$, which is impossible since φ cannot be both positive and negative at a point. Hence φ is not positive anywhere.

(ii) Suppose $\varphi(\alpha) > 0$. If $\varphi_0 = \varphi - \varphi(\alpha)$, then (1), (4), and (3) yield the first of the inequalities (1):

$$\begin{aligned}\varphi_0(x) &\leq F_1(x,\varphi(f_1(x))) - \varphi(\alpha)\\ &\leq F_1(x,\varphi_0(f_1(x))) + F_1(x,\varphi(\alpha)) - \varphi(\alpha)\\ &\leq F_1(x,\varphi_0(f_1(x))) \quad .\end{aligned}$$

Similarly, the second inequality in (1) is satisfied by φ_0. But φ_0 is continuous and $\varphi_0(\alpha) = 0$, whence the first case considered above implies that $\varphi_0 \leq 0$, or $\varphi \leq \varphi(\alpha)$.

As a particular case of Theorem 1, homogeneous linear systems are examined in the following theorem.

THEOREM 2. Let f_1 and f_2 satisfy the condition (H_{12}). Let the functions b_1 and b_2 map $[\alpha,\beta)$ into the closed interval $[0,1]$. If φ is a continuous solution of the system of functional inequalities

$$\varphi(x) \leq b_1(x)\varphi(f_1(x)) \quad , \quad \varphi(f_2(x)) \leq b_2(x)\varphi(x) \quad , \qquad \alpha \leq x < \beta \quad , \tag{6}$$

then (5) holds.

To illustrate Theorem 2, consider the system

$$\varphi(x + w) \leq \varphi(x) \leq b\varphi(cx) \quad , \quad 0 \leq x < \infty \quad ,$$

where b, c, and w are positive, and $b < 1 < c$. Theorem 2 implies that every continuous solution φ satisfies the inequality

$$\varphi(x) \leq \max\{0,\varphi(0)\} \quad , \quad x \geq 0 \quad .$$

But the inequality $\varphi(0) \leq b\varphi(0)$ implies $\varphi(0) \leq 0$, so φ is bounded above by 0. It may be noted that the solutions $\varphi(x) = Kx + L$, where K, $L \leq 0$, do not exceed the upper bound zero.

The results in [5] and [6] are particular cases of the following theorem.

THEOREM 3. Let f_1 and f_2 either satisfy the conditions (H_{12}) and (H_{21}); or satisfy the condition (H_{12}), and the function f_1 have the property that the iterates $f_1^{-n}(x)$ exist and converge to the constant α. If φ is a continuous solution of the system of functional inequalities

$$\varphi(f_2(x)) \leq \varphi(x) \leq \varphi(f_1(x)) \quad , \quad \alpha \leq x < \beta \quad , \tag{7}$$

then φ is constant.

Proof. Without loss of generality, it may be assumed that $\varphi(\alpha) = 0$, so (H_{12}) implies that $\varphi \leq 0$. On the other hand, (H_{21}) implies that $-\varphi \leq 0$. Hence (H_{12}) and (H_{21}) together imply that $\varphi = 0$.

Finally, the inequalities $\varphi(x) \geq \varphi(f_1^{-n}(x))$, in the limit as $n \to \infty$, imply that $\varphi(x) \geq \varphi(\alpha) = 0$, and the extra condition (H_{12}) then implies that $\varphi = 0$.

THEOREM 4. Let f_1 and f_2 satisfy the condition (H_{12}). Let a_1, a_2 be real functions on $[\alpha,\beta)$ and b_1, b_2 map $[\alpha,\beta)$ into the closed interval $[0,1]$. If φ is a continuous solution of the system

$$\begin{aligned} &\varphi(x) \leq a_1(x) + b_1(x)\varphi(f_1(x)) \quad , \\ &\varphi(f_2(x)) \leq a_2(x) + b_2(x)\varphi(x) \quad , \quad \alpha \leq x < \beta \quad , \end{aligned} \tag{8}$$

and ψ is a continuous solution of the system

$$\psi(x) \geq a_1(x) + b_1(x)\psi(f_1(x)) \quad , \quad \psi(f_2(x)) \geq a_2(x) + b_2(x)\psi(x) \quad , \quad \alpha \leq x < \beta \quad , \tag{9}$$

and $\varphi(\alpha) \leq \psi(\alpha)$, then $\varphi \leq \psi$.

Proof. If $\mu = \varphi - \psi$, then (8) and (9) imply that μ satisfies (6). But $\mu(\alpha) \leq 0$, so (5) implies $\mu \leq 0$, and $\varphi \leq \psi$.

THEOREM 5. Let f_1 and f_2 satisfy the conditions of Theorem 3. Let a_1 and a_2 be real functions on $[\alpha,\beta)$. If φ is a continuous solution of the system

$$\varphi(x) \leq a_1(x) + \varphi(f_1(x))$$

(10)

$$\varphi(f_2(x)) \leq a_2(x) + \varphi(x) \quad , \quad \alpha \leq x < \beta \ ,$$

and ψ is a continuous solution of the system

$$\psi(x) \geq a_1(x) + \psi(f_1(x))$$

(11)

$$\psi(f_2(x)) \geq a_2(x) + \psi(x) \quad , \quad \alpha \leq x < \beta \ ,$$

and $\varphi(\alpha) = \psi(\alpha)$, then $\varphi = \psi$.

Proof. If $\mu = \varphi - \psi$, then μ satisfies (7) and so is a constant, which must be zero.

COROLLARY. Let f_1, f_2, a_1, a_2 satisfy the hypotheses of Theorem 5. Then each continuous solution ρ of the system of functional equations

$$\rho(x) = a_1(x) + \rho(f_1(x)) \quad ,$$

(12)

$$\rho(f_2(x)) = a_2(x) + \rho(x) \quad , \qquad \alpha \leq x < \beta \ ,$$

is uniquely determined by its value at α. If (11) has a continuous solution which is not a solution of (12), then (12) has no continuous solution.

Proof. If ρ_1 and ρ_2 are continuous solutions of (12) coincident at α, then $\varphi = \rho_1$ satisfies (10) and $\psi = \rho_2$ satisfies (11), whence $\rho_1 = \rho_2$.

Let ψ be a continuous solution of (11), but not satisfy (12). If (12) has a continuous solution ρ, then $\varphi = \rho - \rho(\alpha) + \psi(\alpha)$ is a continuous solution of (12), and $\varphi(\alpha) = \psi(\alpha)$. Hence $\varphi = \psi$, and ρ cannot exist.

Various authors (Fortet [2], Kac [3], Ciesielski [1]) have examined periodic solutions φ of the functional equation

$$\varphi(x) = f(x) + \varphi(2x) \quad , \quad x \geq 0 \ .$$

In [3] it is shown that continuous, even integrable, periodic solutions do not always exist. In the next theorem, continuous periodic solutions of the corresponding functional inequality are shown not to exist.

THEOREM 6. Let c be a positive real number and f be a nonnegative function which is not identically zero. Then the functional inequality

$$\varphi(x) \geq f(x) + \varphi(cx) \quad , \quad x \geq 0 \quad , \tag{13}$$

has no continuous periodic solution φ.

Proof. Let φ satisfy (13) and have positive period w. Then

$$\varphi(cx) \leq \varphi(x) = \varphi(x + w) \quad , \quad x \geq 0 \quad .$$

If $c > 1$, then Theorem 3 implies that φ is constant, and so cannot satisfy (13).

If $c < 1$, then $1/c > 1$ and the system (13) may be written

$$\varphi(\frac{x}{c}) \leq -\varphi(x) = -\varphi(x + w) \quad , \quad x \geq 0 \quad .$$

Hence the same conclusion is valid.

If $c = 1$, then (13) cannot hold.

REFERENCES

1. Z. Ciesielski, On the functional equation $f(t) = g(t) - g(2t)$, Proc. Amer. Math. Soc. 13 (1962), 388-393.

2. R. Fortet, Sur une suite egalement répartie, Studia Math. 9 (1940), 54-70.

3. M. Kac, On the distribution of values of sums of type $\Sigma f(2^k t)$, Ann. of Math. 47 (1946), 33-49.

4. M. Kuczma and K. Szymiczek, On periodic solutions of a functional equation, Amer. Math. Monthly 70 (1963), 847-850.

5. P. Montel, Sur les propriétés périodiques des fonctions, C. R. Acad. Sci. Paris 251 (1960), 2111-2112.

6. C. Popoviciu, Sur le parallélisme entre des équations différentielles et les équations fonctionnelles, C. R. Acad. Sci. Paris 188 (1929), 763-765.

ALMOST SUBADDITIVE FUNCTIONS

Roman Ger
Department of Mathematics
Silesian University
40-007 Katowice
POLAND

ABSTRACT. Let $(D,+)$ be a subsemigroup of $(\mathbb{R}^n,+)$. This paper is concerned with some properties of functions $f : D \to \mathbb{R}$ fulfilling the subadditivity condition $f(x + y) \leq f(x) + f(y)$ for almost all pairs $(x,y) \in D^2$.

1. INTRODUCTION

Answering a question of P. Erdös [2], N. G. de Bruijn (cf. also W. B. Jurkat [6]) has proved in [1] that every function $f : \mathbb{R} \to \mathbb{R}$ which satisfies the additivity condition almost everywhere in $\mathbb{R}^2$ (almost additive function) must be of the form

$$f(x) = \begin{cases} h(x) & \text{for } x \in \mathbb{R}\setminus W \\ a(x) & \text{for } x \in W \end{cases},$$

where W is a null set on the real line, $h : \mathbb{R} \to \mathbb{R}$ is an additive function, and a is an arbitrary map of W into $\mathbb{R}$. M. Kuczma [7] has shown that the analogous theorem remains true in the class of convex functions (in the sense of Jensen). It is to be expected that a similar result should be obtained for subadditive functions. However, as was pointed out in [5, p. 239], the classes of convex functions and subadditive functions are "rather remotely related" (note that additive functions are simultaneously convex and subadditive). At any rate, our result on almost subadditive functions is not so complete as those quoted above. We have found only a subadditive essential minorant and majorant of a given almost subadditive function; moreover, we present a sufficient condition on the given function in order that these boundary maps coincide almost everywhere. On the other hand, up to now, we have no example of an almost subadditive function which would not be almost everywhere equal to a subadditive function; it is not excluded that such functions do not exist.

2. MAIN RESULTS

In the sequel, the symbols m_n^* and m_n will always denote the n-dimensional outer Lebesgue measure and the n-dimensional Lebesgue measure, respectively. We shall be concerned with real-valued functions f that are defined on a set $D \subset \mathbb{R}^n$ such that $(D,+)$ is a subsemigroup of $(\mathbb{R}^n,+)$, and that fulfill the subadditivity condition

$$f(x + y) \leq f(x) + f(y) \tag{1}$$

for almost all (in the sense of m_{2n}) pairs $(x,y) \in D^2$. For obvious reasons, we shall constantly assume that

$$m_n(D) > 0 . \tag{2}$$

Such functions f will be referred to as almost subadditive functions. We do not restrict our attention to measurable maps.

THEOREM 1. Suppose $f : D \to \mathbb{R}$ to be almost subadditive. The function $\varphi : D \to \mathbb{R}$ given by the formula

$$\varphi(x) = \sup_{h \in D} \text{ess} \, [f(x + h) - f(h)] , \tag{3}$$

$x \in D$, is subadditive and satisfies the condition

$$m_n (\{x \in D \mid \varphi(x) > f(x)\}) = 0 .$$

Proof. Making use of the well-known properties of the sup ess operation, for every $x \in D$, we may find a null set $E_x \subset D$ such that the value $\varphi(x)$ defined by (3) can be represented as follows:

$$\varphi(x) = \sup_{h \in D \setminus E_x} [f(x + h) - f(h)] .$$

Consequently, for $x,y \in D$ arbitrarily fixed, we have obviously the following inequalities:

$$\varphi(x) \geq f(x + h) - f(h) \quad \text{for all } h \in D \setminus E_x$$

and

$$\varphi(y) \geq f(y + k) - f(k) \quad \text{for all } k \in D \setminus E_y .$$

Put $B := E_x \cup (E_y - x)$. Evidently $m_n(B) = 0$. Take a $t \in D \setminus B$. Then $t \notin E_x$ as well as $x + t \notin E_y$, whence, putting $h = t$ and $k = x + t$ in the above inequalities, we get

$$\varphi(x) \geq f(x + t) - f(t) , \quad \varphi(y) \geq f(y + x + t) - f(x + t) ,$$

and hence also

$$\varphi(x) + \varphi(y) \geq f(x + y + t) - f(t) .$$

Thus

$$\varphi(x) + \varphi(y) \geq \sup_{t \in D \setminus B} [f(x + y + t) - f(t)]$$

$$\geq \sup_{t \in D} \text{ess} [f(x + y + t) - f(t)] = \varphi(x + y) ;$$

i.e., φ is subadditive.

Almost subadditivity of f means that there exists a set $M \subset D^2$ such that $m_{2n}(M) = 0$ and (1) is satisfied for all $(x,y) \in D^2 \setminus M$. By Fubini's theorem, there exists a set $U(M) \subset D$ such that $m_n(U(M)) = 0$ and

$$V_x(M) := \{y \in D \mid (x,y) \in M\}$$

is a null set provided $x \in D \setminus U(M)$. Take an $x \in D \setminus U(M)$. Then

$$\varphi(x) = \sup_{h \in D} \text{ess} [f(x + h) - f(h)] \leq \sup_{h \in D \setminus V_x(M)} [f(x + h) - f(h)]$$

$$\leq \sup_{h \in D \setminus V_x(M)} [f(x) + f(h) - f(h)] = f(x) ,$$

which finishes our proof.

For a given $x \in D$, we put

$$\Delta_x := D \cap (x - D) . \tag{4}$$

We shall apply the following:

LEMMA. For every $x,y \in D$, we have $\Delta_y \subset \Delta_{x+y} - h$, provided $h \in \Delta_x$.

<u>Proof</u>. Take a $t \in \Delta_y$ and an $h \in \Delta_x$. Then t, $y - t$, h, and $x - h$ are members of D, whence so are $t + h$ and $y - t + x - h$. Therefore $t \in D - h$ as well as $t \in (x + y - D) - h$, which simply means that $t \in [D \cap (x + y - D)] - h = \Delta_{x+y} - h$.

THEOREM 2. Suppose $f : D \to \mathbb{R}$ to be almost subadditive. If

$$m_n^*(\Delta_x) > 0 \quad \text{for all } x \in D , \tag{5}$$

where Δ_x is defined by (4), then the function $\Phi : D \to \mathbb{R}$ given by the formula

$$\Phi(x) := \inf_{h \in \Delta_x} \text{ess} [f(x - h) + f(h)] , \tag{6}$$

$x \in D$, is subadditive and satisfies the condition

$$m_n(\{x \in D \mid f(x) > \Phi(x)\}) = 0 .$$

<u>Proof</u>. Fix $x, y \in D$ and an $\varepsilon \in (0, \infty)$. Let $M \subset D^2$, $U(M)$, and $V_x(M)$ have the same meaning as in the proof of Theorem 1. We have

$$\begin{aligned} \Phi(x) &\geq \inf_{h \in \Delta_x \setminus [U(M) \cup (x - U(M))]} [f(x-h) + f(h)] \\ &\geq f(x-h) + f(h) - \tfrac{1}{2}\varepsilon \end{aligned} \tag{7}$$

for a certain $h \in \Delta_x$ such that $h \notin U(M)$ and $x - h \notin U(M)$. On account of the preceding Lemma, we infer that $\Delta_y \subset \Delta_{x+y} - h$. Consequently, denoting by E_x a null set such that

$$\operatorname*{inf\,ess}_{h \in \Delta_x} [f(x-h) + f(h)] = \inf_{h \in \Delta_x \setminus E_x} [f(x-h) + f(h)] ,$$

$x \in D$, we get

$$\begin{aligned} &\Delta_y \setminus [V_h(M) \cup (y - V_{x-h}(M)) \cup (E_{x+y} - h)] \\ &\subset (\Delta_{x+y} - h) \setminus [V_h(M) \cup (y - V_{x-h}(M)) \cup (E_{x+y} - h)] \\ &= [(\Delta_{x+y} \setminus E_{x+y}) - h] \setminus [V_h(M) \cup (y - V_{x-h}(M))] . \end{aligned}$$

Hence

$$\begin{aligned} \Phi(y) &\geq \inf_{k \in \Delta_y \setminus [V_h(M) \cup (y - V_{x-h}(M)) \cup (E_{x+y} - h)]} [f(y-k) + f(k)] \\ &\geq \inf_{k \in [(\Delta_{x+y} \setminus E_{x+y}) - h] \setminus [V_h(M) \cup (y - V_{x-h}(M))]} [f(y-k) + f(k)] \\ &\geq f(y-k) + f(k) - \tfrac{1}{2}\varepsilon \end{aligned} \tag{8}$$

for a certain k such that

(i) $h + k \in \Delta_{x+y} \setminus E_{x+y}$;

(ii) $k \notin V_h(M)$, i.e., $(h, k) \notin M$;

(iii) $y - k \notin V_{x-h}(M)$, i.e., $(x - h, y - k) \notin M$.

Observe that $\Phi(x + y) \leq f(x + y - t) + f(t)$ for all $t \in \Delta_{x+y} \setminus E_{x+y}$. In particular, by means of (i), we have

(iv) $\Phi(x + y) \le f(x + y - (h + k)) - f(h + k)$.

Now, recalling (7) and (8), by (iii), (ii), and (iv), we obtain

$$\Phi(x) + \Phi(y) \ge f(x - h) + f(h) - \frac{1}{2}\varepsilon + f(y - k) + f(k) - \frac{1}{2}\varepsilon$$

$$\ge f(x + y - (h + k)) + f(h) + f(k) - \varepsilon$$

$$\ge f(x + y - (h + k)) + f(h + k) - \varepsilon \ge \Phi(x + y) - \varepsilon .$$

Letting ε tend to zero, we get the subadditivity of Φ.

To prove the second part of our assertion, observe, again by Fubini's theorem, that the set

$$W := \{x \in D \mid m_n^x(\{h \in D \mid (x - h,h) \in M\}) > 0\}$$

is of measure zero. Fix, again, an $\varepsilon \in (0,\infty)$, and take an $x \in D \setminus W$. Then $W(x) := \{h \in D \mid (x - h,h) \in M\}$ is a null set, whence

$$\Phi(x) \ge \inf_{h \in \Delta_x \setminus W(x)} [f(x - h) + f(h)] \ge f(x - h) + f(h) - \varepsilon$$

for a certain $h \in \Delta_x \setminus W(x)$. However, for such an h we have $(x - h,h) \notin M$; therefore,

$$\Phi(x) \ge f(x) - \varepsilon ,$$

which completes the proof since ε has been chosen arbitrarily from $(0,\infty)$.

One may ask about the essential distance between φ and Φ. If may happen to be rather great. In the next section, we shall discuss this question more precisely (see Remark 4, below). Here, we are going to give a condition implying that $\varphi = \Phi$ almost everywhere.

THEOREM 3. Let $(D,+)$ be a subgroup of $(\mathbb{R}^n,+)$. Suppose $f : D \to \mathbb{R}$ to be almost subadditive. Given an $\varepsilon > 0$, we put

$$A_f(\varepsilon) := \{h \in D \mid f(h) < \varepsilon \text{ and } f(-h) < \varepsilon\} . \tag{9}$$

If

$$m_n^*(A_f(\varepsilon)) > 0 \quad \text{for all} \quad \varepsilon > 0 \tag{10}$$

and if the functions φ and Φ are defined by (3) and (6), respectively, then $\varphi = \Phi$ almost everywhere in D.

Proof. Evidently, (10) and the fact that $(D,+)$ is a group imply (5) since $\Delta_x = D \supset A_f(1)$ for all $x \in D$. Thus, in particular, φ and Φ are well defined. It follows from Theorems 1 and 2 that the set

$$W := \{x \in D \mid \varphi(x) > \Phi(x)\}$$

is of measure zero. We shall show that $\varphi(x) = \Phi(x)$ for all $x \in D \setminus W$. Assume, for the indirect proof, that $\varphi(x) < \Phi(x)$ for a certain $x \in D \setminus W$. Then there exists an $\varepsilon > 0$ such that $\varphi(x) + 2\varepsilon \leq \Phi(x)$, whence, by means of the definitions of φ and Φ, we infer that

$$Z_0 := \{h \in D \mid f(x+h) - f(h) + 2\varepsilon > f(x-h) + f(h)\}$$

is of measure zero; obviously so also is $Z := -Z_0 \cup Z_0$. By virtue of (10), one may find an $h \in A_f(\varepsilon) \setminus Z$. For such an h, we get

$$f(x+h) < f(x+h) - f(h) + \varepsilon \leq f(x-h) + f(h) - \varepsilon < f(x-h)$$
$$< f(x-h) - f(-h) + \varepsilon \leq f(x+h) + f(-h) - \varepsilon < f(x+h) ,$$

a contradiction. This ends the proof.

3. SOME REMARKS

REMARK 1. Clearly (5) implies (2), but not conversely. However, in the simplest and most important case where D is an open cone in $\mathbb{R}^n$, assumption (5) is easily satisfied. In fact, taking an $x \in D$ and a positive ε such that

$$\{y \in \mathbb{R}^n \mid \|y - x\| < \varepsilon\} =: K(x,\varepsilon) \subset D$$

and $K(-x,\varepsilon) \subset -D$, we infer that $K(0,\varepsilon) \subset x - D$. Therefore, Δ_x contains a nonempty open set $K(0,\varepsilon) \cap D$, which implies (5). However, it may be of interest to observe that (5) holds for rather pathological semigroups, too. For, take a saturated Hamel base of the reals, i.e., such a base H that

$$m_1^*(H \cap (\alpha,\beta)) = m_1((\alpha,\beta))$$

for every interval $(\alpha,\beta) \subset \mathbb{R}$. Such bases do exist (cf., for instance, [7]). Now put

$$a_0(h) := \begin{cases} h & \text{for } h \in H \cap (0,\infty) , \\ 0 & \text{for } h \in H \cap (-\infty,0] , \end{cases}$$

and extend a_0 to an additive function $a : \mathbb{R} \to \mathbb{R}$. Obviously, a is discontinuous; both of the sets $D := \{t \in \mathbb{R} \mid a(t) > 0\}$ and $\mathbb{R} \setminus D$ are nonmeasurable. However, it turns out that $(D,+)$ yields a subsemigroup of $(\mathbb{R},+)$ fulfilling (5). Indeed, take an $x \in D$ and write $\beta := a(x) > 0$. Now,

$$x - D = \{x - t \in \mathbb{R} \mid a(t) > 0\} = \{t \in \mathbb{R} \mid \beta > a(t)\}$$

and

$$\Delta_x = D \cap (x - D) = \{t \in \mathbb{R} \mid 0 < a(t) < \beta\} \supset \{h \in H \mid 0 < a(h) < \beta\} = H \cap (0,\beta).$$

Consequently,

$$m_1^*(\Delta_x) \geq m_1^*(H \cap (0,\beta)) = \beta > 0.$$

REMARK 2. If we admit almost-everywhere-finite functions, then (5) may be weakened; namely, it suffices to postulate (5) almost everywhere in D. To show this, write $W := \{x \in D \mid m_n(\Delta_x) = 0\}$ and assume that $m_n(W) = 0$. Note that, for all $x,y \in D$,

(11) $\qquad x + y \in W$ implies $x \in W$ and $y \in W$.

Indeed, suppose that $x \notin W$, for instance. Then $m_n^*(\Delta_x) > 0$ and, taking an $h \in \Delta_x$, in view of our Lemma we obtain

$$0 < m_n^*(\Delta_x) \leq m_n^*(\Delta_{x+y} - h) = m_n^*(\Delta_{x+y}) ,$$

i.e., $x + y \notin W$.

Now, define $\Phi(x)$ by formula (6) for all $x \in D \setminus W$ and put

$$\tilde{\Phi}(x) := \begin{cases} \Phi(x) & \text{for } x \in D \setminus W, \\ \infty & \text{for } x \in W . \end{cases}$$

With the aid of (11), one can easily check that $\tilde{\Phi}$ is subadditive. Clearly, $\tilde{\Phi}$ is almost everywhere finite and $f \leq \tilde{\Phi}$ almost everywhere in D.

REMARK 3. Condition (10) is trivially satisfied provided $D = \mathbb{R}^n$, the limit $\lim_{h \to 0} f(h)$ does exist, and it is nonpositive. However, (10) may hold for completely irregular almost subadditive functions whose domain does not contain interior points. To present a suitable example, consider a measurable Hamel base H of the reals; it is known (see [9]) that we then have $m_1(H) = 0$. Let $(H^*,+)$ be the group generated by H, and let

$$D := \bigcup_{n=0}^{\infty} \frac{1}{2^n} H^* .$$

Evidently, $(D,+)$ is also a group, and $H^* \subset D$. Both H^* and D are non-measurable (cf. [3] and [8], respectively). Now, put

$$f(x) = \begin{cases} 0 & \text{for } x \in H^* \setminus H , \\ \alpha(x) & \text{for } x \in H , \\ 1 & \text{for } x \in D \setminus H^* ; \end{cases}$$

here $\alpha : H \to \mathbb{R}$ is quite arbitrary. It is easily seen that

$$M := (H \times D) \cup (D \times H) \cup \{(x,y) \in \mathbb{R}^2 \mid x + t \in H\}$$

is of planar Lebesgue measure zero, and that f satisfies (1) for all $(x,t) \in D^2 \setminus M$. Fix an $\varepsilon > 0$. Then (see (9)),

$$A_f(\varepsilon) \supset (H^* \setminus H) \cap [-(H^* \setminus H)] = (H^* \setminus H) \cap [H^* \setminus (-H)] = H^* \setminus [H \cup (-H)] .$$

The latter set is of positive outer measure since $m_1(H \cup (-H)) = 0$, whereas H^* is nonmeasurable.

Clearly, f is almost everywhere equal to the subadditive function $f : D \to \mathbb{R}$ given by the formula

$$\tilde{f}(x) = \begin{cases} 0 & \text{for } x \in H^* , \\ 1 & \text{for } x \in D \setminus H^* . \end{cases}$$

REMARK 4. Assuming (5), suppose $f : D \to \mathbb{R}$ to be almost subadditive, and define φ and Φ by (3) and (6), respectively. The estimation

$$\varphi(x) \leq f(x) \leq \Phi(x) \quad \text{a.e. in } D \tag{12}$$

in general is not sharp. To see this, take the function $f : \mathbb{R} \to \mathbb{R}$ given by the formula

$$f(x) = 3 + \sin x, \quad x \in \mathbb{R} .$$

Since f is subadditive, it is, in particular, almost subadditive. A simple calculation shows that here we have

$$\varphi(x) = 2\left|\sin \frac{x}{2}\right| \quad \text{and} \quad \Phi(x) = 6 - 2\left|\sin \frac{x}{2}\right|, \quad x \in \mathbb{R} .$$

Thus, φ and Φ do not both coincide with f.

In general, the left-hand estimation in (12) may be improved by putting

$$\tilde{\varphi}(x) := \varphi(x) + \operatorname*{inf\,ess}_{h \in D} [f(h) - \varphi(h)], \quad x \in D,$$

instead of φ. Similarly, in the case where

$$c := \min\left\{\operatorname*{inf\,ess}_{h \in D} [\Phi(h) - f(h)], \inf_{(x,y) \in D^2} [\Phi(x) + \Phi(y) - \Phi(x + y)]\right\}$$

is positive, $\tilde{\Phi} := \Phi - c$ improves the right-hand estimation in (12).

Finally, observe that if, for a given null set $W \subset D$, the class

$$(13)\quad \mathfrak{F}_W := \{\Psi : D \to \mathbb{R} \mid \Psi \text{ is subadditive and } \Psi(x) \leq f(x) \text{ for all } x \in D \setminus W\} \neq \emptyset ,$$

then

$$\varphi_0(x) := \sup_{\Psi \in \mathfrak{F}_W} \Psi(x) , \quad x \in D ,$$

belongs to $\mathfrak{F}_W$, and φ_0 is the best of its kind. Indeed, it suffices to prove only that φ_0 is finite. Obviously $\varphi_0(x) \leq f(x)$ for $x \in D \setminus W$. Take an $x \in W$ and an $h \in \Delta_x \setminus [W \cup (x - W)]$ (cf. (4)); then

$$\Psi(x) \leq \Psi(x - h) + \Psi(h) \leq f(x - h) + f(h) \quad \text{for every } \Psi \in \mathfrak{F}_W ,$$

whence $\varphi_0(x) \leq f(x - h) + f(h)$.

Note that, with the aid of the symbols used in the proof of Theorem 1, we have (13) for $W = U(M)$.

REFERENCES

1. N. G. de Bruijn, On almost additive functions, Colloq. Math. 15 (1966), 59-63.
2. P. Erdös, P 310, Colloq. Math. 7 (1960), 311.
3. P. Erdös, On some properties of Hamel bases, Colloq. Math. 10 (1963), 267-269.
4. I. Halperin, Nonmeasurable sets and the equation $f(x + y) = f(x) + f(y)$, Proc. Amer. Math. Soc. 2 (1951), 221-224.
5. E. Hille and R. S. Phillips, Functional analysis and semigroups, Amer. Math. Soc. Coll. Pub. 31, 1957.
6. W. B. Jurkat, On Cauchy's functional equation, Proc. Amer. Math. Soc. 16 (1965), 683-686.
7. M. Kuczma, Almost convex functions, Colloq. Math. 21 (1970), 279-284.
8. J. Smital, A necessary and sufficient condition for continuity of additive functions, Czechoslovak Math. J. 26 (1976), 171-173.
9. H. Steinhaus, Sur les distances des points des ensembles de mesure positive, Fund. Math. 1 (1920), 93-104.

ADDITIVE FUNCTIONS AND THE EGOROV THEOREM

Marek Kuczma
Department of Mathematics
Silesian University
40-007 Katowice
POLAND

ABSTRACT. For discontinuous additive functions $f : \mathbb{R}^N \to \mathbb{R}$, the known result for $N = 1$, that f assumes on every set of a positive measure values from every interval, is shown to hold for arbitrary N. An application is made to the Egorov theorem.

1. INTRODUCTION

A function $f : \mathbb{R}^N \to \mathbb{R}$ is called <u>additive</u> whenever the relation

$$f(x + y) = f(x) + f(y) \tag{1}$$

holds for all $x, y \in \mathbb{R}^N$. As was shown by Hamel [4] (with the aid of the axiom of choice), there exist discontinuous additive functions. Such functions display many pecular properties; in particular, if f is a discontinuous additive function, then the graph of f is dense in $\mathbb{R}^{N+1}$. This fact has repeatedly been observed by various authors, starting with Hamel [4] (in the case $N = 1$). It is an easy consequence of the formula

$$f(qx) = qf(x) \quad , \tag{2}$$

valid for any additive function f (cf., e.g., [1]) with arbitrary <u>rational</u> q, and of the theorem of Ostrowski ([8] for $N = 1$, [7] for arbitrary N) stating that a discontinuous additive function cannot be bounded from above, nor from below, on any set of a positive Lebesgue measure. (In the sequel, measure always refers to Lebesgue measure.)

Thus a discontinuous additive function assumes on every open set values from every interval. In this statement, "open set" may be replaced by "set of a positive measure." This fact was proved by A. M. Ostrowski [8] in the case $N = 1$. In this note, we shall present a proof for arbitrary N.

2. DISCONTINUOUS ADDITIVE FUNCTIONS

A set A contained in $\mathbb{R}^N$ is called <u>saturated nonmeasurable</u> whenever the inner measure both of A and of its complement is zero (cf. [3]). Such a set has the property that, for every measurable set $B \subset \mathbb{R}^N$, the outer measure of $A \cap B$ equals the measure of B:

$$m_e(A \cap B) = m(B) \quad .$$

Thus our aim will be achieved if we show that, for every discontinuous additive function f and for every (nondegenerated) interval $J \subset \mathbb{R}$, the set $f^{-1}(J)$ is saturated nonmeasurable. The proof will be based on a lemma; for $N = 1$ this lemma was proved in [6], but similar ideas are already found in [8] and [2].

LEMMA. Let B, D be subsets of $\mathbb{R}^N$, where D is dense and B has positive outer measure, and write

$$A = B + D = \{a \mid a = b + d, b \in B, d \in D\} \quad .$$

Then the complement of A has inner measure zero.

<u>Proof</u>. Let x_o be a point of outer density of B. Fix a $c \in (0,1)$. Then there exists an $\eta > 0$ such that, for every cube K with edge not exceeding η, we have

$$\text{(3)} \qquad x_o \in K \text{ implies } m_e(K \cap B) > cm(K) \quad .$$

Take an arbitrary $y \in D$ and write $x = x_o + y$. Then (3) yields:

$$\text{(4)} \qquad x \in K \text{ implies } m_e(K \cap (B + y)) > cm(K) \quad .$$

Since $B + y \subset A$, we infer from (4) that, for every $x \in D + x_o$,

$$\text{(5)} \qquad x \in K \text{ implies } m_e(K \cap A) > cm(K) \quad .$$

Fix an arbitrary N-dimensional open interval P. In virtue of (5), the family $\mathfrak{K}$ of N-dimensional cubes $K \subset P$ fulfilling the condition

$$m_e(K \cap A) > cm(K)$$

forms a Vitali covering of P. Thus there exists a sequence $K_n \in \mathfrak{K}$ of disjoint cubes such that $m(P \setminus \bigcup K_n) = 0$. Hence

$$m(P) \geq m_e(P \cap A) = m_e(\bigcup K_n \cap A) = \sum m_e(K_n \cap A)$$

$$\geq c \sum m(K_n) = cm(\bigcup K_n) = cm(P) \quad .$$

Letting $c \to 1$, we obtain hence the relation $m_e(P \cap A) = m(P)$, valid for every open interval P. This easily implies that the inner measure of the complement of A is zero.

THEOREM. If $f : \mathbb{R}^N \to \mathbb{R}$ is a discontinuous additive function, then for every interval $J \subset \mathbb{R}$, the set $f^{-1}(J)$ is saturated nonmeasurable.

Proof. Since f is odd and we may always replace J by a smaller interval, there is no loss of generality in assuming that $J = [c,d]$, $0 < c < d$, and $q = d/c$ is rational. Then (2) holds for all $x \in \mathbb{R}^N$.

Suppose that $f^{-1}(J)$ has measure zero. By (2), also

$$f^{-1}([c,\infty)) = f^{-1}\left(\bigcup_{n=0}^{\infty} q^n J\right) = \bigcup_{n=0}^{\infty} f^{-1}(q^n J) = \bigcup_{n=0}^{\infty} q^n f^{-1}(J)$$

has measure zero, which means that f is bounded from above by c on a set of a positive measure. But this is impossible in view of the theorem of Ostrowski mentioned above, since f is a discontinuous additive function.

Consequently, for every interval $J \subset \mathbb{R}$, the set $f^{-1}(J)$ has positive outer measure, and, as pointed out at the beginning of this note, is dense. For $J = [c,d]$, write $J_1 = [c,c + r]$, $J_2 = [0,r]$, where $r = (d - c)/2$. Then $f^{-1}(J) = f^{-1}(J_1) + f^{-1}(J_2)$, and in virtue of the Lemma the complement of $f^{-1}(J)$ has inner measure zero. The set $f^{-1}(J)$ itself also has inner measure zero, for otherwise f would be bounded on a set of a positive measure. Thus $f^{-1}(J)$ is saturated nonmeasurable.

REMARK. The result obtained may also be formulated as follows: Let

$$I_f = \{(x,t) \in \mathbb{R}^{N+1} | t = f(x)\}$$

be the graph of f. If $f : \mathbb{R}^N \to \mathbb{R}$ is a discontinuous additive function, then, for every measurable set $A \subset \mathbb{R}^N$ of a positive measure and for every interval $J \subset \mathbb{R}$, we have

$$I_f \cap (A \times J) \neq \emptyset \quad . \tag{6}$$

Now, let $f : \Delta \to \mathbb{R}$, where $\Delta \subset \mathbb{R}^N$ is an open convex domain, be a discontinuous convex function:

$$f\left(\frac{x + y}{2}\right) \leq \frac{f(x) + f(y)}{2} \quad \text{for all} \quad x, y \in \Delta \quad .$$

It is an open problem (cf. [5]) whether (6) holds for every measurable set $A \subset \Delta$ of a positive measure and for every interval $J = [c,d] \subset \mathbb{R}$ such that $\inf_{x \in A} f(x) < c$. In the case $N = 1$, A. M. Ostrowski [9] proved that $f^{-1}(J)$

has postive outer measure for $J = [c,d]$ such that $\inf_{x\in\Delta} f(x) < c$, but in the case of $N > 1$ even this is not known.

3. APPLICATION TO THE EGOROV THEOREM

At the 13th International Symposium of Functional Equations, A. M. Ostrowski raised the question whether or not the famous Egorov theorem will remain valid if instead of a sequence of functions we take a family depending on a continuous parameter. The answer turns out to be in the negative; various counterexamples may be found in [11], [12], [13]. Here we give another counterexample, based on the results of the preceding section, and thus essentially on the ideas stemming from A. M. Ostrowski.

Let $f : \mathbb{R}^N \to \mathbb{R}$ be an invertible, discontinuous additive function, and let $A \subset \mathbb{R}^N$ be an arbitrary measurable set with a finite positive measure. Define the function $g : A \times (0,1) \to \mathbb{R}$ by

$$g(x,t) = \begin{cases} 1 & \text{if } t = f(x) \quad , \\ 0 & \text{otherwise} \quad . \end{cases}$$

Clearly, for every fixed $x \in A$ the function $g(x,\cdot)$ is zero except at at most one point, viz., $t = f(x)$, and similarly, for every fixed $t \in (0,1)$ the function $g(\cdot,t)$ is zero except at at most one point, viz. $x = f^{-1}(t)$. Therefore both sections $g(x,\cdot)$ and $g(\cdot,t)$ are measurable, and

$$\lim_{t\to 0} g(x,t) = 0 \quad \text{for all } x \in A \quad . \tag{7}$$

However, the convergence in (7) cannot be uniform on any set $F \subset A$ of a positive measure. In fact, suppose that (7) holds uniformly on a set $F \subset A$. Then there is a $\delta > 0$ such that for all $x \in F$ and all $t \in (0,\delta)$ we have $g(x,t) < \frac{1}{2}$. Thus F must be disjoint with the set

$$B = \{x \in A \mid g(x,t) = 1 \text{ for some } t \in (0,\delta)\} = A \cap f^{-1}((0,\delta)) \quad .$$

But since $f^{-1}((0,\delta))$ is saturated nonmeasurable (cf. the Theorem above), the inner measure of F must be zero.

The counterexamples given in [11] and [13] also have this extremal property, that the convergence cannot be uniform on any set of a positive measure. On the other hand, G. H. Sindalovskiĭ [10] gave some additional conditions under which the continuous version of the Egorov theorem holds true.

ACKNOWLEDGMENT. The author is indebted to Professor J. S. Lipiński for calling his attention to papers [10], [11], [13].

REFERENCES

1. J. Aczél, Lectures on functional equations and their applications, Academic Press, New York - London, 1966.

2. P. Erdös, On some properties of Hamel bases, Colloq. Math. 10 (1963), 267-269.

3. I. Halperin, Non measurable sets and the equation $f(x + y) = f(x) + f(y)$, Proc. Amer. Math. Soc. 2 (1951), 221-224.

4. G. Hamel, Eine Basis aller Zahlen und die unstetigen Lösungen der Funktionalgleichung $f(x + y) = f(x) + f(y)$, Math. Ann. 60 (1905), 459-462.

5. M. Kuczma, Problem (P 161), Aequationes Math. 14 (1976), 235-236.

6. M. Kuczma and J. Smítal, On measures connected with the Cauchy equation, Aequationes Math. 14 (1976), 421-428.

7. S. Marcus, Généralisation, aux fonctions de plusieurs variables, des théorèmes de Alexander Ostrowski et Masuo Hukuhara concernant les fonctions convexes (J), J. Math. Soc. Japan 11 (1959), 171-176.

8. A. M. Ostrowski, Über die Funktionalgleichung der Exponentialfunktion und verwandte Funktionalgleichungen, Jahresber. Deutsch. Math. Verein. 38 (1929), 54-62.

9. A. M. Ostrowski, Zur Theorie der konvexen Funktionen, Comment. Math. Helv. 1 (1929), 157-159.

10. G. H. Sindalovskiĭ, On the uniform convergence of a family of functions depending on a continuously varying parameter, (Russian), Vestnik Moskov. Univ. 5 (1960), 14-18.

11. G. P. Tolstov, A remark on a theorem of D. F. Egorov, (Russian), Dokl. Akad. Nauk SSSR 6 (1939), 309-311.

12. W. Walter, A remark on Egorov's theorem, Amer. Math. Monthly (to appear).

13. J. D. Weston, A counter-example concerning Egoroff's theorem, J. London Math. Soc. 34 (1959), 139-140.

NONNEGATIVE CONTINUOUS SOLUTIONS OF A FUNCTIONAL INEQUALITY IN A SINGLE VARIABLE*

Marek Kuczma
Department of Mathematics
Silesian University
40-007 Katowice
POLAND

ABSTRACT. Asymptotic properties of nonnegative continuous solutions of a functional inequality in one variable are presented, and examples are given.

1. INTRODUCTION

Let $I = [0,a]$ or $[0,a)$, where $0 < a \leq \infty$, be a real interval. Suppose that a function $\Phi : I \to X$, where X is a normed space, is a continuous solution of the functional equation

$$\Phi(x) = H(x,\Phi[f(x)]).$$

If the function H fulfills a Lipschitz condition,

$$\|H(x,y) - H(x,0)\| \leq g(x)\|y\|,$$

then $\varphi(x) = \|\Phi(x)\|$ is a nonnegative continuous solution of the functional inequality

$$\varphi(x) \leq g(x)\varphi[f(x)] + h(x), \tag{1}$$

where $h(x) = \|H(x,0)\|$. This motivates the investigation of the properties of nonnegative continuous solutions φ of inequality (1). Our aim is to obtain some information about the asymptotic properties of φ for $x \to 0 + 0$.

*Prepublication announcement.

We shall consider, along with inequality (1), the associated functional equation

$$\varphi(x) = g(x)\varphi[f(x)] + h(x). \tag{2}$$

The asymptotic behaviour of continuous solutions of equation (2) has been extensively investigated by B. Choczewski [1]. However, our results will also yield some new information concerning equation (2).

In the sequel, we shall use the notation $I^* = I\setminus\{0\}$. All the asymptotic symbols will refer to $x \to 0 + 0$.

2. ASSUMPTIONS

The functions f, g, h occurring in (1) and (2) will usually be subjected to some of the following general assumptions.

(i) $f : I \to I$, $g : I \to \mathbb{R}$, $h : I \to \mathbb{R}$ are continuous in I. Moreover, $0 < f(x) < x$ in I^*, $0 < g(x) < 1$ in I^*, $h(x) \geq 0$ in I.

(ii) $\lim_{n\to\infty} G_n(x) = 0$ in I^*, where

$$G_n(x) = \prod_{i=0}^{n-1} g[f^i(x)],$$

and f^i denotes the i-th iterate of f.

(iii) $A(x) = o(1)$, where $A(x) = h(x)/(1 - g(x))$ for $x \in I^*$.

(iv) $A(x)$ is monotonic in I^*.

Assumptions (i) and (ii) imply that equation (2) may have at most one continuous solution in I (cf. [2], [3]). On the other hand, every continuous function $\varphi : I \to \mathbb{R}$ fulfilling the condition $0 \leq \varphi(x) \leq h(x)$ yields a nonnegative continuous solution of (1). Consequently, there is no nontrivial problem of the existence and/or uniqueness of nonnegative continuous solutions of inequality (1): we have always the existence and never (except when $h \equiv 0$) the uniqueness.

3. THEOREMS AND EXAMPLES

THEOREM 1. Let conditions (i), (ii), (iii) be fulfilled, and let $\varphi : I \to \mathbb{R}$ be a nonnegative continuous solution of (1). Then $\varphi(x) = o(1)$.

THEOREM 2. Let conditions (i), (ii) be fulfilled, let $\varphi : I \to \mathbb{R}$ be a nonnegative continuous solution of (1), and let $\varphi_0 : I \to \mathbb{R}$ be a continuous solution of (2). Then $\varphi(x) = O(\varphi_0(x))$.

THEOREM 3. Let conditions (i) through (iv) be fulfilled. Then equation (2) actually has a unique continuous solution $\varphi_0 : I \to \mathbb{R}$, and

$$\varphi_0(x) = \sum_{n=0}^{\infty} G_n(x)h[f^n(x)]. \tag{3}$$

THEOREM 4. Let conditions (i) through (iv) be fulfilled, and let $\varphi : I \to \mathbb{R}$ be a nonnegative continuous solution of (1). Then $\varphi(x) = O(A(x))$.

Clearly, the estimation given in Theorem 4 is less sharp than that given in Theorem 2. However, due to the simple form of the function $A(x)$ and a fairly complicated form of formula (3) defining φ_0, it may be much more convenient in practice. On the other hand, the estimation in Theorem 4 may still be quite good, as may be seen from the following examples. (In all these examples, $I = [0,a)$, where $a < 1$ is so small that the function $A(x)$ is monotonic in I.)

EXAMPLE 1. Take the equation

$$\varphi(x) = (1 - x)\varphi(x - x^3) + x^3 + 2x^4 - 2x^5 - x^6 + x^7. \tag{4}$$

Here $\varphi_0(x) = x^2$ and $A(x) = x^2 + 2x^3 - 2x^4 - x^5 + x^6$, and so φ_0 and A differ only in terms of higher order.

EXAMPLE 2. Take the equation

$$\varphi(x) = \left(1 + \frac{1}{\log x}\right)\varphi\left(\frac{x}{x+1}\right) + \frac{x^2 \log x - x}{(x+1)\log x}.$$

Here $\varphi_0(x) = x$ and

$$A(x) = \frac{x - x^2 \log x}{x + 1} = x - x^2 \log x - x^2 + x^3 \log x + x^3 - \cdots .$$

Again φ_0 and A differ only in terms of higher order.

Examples 1 and 2 are particular cases of a more general situation. Write $p(x) = \inf_{[0,x]} h[f(t)]/h(t)$, $x \in I^*$.

THEOREM 5. Let conditions (i) through (iv) be fulfilled, and assume that $g(x)$ is decreasing, $h(x) > 0$ in I^*, and $1 - p(x) = o(1 - g(x))$. Then $\lim_{x\to 0+0} \varphi_0(x)/A(x) = 1$, where φ_0 is given by (3).

On the other hand, if the assumptions of Theorem 5 are not fulfilled, then it may happen that the estimation given in Theorem 2 is essentially better that that given in Theorem 4.

EXAMPLE 3. Take the equation

$$\varphi(x) = \left(1 + \frac{1}{\log x}\right)\varphi\left(\frac{x}{2}\right) + \frac{x}{2}\left(1 - \frac{1}{\log x}\right).$$

Here $\varphi_0(x) = x$ and $A(x) = \frac{x}{2}(1 - \log x) = O(x \log x)$.

This again is a particular case of a more general situation.

THEOREM 6. Let conditions (i) through (iv) be fulfilled, and assume that $g(0) = 1$, $h(x) > 0$ in I^*, and

$$\limsup_{x\to 0+0} h[f(x)]/h(x) < 1.$$

Then $\varphi_0(x) = o(A(x))$, where φ_0 is given by (3).

The results obtained so far allow us to improve on a theorem of B. Choczewski [1].

THEOREM 7. Let the functions $f : I \to I$, $g : I \to \mathbb{R}$, and $h : I \to \mathbb{R}$ be continuous in I, with $0 < f(x) < x$ in I^* and $g(x) \neq 0$ in I. Assume further that

$$f(x) = x - x^{m+1}u(x), \quad g(x) = 1 - x^k v(x), \quad \text{and} \quad h(x) = cv(x)x^k + x^q w(x),$$

where the functions u and w are bounded in I,

$$v_o = \liminf_{x \to 0+0} v(x) > 0,$$

c is a real constant, and the real constants k, m, q fulfill $0 < k \leq m \leq q$, $k < q$. Then equation (2) actually has in I a unique continuous solution φ_0 (given by (3)), and $\varphi_0(x) = c + 0(x^{q-k})$.

B. Choczewski, under slightly stronger assumptions, obtained only the weaker estimation $\varphi_0(x) = c + 0(x^{q-m})$. In particular, in the case of equation (4), Theorem 7 above yields a fairly sharp estimation $\varphi_0(x) = 0(x^2)$, whereas Choczewski's theorem gives only $\varphi_0(x) = 0(x)$. In fact, for equations like (4) we are able to deduce a still better estimation.

THEOREM 8. Let the assumptions of Theorem 7 be fulfilled, and assume, moreover, that there exist the limits

$$v_o = \lim_{x \to 0+0} v(x) > 0, \quad w_o = \lim_{x \to 0+0} w(x) > 0.$$

Then

$$\varphi_0(x) = c + \frac{w_0}{v_0} x^{q-k} + o(x^{q-k}),$$

where φ_0 is given by (3).

The full version of this paper will appear in Annales Polonici Mathematici 36 (1978).

REFERENCES

1. B. Choczewski, Przebieg asymptotyczny rozwiazań ciagłych pewnych równań funkcyjnych, Zeszyty Naukowe Akademii Górniczo-Hutniczej w Krakowie, Nr. 274, Mat. Fiz. Chem. 4 (1970), 66 pp.

2. B. Choczewski and M. Kuczma, On the "indeterminate case" in the theory of a linear functional equation, Fund. Math. 58 (1966), 163-175.

3. M. Kuczma, Functional equations in a single variable, Monografie Mat. 46, PWN - Polish Scientific Publishers, Warszawa, 1968.

NONLINEAR FUNCTIONAL INEQUALITIES IN A SINGLE VARIABLE

Dobiesław Brydak
Institute of Mathematics
Pedagogical University
30-011 Kraków
POLAND

ABSTRACT. Under suitable hypotheses, solutions of functional inequalities are studied in relation to solutions of corresponding functional equations.

1. INTRODUCTION

In the present paper, we shall deal with the functional inequalities

$$\psi[f(x)] \leq g[x,\psi(x)] \tag{1}$$

and

$$\psi[f(x)] \geq g[x,\psi(x)] \ , \tag{2}$$

related with the functional equation

$$\varphi[f(x)] = g[x,\varphi(x)] \ , \tag{3}$$

where f and g are given functions, and ψ and φ are unknown functions.

These inequalities have been studied in [1], where some general results have been obtained. Here we are going to restrict our attention to the case where equation (3) has a one-parameter family of continuous solutions.

2. HYPOTHESES

In the sequel, the given functions will be subjected to the following:

HYPOTHESIS H_1. The function f is defined, continuous, and strictly increasing in an interval $I=[\xi,b)$, $f(\xi)=\xi$, and $\xi<f(x)<x$ for $x \in I_0 = (\xi,b)$. The function g is defined and continuous in a set $\Omega = I \times E$, where E is an open interval in the set of real numbers. Moreover, for every $x \in I$, g is strictly increasing with respect to the second variable; further, $\Gamma_x \subset \Omega_{f(x)}$ for $x \in I$, where $\Omega_x = \{y : x,y \in \Omega\}$ and Γ_x is the set of values of g for $y \in \Omega_x$.

REMARKS. (i) The number ξ as well as b may be infinite. We say that a function ψ is continuous at infinity if there exists a finite limit of ψ at infinity. This limit will be called the value of ψ at infinity.

(ii) Hypothesis H_1 implies that ξ is an attractive fixed point of f in I; that is,

$$\lim_{n\to\infty} f^n(x) = \xi \quad \text{for} \quad x \in I \tag{4}$$

(see [2]). Here $f^n(x)$ denotes the n-th iterate of the function f; that is,

$$f^0(x) = x, \quad f^{n+1}(x) = f[f^n(x)], \quad \text{for} \quad n = 0,1,\dots .$$

Hypothesis H_1 also implies that ξ is the only fixed point of f in I.

(iii) The results obtained in this paper can also be applied to the case where the point ξ is the right-hand endpoint of I, after a slight modification: We have to assume that $f(x) > x$ for $x \in I_0$. This is the case of difference inequalities.

HYPOTHESIS H_2. For every point $(x,y) \in \Omega$, there exists exactly one continuous solution of equations (3) in I passing through this point.

Hypotheses H_1 and H_2 imply that

$$\eta_0 = g(\xi,\eta_0) \quad \text{for} \quad \eta_0 \in \Omega_\xi \tag{5}$$

(see [2]).

DEFINITION 1. If hypotheses H_1 and H_2 are fulfilled, we define the function $R : \Omega \to \Omega_\xi$ by the formula

$$R(x,y) = \varphi(\xi) \quad \text{for} \quad (x,y) \in \Omega , \tag{6}$$

where φ is a continuous solution of equation (3) in I satisfying the condition

$$\varphi(x) = y . \tag{7}$$

It is easy to see that the function R is a prime integral of equation (3); that is, R is constant on every solution of (3).

3. LEMMAS

We are going to list here some properties of R as the following lemmas.

LEMMA 1. If hypotheses H_1 and H_2 are fulfilled, then the function R is continuous in Ω.

LEMMA 2. Let hypotheses H_1 and H_2 be fulfilled, let ψ be a function defined in I, let the graph of ψ lie in Ω, and let η be a function defined in I by the formula

$$\eta(x) = R[x,\psi(x)] = \varphi(\xi) \quad \text{for} \quad x \in I \; , \tag{8}$$

where φ is a continuous solution of (3) in I satisfying condition (7) with $y = \psi(x)$. Then the function η is continuous in I if and only if the function ψ is continuous in I.

These two lemmas have been proved in [1].

LEMMA 3. If hypotheses H_1 and H_2 are fulfilled, then the function R is strictly increasing with respect to the second variable for every $x \in I$.

Proof. Let $x \in I$, $z,y \in \Omega_x$, and

$$z < y \; . \tag{9}$$

It follows from Definition 1 that

$$R(x,z) = \varphi_1(\xi), \quad R(x,y) = \varphi_2(\xi) \; ,$$

where φ_1 and φ_2 are continuous solutions of (3) in I satisfying the conditions:

$$\varphi_1(x) = z, \quad \varphi_2(x) = y \; .$$

Since the family of continuous solutions of equation (3) in I is a one-parameter family, in view of hypothesis H_2, the inequality (9) implies that $\varphi_1(\xi) < \varphi_2(\xi)$. This completes the proof of the lemma.

DEFINITION 2. A function η defined in an interval I is called $\{f\}$-decreasing ($\{f\}$-increasing) in I if and only if the inequality

$$\eta[f(x)] \leq \eta(x) \quad \text{for} \quad x \in I \tag{10}$$

holds (the inequality

$$\eta[f(x)] \geq \eta(x) \quad \text{for} \quad x \in I \tag{11}$$

holds).

The relation between solutions of inequalities (1) and (2) and $\{f\}$-monotonic functions gives the following:

LEMMA 4. Let hypotheses H_1 and H_2 be fulfilled. The function ψ defined in I having its graph in Ω satisfied the inequality (1) (the inequality (2)) if and only if the function η, defined by the formula (8), is an $\{f\}$-decreasing (an $\{f\}$-increasing) function in I.

One can find a proof of Lemma 4 in [1].

DEFINITION 3. Let hypotheses H_1 and H_2 be fulfilled. We denote: by L the family of all continuous solutions of inequality (1) in I having their graphs in Ω; by L' the family of all continuous solutions of inequality (2) in I having their graphs in Ω; by L_0 the family of all continuous solutions of equation (3) in I having their graphs in Ω.

Let $\psi_1, \psi_2 \in L \cup L'$. We denote:

$$(\psi_1 \vee \psi_2)(x) = \max[\psi_1(x), \psi_2(x)] \quad \text{for} \quad x \in I \ ;$$

$$(\psi_1 \wedge \psi_2)(x) = \min[\psi_1(x), \psi_2(x)] \quad \text{for} \quad x \in I \ .$$

The following has been proven in [1]:

LEMMA 5. If hypotheses H_1 and H_2 are fulfilled, then the families L, L' and L_0 are distributive lattices with respect to the operations $\vee$ and $\wedge$.

DEFINITION 4. Let us denote by K the family of all continuous functions $\{f\}$-decreasing in I and taking values in $I \times \Omega_\xi$; by K' the family of all continuous functions $\{f\}$-increasing in I and taking values in Ω_ξ; by K_0 the family of all constant functions in I taking values in Ω_ξ.

Since the inequalities (10) and (11) are singular cases of inequalities (1) and (2), respectively, and the equation

$$\eta[f(x)] = \eta(x) \quad \text{for} \quad x \in I \tag{12}$$

is a singular case of equation (3), we are able to obtain, as an immediate consequence of Lemma 5, the following:

LEMMA 6. If hypothesis H_1 is fulfilled, then the families K, K', and K_0 are distributive lattices with respect to the operations $\vee$ and $\wedge$.

Denoting by T the map $T : L \vee L' \to K \vee K'$, defined by formula (8), we obtain (see [1]) the following:

LEMMA 7. If hypotheses H_1 and H_2 are fulfilled, then the map T is an isomorphism between L and K, L' and K', L_0 and K_0, respectively.

LEMMA 8. The family of all functions defined and increasing (decreasing) in a set $A \in \mathbb{R}$, taking values in a set $B \subset \mathbb{R}$ is a distributive lattice with respect to the operations $\vee$ and $\wedge$.

Proof. We are going to prove the lemma for increasing functions only, because the proof is similar in both cases. Let η_1, η_2 be increasing functions in A, and let $x,y \in A$, $x < y$. We are going to consider one of four possible cases, because the proof is very similar in each of them: Let us assume that

$$\eta_1(x) \leq \eta_2(x) \quad \text{and} \quad \eta_1(y) \geq \eta_2(y) .$$

Then, in view of the definition of operations $\vee$ and $\wedge$, we have

$$(\eta_1 \vee \eta_2)(x) = \eta_2(x) \leq \eta_2(y) \leq \eta_1(y) = (\eta_1 \vee \eta_2)(y) ,$$

$$(\eta_1 \wedge \eta_2)(x) = \eta_2(x) \leq \eta_2(y) \leq \eta_1(y) = (\eta_1 \wedge \eta_2)(y) ,$$

because the functions η_1 and η_1 are increasing in A.

4. THEOREMS

Now we are going to prove a comparison theorem for the inequalities (1) and (2), similar to the comparison theorems proved in [1]. It turns out that the condition of Lipschitzian type, assumed in [1], can be replaced by hypothesis H_2.

THEOREM 1. Let hypotheses H_1 and H_2 be fulfilled, let $\psi \in L$ ($\psi \in L'$), and let $\varphi \in L_0$. If

$$\psi(\xi) \geq \varphi(\xi) \tag{13}$$
$$(\psi(\xi) \leq \varphi(\xi)) ,$$

then

$$\psi(x) \geq \varphi(x) \quad \text{for} \quad x \in I \tag{14}$$
$$(\psi(x) \leq \varphi(x) \quad \text{for} \quad x \in I) .$$

Proof. We are going to prove the theorem in the case where $\psi \in L$. In the other case, the proof is similar. Let $\psi \in L$ and $\varphi \in L_0$. Moreover, let us assume (13) to hold. It is very easy to see, in view of Definition 3, that $L_0 \subset L$; thus both functions ψ and φ belong to L. Hence, by Lemmas 3 and 7, there exist a function $\eta \in K$ and a function $\rho \in K_0$ such that $T(\psi) = \eta$ and $T(\varphi) = \rho$. It follows from (13) that

$$\eta(\xi) \geq \rho(\xi) , \tag{15}$$

because Hypothesis H_1 and Lemma 2 imply that (8) holds, whence $\eta(\xi) = \psi(\xi) \geq \varphi(\xi) = \rho(\xi)$. Therefore, inequality (15) implies the inequality

$$\eta(x) \geq \rho(x) \quad \text{for} \quad x \in I , \tag{16}$$

because of the comparison theorem for the homogeneous inequality (10), proved in [1]. It follows from Definition 4 that $K_0 \subset K$, so that $\eta, \rho \in K$. Therefore, inequality (16) means that $\eta \geq \rho$ in terms of lattice theory. Since T is an isomorphism between the lattices L and K, in view of Lemma 7, the inequality $\psi \geq \varphi$ holds in terms of lattice theory, too, whence inequality (14) follows. This completes the proof of the theorem.

DEFINITION 5. Let hypotheses H_1 and H_2 be fulfilled. We say that the function ψ, defined in I and having its graph in Ω, is increasing with respect to equation (3) in I (decreasing with respect to equation (3) in I), if, for every $x_0 \in I$ and every $\varphi \in L_0$, the inequality

$$\psi(x_0) \geq \varphi(x_0) \tag{17}$$

(the inequality

$$\psi(x_0) \leq \varphi(x_0) \;)$$

implies the inequality

$$\psi(x) \geq \varphi(x) \quad \text{for} \quad x \in I, \quad x \geq x_0 \tag{18}$$

(the inequality $\psi(x) \leq \varphi(x)$ for $x \in I$, $x \geq x_0$).

We say that ψ is $\{f\}$-increasing with respect to equation (3) in I ($\{f\}$-decreasing with respect to equation (3) in I), if (17) implies

$$\psi[f^{-1}(x_0)] \geq \varphi[f^{-1}(x_0)] \tag{19}$$

$$(\psi[f^{-1}(x_0)] \leq \varphi[f^{-1}(x_0)]) ,$$

whenever $f^{-1}(x_0)$ is defined and belongs to I.

It is easy to see that if ψ is increasing (decreasing) with respect to (3) in I, then ψ is $\{f\}$-increasing ($\{f\}$-decreasing) with respect to (3) in I.

THEOREM 2. Let hypotheses H_1 and H_2 be fulfilled. A function ψ is $\{f\}$-increasing ($\{f\}$-decreasing) with respect to equation (3) in I if and only if ψ satisfies the inequality (1) (the inequality (2)) in I.

Proof. Let ψ be $\{f\}$-increasing with respect to (3) in I. It follows from hypothesis H_2 that there exists a solution φ of equation (3) in I such that $\varphi[f(x)] = \psi[f(x)]$ for an $x \in I$. Putting $f(x) = x_0$, we see, by virtue of (3), (19), and hypothesis H_1, that

$$\psi[f(x)] = \varphi(x_0) = \varphi[f(x)] = g[x,\varphi(x)] \le g[x,\psi(x)] .$$

Conversely, if ψ satisfies (1) in I, then (17) implies (19); this has been proved in [1].

The proof for inequality (2) is similar.

An immediate consequence of Theorem 2 is the following:

COROLLARY 1. Let hypotheses H_1 and H_2 be fulfilled. If a function ψ is increasing (decreasing) with respect to equation (3) in I, then ψ satisfies (1) (ψ satisfies (2)) in I.

Let us notice that if ψ satisfies (1) in I, it does not have to be an increasing function with respect to equation (3) in I.

Now we are going to prove a criterion for ψ to be an increasing function with respect to (3) in I:

THEOREM 3. Let hypotheses H_1 and H_2 be fulfilled. The function ψ is an increasing (a decreasing) function with respect to equation (3) in I if and only if the function η, defined by formula (8), is an increasing (a decreasing) function in I.

Proof. Let ψ be an increasing function with respect to equation (3) in I. Let $x_0, x \in I$, $x_0 < x$. It follows from hypothesis H_2 that there exists exactly one $\varphi_1 \in L_0$ and exactly one $\varphi_2 \in L_0$ such that

$$\psi(x_0) = \varphi_1(x_0), \quad \psi(x) = \varphi_2(x) . \tag{20}$$

Since ψ is increasing with respect to (3) in I, equalities (20) imply that inequality (18) holds, whence

$$\eta(x) = \varphi_2(\xi) = R[x,\psi(x)] \ge R[x,\varphi_1(x)] = \varphi_1(\xi) = R[x_0,\psi(x_0)] = \eta(x_0) ,$$

by virtue of (8), Definition 1, (18), Lemma 4, and (20).

Conversely, let ψ be a function defined in I and having its graph in Ω; let η, defined by formula (8), be an increasing function in I; and let $x_0, x \in I$, $x_0 < x$. Moreover, let $\varphi_1, \varphi_2 \in L_0$ satisfy (20). Thus

$$R[x_0,\psi(x_0)] = \eta(x_0) \leq \eta(x) = R[x,\psi(x)] ,$$

by virtue of (8), whence (18) follows, in view of Lemma 4.

The proof for decreasing functions is similar.

COROLLARY 2. Under the assumptions of Theorem 3, the family M of all continuous functions increasing (decreasing) with respect to equation (3) in I is a sublattice of the lattice L (of the lattice L').

<u>Proof</u>. Since the family M_0 of all continuous functions increasing in I and having their graphs in Ω is a sublattice of K, in view of Definition 2 and Lemma 8, M is a sublattice of L, as an isomorphic image of M_0, in view of Lemma 7. The proof for decreasing functions is similar.

5. A FINAL RESULT

In the sequel we shall need the following:

HYPOTHESIS H_3. For every $y \in E$, the set $I_y = \{x : (x,y) \in \Omega\}$ is an open interval.

THEOREM 4. Let hypotheses $H_1 - H_3$ be fulfilled. The following conditions are equivalent:

(i) All functions belonging to L_0 are decreasing (increasing) in I.

(ii) For every $y \in E$, the function R is an increasing (a decreasing) function with respect to the first variable in I.

(iii) Every constant function $\psi(x) = y$, where $y \in E$, is an increasing (a decreasing) function with respect to equation (3) in I_y.

<u>Proof</u>. Let us assume (iii) to hold. If ψ is an increasing function with respect to (3) in I_y, then the function η, defined by (8), is increasing in I_y, in view of Theorem 3. Hence R is an increasing function with respect to the first variable, in view of Definition 1, in I_y for every $y \in E$, whence R is also increasing with respect to the first variable in I, for every $y \in E$. The proof is similar for decreasing ψ with respect to (3). Thus (iii) implies (ii).

It follows from Definition 1 that if $\varphi \in L_0$, then $R[x,\varphi(x)] = \text{const}$. Therefore (ii) implies (i), because of Lemma 3.

Let us assume (i) to hold. If every $\varphi \in L_0$ is decreasing in L, then for every $x_0 \in I$, we have

$$\varphi(x) \leq y = \varphi(x_0) \quad \text{for} \quad x \geq x_0, \quad x \in I ,$$

whence every constant function is increasing with respect to equation (3) in I, in view of Definition 5. Thus (i) implies (iii). The proof is similar in case of increasing φ. The last implication ends the proof of the theorem.

REFERENCES

1. D. Brydak, On functional inequalities in a single variable, Dissertationes Math. (to appear).

2. M. Kuczma, Functional equations in a single variable, PWN, Warszawa, 1968.

APPLICATION OF FUNCTIONAL INEQUALITIES TO DETERMINING ONE-PARAMETER FAMILIES OF SOLUTIONS OF A FUNCTIONAL EQUATION

Dobiesław Brydak
Institute of Mathematics
Pedagogical University
30-011 Kraków
POLAND

Bogden Choczewski
Institute of Mathematics
University of Mining and Metallurgy
30-059 Kraków
POLAND

ABSTRACT. One-parameter families of solutions φ of the functional equation $\varphi(f(x)) = g(x)\varphi(x)$, belonging to a class U_0 of functions asymptotically comparable at the fixed point of the function f with a solution of a corresponding functional inequality, are determined in the case where the continuous solutions φ of the given functional equation depend on an arbitrary function.

1. INTRODUCTION

In the present paper, we shall deal with the functional equation

$$\varphi(f(x)) = g(x)\varphi(x) \tag{1}$$

in the case where the sequence

$$G_n(x) := \prod_{i=0}^{n-1} g(f^i(x)), \; n = 1,2,\ldots, \tag{2}$$

converges uniformly to the zero function in an interval $[f(x_0),x_0] \subset [0,b)$, where $f : [0,b) \to [0,b)$ is a continuous function satisfying the condition $0 < f(x) < x$ for $x \in (0,b)$. Here and in the sequel, upper indices denote functional iterates.

In the considered case, continuous solutions $\varphi : [0,b) \to \mathbb{R}$ of equation (1) depend on an arbitrary function, and, moreover, every solution of (1) which is continuous at the origin takes on the value zero there.

We shall be interested in solutions of (1) belonging to the class U_0 of functions, which is defined as follows.

DEFINITION. Let $u_0 : [0,b) \to [0,\infty)$ be a continuous function, vanishing only at the origin. A function $p \in C[0,b)$ is said to belong to the class U_0 of functions if and only if it is positive in $(0,b)$ and there exists a positive limit of the quotient $p(x)/u_0(x)$ as $x \to 0+$.

In the case where $u_0(x) = x^a$, $a > 0$, the class U_0 reduces to the class U^a introduced by M. Kuczma [6].

2. LEMMA

We shall be concerned with functions u_0 that fulfill the functional inequality

$$u(f(x)) \leq q(x)u(x), \quad x \in [0,b). \tag{3}$$

Let us note some results obtained by D. Brydak [1] (cf. also [2]) for inequalities of type (3).

LEMMA. Assume that the function $f \in C[0,b)$ has the property $0 < f(x) < x$ in $(0,b)$; the function $q : [0,b) \to [0,\infty)$ is also continuous in $[0,b)$, $q(x) > 0$ in $(0,b)$; and the sequence

$$Q_n(x) := \prod_{i=0}^{n-1} q(f^i(x)), \quad x \in [0,b), \ n = 1,2,\ldots, \tag{4}$$

converges uniformly to zero in an interval $[f(x_0),x_0] \subset (0,b)$. Then for any continuous solution $u_0 : [0,b) \to [0,\infty)$ of inequality (3), vanishing only at the origin, there exists the limit

$$w(x) := \lim_{n\to\infty} u_0(f^n(x))/Q_n(x), \quad x \in (0,b); \tag{5}$$

further, the function

$$v_0(x) := w(x), \quad x \in (0,b); \quad v_0(0) := 0, \tag{6}$$

is upper semi-continuous in $[0,b)$, is continuous at the origin, and is a solution of the equation

$$v(f(x)) = q(x)v(x), \quad x \in [0,b). \tag{7}$$

Moreover, if there is a solution $v \in U_0$ of equation (7) that is positive in $(0,b)$, then the function v_0 is continuous in $[0,b)$ and positive in $(0,b)$, and there exists exactly one function d, continuous and $\{f\}$ - decreasing in $[0,b)$ (this means that

$$d(f(x)) \leq d(x) \tag{8}$$

in the interval), such that

$$u_0(x) = d(x)v_0(x), \quad x \in [0,b), \tag{9}$$

and

$$d(0) = 1. \tag{10}$$

3. THEOREM

Now we are going to prove the following:

THEOREM. Suppose that the functions f and q fulfill the assumptions of the foregoing Lemma, and that the function $g : [0,b) \to [0,\infty)$ can be written in the form

$$g(x) = q(x)r(x),\ x \in [0,b), \tag{11}$$

where $r \in C[0,b)$. Let the sequence (Q_n), defined by (4), converge uniformly to zero in $[f(x_0),x_0] \subset (0,b)$, and let the sequence

$$R_n(x) := \prod_{i=0}^{n-1} r(f^i(x)) \tag{12}$$

possess in the interval $[0,b)$ a limit

$$R(x) := \lim_{n\to\infty} R_n(x) \tag{13}$$

which is continuous and positive. If u_0 is a solution of inequality (3) satisfying the conditions of the foregoing Definition, then the solutions of equation (1) that belong to the class U_0 constitute at most a one-parameter family of functions. Solutions actually exist if $v_0 \in C[0,b)$, $v_0(x) > 0$ in $(0,b)$ (cf. (6)). On the other hand, if $\varphi \in C[0,b)$ is a solution of (1) such that

$$\lim_{x\to 0+} \varphi(x)/u_0(x) = 0, \tag{14}$$

then

$$\varphi(x) = 0 \quad \text{for} \quad x \in [0,b). \tag{15}$$

Proof. Let u_0 be a solution of (3) satisfying the conditions of the Definition. Take a solution $\varphi \in U_0$ of equation (1). If we write

$$\varphi(x) = \alpha(x)u_0(x),\ x \in [0,b), \tag{16}$$

then, by the Definition, α will be a function continuous and positive in $[0,b)$. As a consequence of (1), we get

$$\alpha(f(x)) = k(x)\alpha(x),\ x \in [0,b), \tag{17}$$

where we put

$$k(x) := g(x)u_0(x)/u_0(f(x)),\ x \in (0,b);\ k(0) := 1. \tag{18}$$

We claim that the function k is continuous in $[0,b)$.

This is obvious for $x \in (0,b)$. To prove the continuity of k at the origin, we first verify that relations (9) and (10) hold true for u_0.

According to the Lemma, it is enough to find a solution $v \in U_0$ of equation (7) that is positive in $(0,b)$. To this end, we consider the equation

$$\lambda(f(x)) = \lambda(x)/r(x),\ x \in [0,b). \tag{19}$$

Continuous solutions of (19) can be determined with the aid of the sequence $(1/R_n(x))$. Since the sequence tends in $[0,b)$ to the continuous and positive function $1/R$ (cf. (13)), it follows that the functions

$$\lambda_t(x) := t\,R(x)$$

are the only solutions of (19) that are continuous in $[0,b)$ (cf. [4] and also [5], p. 48). Take $t = 1$, and put

$$v(x) := R(x)\varphi(x),\ x \in [0,b).$$

This is a function continuous in $[0,b)$ and positive in $(0,b)$. Moreover, for $x \in [0,b)$, we have

$$v(f(x)) = R(f(x))\varphi(f(x)) = [g(x)/r(x)]R(x)\varphi(x) = q(x)v(x);$$

i.e., the function v satisfies equation (7). We check that $v \in U_0$ by calculating

$$\lim_{x\to 0+} v(x)/u_0(x) = \lim_{x\to 0+} [\varphi(x)/u_0(x)]R(x).$$

We see that the limit is positive since $\varphi \in U_0$ and $R(0) = 1$, and therefore that $v \in U_0$.

Thus formulas (9) and (10) for u_0 hold true. For $x \in (0,b)$, they yield, when used in (18) together with (11) and (7),

$$k(x) = r(x)d(x)/d(f(x)). \tag{20}$$

By (10) and (20), we get

$$\lim_{x\to 0+} k(x) = r(0).$$

On the other hand,

$$r(0) = \lim_{i\to\infty} r(f^i(x)) = 1,$$

since the sequence (12) is convergent. Thus $k \in C[0,b)$.

Now we turn to the examination of continuous solutions of equation (17), with the aid of the sequence

$$K_n(x) := \prod_{i=0}^{n-1} k(f^i(x)), \ x \in [0,b),$$

which by (20) can be written in the form

$$K_n(x) = R_n(x)d(x)/d(f^n(x)).$$

Passing to the limit as $n \to \infty$, we obtain, on account of (13) and (10) (note that $f^n(x) \to \infty$),

$$K(x) := \lim_{n\to\infty} K_n(x) = R(x)d(x), \ x \in [0,b). \tag{21}$$

Because of the continuity of the functions R and d, we have $K \in C[0,b)$. Moreover, $K(x) > 0$ in $[0,b)$ since so is the function R and

$$d(x) \geq d(f^n(x)) \geq d(0) = 1$$

(cf. (8) and (10)). This shows, as for equation (19), that all the solutions of (17) that are continuous in $[0,b)$ are given by the formula

$$\alpha_t(x) = t/K(x), \ x \in [0,b). \tag{22}$$

As $\alpha(0) > 0$, we have here $t > 0$. Taking into account (16) and (21), we get

$$\varphi_t(x) = t\ u_0(x)/K(x) = t\ u_0(x)/d(x)R(x),$$

and finally, by (9),

$$\varphi_t(x) = t\ v_0(x)/R(x), \ t > 0, \ x \in [0,b). \tag{23}$$

We have proved that in the class U_0 there exists at most a one-parameter family (23) of solutions of equation (1).

Now we shall prove that all the functions (23) actually fulfill (1) and belong to U_0, if the $v_0 \in C[0,b)$, $v_0(x) > 0$ in $(0,b)$.

First observe that $\varphi_t \in C[0,b)$ by the assumptions, and that the functions $\rho(x) := t/R(x)$ fulfill the equation

$$\rho(f(x)) = r(x)\rho(x), \quad x \in [0,b)$$

(cf. the argument for equation (19)). Thus, by (23), (7), and (11), we get

$$\varphi_t(f(x)) = t\, q(x)r(x)v_0(x)/R(x) = g(x)\varphi_t(x);$$

i.e., the functions φ_t fulfill equation (1) for $x \in [0,b)$. We also see that $\varphi_t \in U_0$, since, by (9), (10), and the equalities $R(0) = r(0) = 1$, for $x \to 0+$ we have

$$\lim \varphi_t(x)/u_0(x) = t \lim[v_0(x)/u_0(x)] \cdot \lim[1/R(x)] = t,$$

and the limit is positive.

Finally, let $\varphi \in C[0,b)$ be a solution of (1) fulfilling (14). Then, by (16), we see that $\alpha \in C[0,b)$ and $\alpha(0) = 0$. As the function α is given by (22), we have $t = 0$ there, and so $\alpha(x) = 0$ for $x \in [0,b)$. By (16), we get (15). Thus the proof of the theorem is complete.

4. CONCLUSION

We conclude the paper with two remarks and an example.

REMARKS. (i) The sets U_0 are equivalence classes of the relation L defined in the class of functions h that are continuous and positive in $(0,b)$ in the following way: $h_1 \, L \, h_2$ if the limit of their quotient exists at the origin and is positive.

(ii) In the paper [3], we dealt with the solutions of the inequality

$$\psi(f(x)) \leq g(x)\psi(x), \quad x \in [0,b), \tag{24}$$

belonging to the class U^a of functions (cf. [6]). Here we show the following generalization of a result from [3]:

PROPOSITION. Let the functions f, g, and u_0 satisfy the assumptions of the foregoing Theorem. If ψ is a solution of inequality (24) for which the function

$$\varphi_0(x) := \lim_{n\to\infty} \psi(f^n(x))/G_n(x), \quad x \in (0,b); \quad \varphi_0(0) := 0,$$

is defined and continuous in $[0,b)$ and positive in $(0,b)$, then $\psi \in U_0$ if and only if $\varphi_0 \in U_0$.

<u>Proof</u>. On account of the Lemma applied to inequality (24) instead of (3), we can write

$$\psi(x) = d_0(x)\varphi_0(x)$$

for $x \in [0,b)$, where $d_0 \in C[0,b)$, and d_0 satisfies conditions (8) and (10). Thus the assertion results from the definition of the class U_0.

EXAMPLE (cf. [1]). Consider the equation

$$\varphi(x/2) = e^{-1/x}\cos(x/2)\varphi(x),\ x \in (0,\pi),\ \varphi(0) = 0.$$

Put

$$q(x) := e^{-1/x} \quad \text{for} \quad x \in (0,\pi),\ q(0) := 0.$$

Then, according to (11), we have

$$r(x) = \cos(x/2) \quad \text{for} \quad x \in [0,\pi),$$

$$R(x) = (\sin x)/x \quad \text{for} \quad x \in (0,\pi),\quad R(0) = 1.$$

The function

$$u_0(x) := q(x) \quad \text{for} \quad x \in [0,\pi)$$

fulfills inequality (3), and $v_0(x) = u_0(x)$. Now the Theorem implies that the only solutions of our equation in the class U_0 (determined by the u_0) are of the form

$$\varphi_t(x) = t\, x\, e^{-1/x}/\sin x \quad \text{for} \quad x \in (0,\pi),\ \varphi_t(0) = 0,\ t > 0.$$

REFERENCES

1. D. Brydak, On functional inequalities in a single variable, Dissertationes Math. (to appear).

2. D. Brydak, On the homogeneous functional inequality, Rocznik Nauk.-Dydak. WSP w Krakowie (to appear).

3. D. Brydak and B. Choczewski, Classification of continuous solutions of a functional inequality, Zeszyty Nauk. Uniw. Jagiello. Prace Mat. 17 (1975), 33-40.

4. B. Choczewski and M. Kuczma, On the "indeterminate case" in the theory of a linear functional equation, Fund. Math. 58 (1966), 163-175.

5. M. Kuczma, Functional equations in a single variable, Polish Scientific Publishers, Warszawa, 1968.

6. M. Kuczma, Sur l'équation fonctionnelle de Böttcher, Mathematica (Cluj) 8(31) (1966), 279-285.

COMPARISON THEOREMS FOR A FUNCTIONAL INEQUALITY

Erwin Turdza
Institute of Mathematics
Pedagogical University
30-011 Kraków
POLAND

ABSTRACT. Solutions ψ of the inequality $\psi^n(x) \leq g(x)$ are compared with solutions φ of the equation $\varphi^n(x) = g(x)$ in the cases in which the solutions are (a) continuous, (b) continuously differentiable, and (c) of a given asymptotic behaviour. (Here upper indices denote functional iterates.)

1. INTRODUCTION

In this paper, we shall deal with the functional inequality of the n-th order

$$\psi^n(x) \leq g(x) \ , \tag{1}$$

where g is a given function, and where ψ^n denotes the n-th iterate of the unknown function ψ. We shall be interested in continuous solutions of (1).

The inequality (1) was examined by G. Brauer [1] under the assumption that the function $g : \mathbb{R} \to \mathbb{R}$ is increasing in $\mathbb{R}$.

For inequalities of the first order, some comparison theorems have been proved by D. Brydak [2]. In the present paper, we prove some theorems for inequality (1) in which we compare solutions ψ of inequality (1) with solutions φ of the functional equation

$$\varphi^n(x) = g(x) \ . \tag{2}$$

In the sequel, we shall assume that the following hypothesis is fulfilled:

(H) The function $g : [a,b] \to [a,b]$ is continuous and strictly increasing in $[a,b]$ and

$$g(x) < x \quad \text{for} \quad x \in (a,b), \quad g(a) = a, \quad g(b) = b \ .$$

We shall consider solutions ψ of (1) that are commutable with g, i.e.,

$$\psi(g(x)) = g(\psi(x)) \ . \tag{3}$$

Note that the class of functions commuting in $[a,b]$ with a function g fulfilling the hypothesis (H) is large enough. Namely, equation (3) has then in the interval $[a,b]$ a continuous solution depending on an arbitrary function ([4], Theorem 10.1, p. 213). More precisely, for any $x_0 \in (a,b)$ and any continuous function ψ_0 defined on $[g(x_0),x_0] \subset (a,b)$ and fulfilling the condition

$$\psi_0(g(x_0)) = g(\psi_0(x_0)) , \tag{4}$$

there exists exactly one continuous function ψ fulfilling equation (3) and the conditions

$$\psi(x) = \psi_0(x) \quad \text{for} \quad x \in [g(x_0),x_0] , \tag{5}$$

$$\psi(a) = a, \quad \psi(b) = b . \tag{6}$$

In particular, any continuous solution of equation (2) is commutable with g.

Now we shall give the construction of the general continuous solution of equation (2), due to M. Kuczma [3]. We shall use it in proofs of our theorems.

CONSTRUCTION. Take an arbitrary $x_0 \in (a,b)$, and let $x_{n-1},x_{n-2},\dots,x_1 \in (a,b)$ fulfill the inequalities

$$g(x_0) < x_{n-1} < x_{n-2} < \cdots < x_1 < x_0 . \tag{7}$$

Put

$$x_{\nu+n} = g(x_\nu) \quad \text{for} \quad \nu = 0,\pm1,\pm2,\cdots . \tag{8}$$

(The sequence x_ν $(\nu \geq 0)$ is decreasing, and it converges to a; the sequence $x_{-\nu}$ $(\nu \geq 0)$ is increasing, and it converges to b.) Let $\varphi_1,\varphi_2,\dots,\varphi_{n-1}$ be arbitrary, continuous, and strictly increasing functions defined on the intervals $[x_1,x_0],[x_2,x_1],\dots,[x_{n-1},x_{n-2}]$, respectively, and fulfilling the following conditions:

$$\varphi_i(x_{i-1}) = x_i , \quad i = 1,2,\dots,n . \tag{9}$$

Put

$$\varphi_{\nu+n}(x) = g(\varphi_{\nu+1}^{-1}(\varphi_{\nu+2}^{-1}(\cdots(\varphi_{\nu+n-1}^{-1}(x))\cdots))) \tag{10}$$
$$\text{for} \quad x \in [x_{\nu+n},x_{\nu+n-1}], \quad \nu = 0,1,2,\dots .$$

The function $\varphi\colon [a,b] \to \mathbb{R}$ defined by

$$\varphi(x) = \varphi_\nu(x) \quad \text{for} \quad x \in [x_\nu,x_{\nu-1}] , \quad \nu = 0,\pm1,\pm2,\dots, \tag{11}$$
$$\varphi(a) = a, \quad \varphi(b) = b ,$$

is a continuous solution of equation (2) in $[a,b]$.

Taking all possible systems of points $x_1, x_2, \dots, x_{n-1}$ fulfilling conditions (7), and all possible systems of continuous and strictly increasing functions $\varphi_1, \varphi_2, \dots, \varphi_{n-1}$ fulfilling conditions (9), we get all continuous solutions of equation (2).

2. CONTINUOUS SOLUTIONS

In this section, we deal with continuous solutions of (1). It will be convenient to accept the following:

DEFINITION 1. A function f is said to belong to the class C_a^b of functions if and only if it is a continuous map of the interval $[a,b]$ into itself.

First we prove two lemmas.

LEMMA 1. Let the function g fulfill hypothesis (H), x_0 be any point of the interval (a,b), and the function $\varphi \in C_a^b$ be commutable with g (i.e., (3) holds for $x \in [a,b]$). If a function $\psi \in C_a^b$ is also commutable with g, and it fulfills the inequality

$$\psi(x) \le \varphi(x) \quad \text{for} \quad x \in [g(x_0), x_0] , \tag{12}$$

then

$$\psi(x) \le \varphi(x) \quad \text{for} \quad x \in [a,b] . \tag{13}$$

<u>Proof</u>. We shall prove separately that

(a) $\psi(x) \le \varphi(x)$ for $x \in [a, x_0]$,

and

(b) $\psi(x) \le \varphi(x)$ for $x \in [g(x_0), b]$.

<u>Part (a)</u>. It follows from hypothesis (H) that the sequence $(g^n(x_0))$ is decreasing and

$$\lim_{n\to\infty} g^n(x_0) = a .$$

We claim that

$$\psi(x) \le \varphi(x) \quad \text{for} \quad x \in [g^n(x_0), x_0], \quad n = 1,2,\dots . \tag{14}$$

For $n = 1$, inequality (14) reduces to (12). Let (14) be fulfilled for a $k \ge 1$, and let

$$x \in [g^{k+1}(x_0), g^k(x_0)) .$$

It follows from (H) that there exists a

$$t \in [g^k(x_0), g^{k-1}(x_0)) \subset [g^k(x_0), x_0]$$

such that $x = g(t)$. Hence the commutability of φ with g and monotonicity of g imply that

$$\varphi(x) = \varphi(g(t)) = g(\varphi(t)) \geq g(\psi(t)) = \psi(g(t)) = \psi(x) , \tag{15}$$

which ends the inductive proof of inequality (14). Inequality (13) is then fulfilled in $(a, x_0]$. But it is also fulfilled at the point a, since φ and ψ are continuous at this point.

Part (b). We are going to prove by induction that

$$\psi(x) \leq \varphi(x) \quad \text{for} \quad x \in [g(x_0), g^{-k}(x_0)) , \quad k = 0,1,\dots . \tag{16}$$

For $k = 1$, inequality (16) reduces to (12), so it is true. Let (16) be fulfilled for a $k \geq 1$, and let

$$x \in [g^{-k}(x_0), g^{-k-1}(x_0)) .$$

Then

$$g(x) \in [g^{-k+1}(x_0), g^{-k}(x_0)) ,$$

whence

$$g(\psi(x)) = \psi(g(x)) \leq \varphi(g(x)) = g(\varphi(x)) ,$$

by virtue of the commutability of ψ with g. This inequality implies $\psi(x) \leq \varphi(x)$, since g is strictly increasing. This ends the inductive proof. The inequality (16) is also fulfilled at the point b, because φ and ψ are continuous in $[a,b]$.

Together, (a) and (b) imply (13), and the proof of Lemma 1 is complete.

LEMMA 2. Every solution $\psi \in C_a^b$ of the inequality (1) fulfills the inequality

$$\psi(x) < x \quad \text{for} \quad x \in (a,b) . \tag{17}$$

Proof. Let us suppose that (17) does not hold. Then either $\psi(x) > x$ for every $x \in (a,b)$ or there exists a point $x_0 \in (a,b)$ such that $\psi(x_0) = x_0$. In the first case, $\psi^k(x) > x$ for $x \in (a,b)$, and $k = 1,2,\dots$, whence $\psi^n(x) > x > g(x)$, which is impossible by virtue of (1). In the other case, $\psi^k(x_0) = x_0$ for $k = 1,2,\dots$, whence $\psi^n(x_0) = x_0 > g(x_0)$, which is also impossible. Thus Lemma 2 is true.

THEOREM 1. Assume that the function g fulfills hypothesis (H),

$\psi \in C_a^b$ is a solution of the inequality (1), and there exists a point $x_0 \in (a,b)$ such that

$$(18) \qquad \psi^n(x_0) = g(x_0),$$

and ψ is strictly increasing in $[\psi^{n-1}(x_0), g(x_0)]$. If the function ψ is commutable with g, then there exists a continuous solution φ of equation (2) such that inequality (13) holds. The solution φ is unique, and $\psi(x) = \varphi(x)$ for $x \in [\psi^{n-1}(x_0), x_0]$.

Proof. Put

$$(19) \qquad x_k = \psi^k(x_0) \quad \text{for} \quad k = 1,2,\dots,n-1 .$$

It follows from Lemma 2 that the function ψ fulfills inequality (17), so that by (18) the x_k fulfill (7). Now we define the sequence φ_ν of functions in the following way:

$$(20) \qquad \varphi_\nu(x) = \psi(x) \quad \text{for} \quad x \in [x_\nu, x_{\nu-1}], \quad \nu = 1,2,\dots,n-1 .$$

The functions φ_ν are strictly increasing in their domains of definition, because ψ is assumed to be strictly increasing in the interval $[x_{n-1}, x_0]$. Moreover, $\varphi_\nu([x_\nu, x_{\nu-1}]) = [x_{\nu+1}, x_\nu]$ for $\nu = 1,2,\dots,n-1$. The foregoing Kuczma Construction then yields the existence of exactly one continuous and strictly increasing solution φ of equation (2) given by (11) with (10), and (8) with (19).

Formulas (10) and (20) imply

$$(21) \qquad \psi(x) \le \varphi(x) \quad \text{for} \quad x \in [x_{n-1}, x_0] .$$

We shall prove that also

$$\psi(x) \le \varphi(x) \quad \text{for} \quad x \in [g(x_0), x_{n-1}] .$$

In this interval, we have $\varphi(x) = \varphi_n(x)$ (cf. (11), (10) and (8)). Since the function φ_ν is a bijection of the interval $[x_\nu, x_{\nu-1}]$ onto $[x_{\nu+1}, x_\nu]$, it follows from (18) and (19) that, for each

$$x \in [g(x_0), x_{n-1}] = [\psi(x_{n-1}), x_{n-1}] ,$$

there exists a number $t \in [x_1, x_0]$ such that

$$\psi^{n-1}(t) = x = \varphi_{n-1}(\dots(\varphi_1(t))\dots) .$$

Thus, for $x \in [g(x_0), x_{n-1}]$, we have

$$\varphi(x) = \varphi_n(x) = g(\varphi_1^{-1}(\ldots(\varphi_{n-1}^{-1}(x))\ldots))$$
$$= g(\varphi_1^{-1}(\ldots(\varphi_{n-1}^{-1}(\varphi_{n-1}(\ldots\varphi_1(t)\ldots)))\ldots)))\ldots))$$
$$= g(t) \geq \psi^n(t) = \psi(\psi^{n-1}(t)) = \psi(x) .$$

The last inequality, together with (21), gives (12), and the asserted inequality (13) follows from Lemma 1.

Continuous solutions of equation (2) intersect each other in the interval $[x_{n-1},x_0]$ (see the Kuczma Construction, above). Thus the solution φ is unique in the sense that for any other solution $\overline{\varphi}$ of equation (2) there exists a point $\overline{x} \in [x_{n-1},x_0]$ such that $\overline{\varphi}(\overline{x}) < \varphi(\overline{x}) = \psi(\overline{x})$. This completes the proof of Theorem 1.

3. DIFFERENTIABLE SOLUTIONS

It turns out that if a ψ fulfilling (1) is of class C^1, then we can find a C^1 solution φ of (2) fulfilling inequality (13).

We start with the following simple result:

LEMMA 3. If the function f defined in a neighborhood of x_0, and continuous at the point x_0, is differentiable in a set $(c,x_0) \cup (x_0,d)$, and if

$$\lim_{x \to x_0^+} f'(x) = \lim_{x \to x_0^-} f'(x) = d_0 , \tag{22}$$

then f is differentiable at x_0, and $f'(x_0) = d_0$.

Proof. Let $x \in (x_0,d)$. It follows from the mean-value theorem that there exists a point $\xi(x) \in (x_0,x)$ such that

$$f(x) - f(x_0) = f'(\xi(x))(x - x_0) .$$

Since $\xi(x)$ converges to x_0^+ as x does, it follows that

$$f'(x_0^+) = \lim_{x \to x_0^+} f'(\xi(x)) = d_0 .$$

Similarly, we prove that $f'(x_0^-) = d_0$.

THEOREM 2. Let the following assumptions be fulfilled:

(i) the function $g \in C^1(a,b)$ fulfills hypothesis (H), and $g'(x) > 0$ for $x \in (a,b)$;

(ii) the function $\psi \in C_a^b$ is a $C^1(a,b)$-solution of inequality (1);

(iii) there exists a point $x_0 \in (a,b)$ such that

$$\psi^n(x_0) = g(x_0), \quad \psi'(x) > 0 \quad \text{for} \quad x \in [x_{n-1}, x_0] ,$$

and

$$\psi'(x_{n-1}) = g'(x_0) \prod_{i=0}^{n-2} [\psi'(x_i)]^{-1} , \tag{23}$$

where $x_i = \psi^i(x_0)$ for $i = 1,2,\dots,n-1$.

If the function ψ commutes with g, then there exists a $C^1(a,b)$-solution φ of equation (2) for which (13) holds. The solution φ is unique, and $\psi(x) = \varphi(x)$ for $x \in [x_{n-1}, x_0]$.

Proof. Let us define further terms of the sequence (x_n) by (8), and take the function φ we have found in the proof of Theorem 1. It is a continuous solution of (2) satisfying inequality (13).

We are going to prove that $\varphi \in C^1(a,b)$. It follows from the definition of φ that it is a C^1-function in the set

$$\bigcup_{\nu=-\infty}^{+\infty} (x_\nu, x_{\nu-1}) ,$$

and $\varphi'(x) > 0$ there. We shall prove that φ is differentiable at the point x_{n-1}. By (20), we have

$$\lim_{x \to x_{n-1}^+} \varphi'(x) = \lim_{x \to x_{n-1}^+} \psi'(x) = \psi'(x_{n-1}) .$$

Let $x \in (x_n, x_{n-1})$. Then from (10) and (23), we have

$$\begin{aligned}
\lim_{x \to x_{n-1}^-} [\varphi(x)]' &= \lim_{x \to x_{n-1}^-} [g(\varphi_1^{-1}(\dots(\varphi_{n-1}^{-1}(x))\cdots))]' \\
&= \lim_{x \to x_{n-1}^-} [g(\psi^{-n+1}(x))]' \\
&= \lim_{x \to x_{n-1}^-} g'(\psi^{-n+1}(x)) \cdot \prod_{i=0}^{n-2} (\psi'(\psi^{-i}(x)))]^{-1} \\
&= g'(x_0) \cdot \prod_{i=0}^{n-2} [\psi'(x_i)]^{-1} = \psi'(x_{n-1}) .
\end{aligned}$$

The two limits prove, by Lemma 3, that

$$\varphi \in C^1(g(x_0), x_0) \quad \text{and} \quad \varphi'(x) > 0 \quad \text{in} \quad (g(x_0), x_0) . \tag{24}$$

To prove that $\varphi \in C^1(a,x_0)$ and $\varphi'(x) > 0$ in (a,x_0), we prove the following implication:

(*) If φ is a C^1-function and $\varphi'(x) > 0$ in $(x_{\nu+n}, x_0)$, then it has the same properties in the interval $(x_{\nu+n+1}, x_0)$, $\nu = 0,1,\dots$.

For, first note that the function φ is a one-to-one map of any interval $(x_\nu, x_{\nu-1})$ onto $(x_{\nu+1}, x_\nu)$ (cf. the Kuczma Construction). Put

$$I_{\nu+n} = (x_{\nu+n+1}, x_{\nu+1}) .$$

The iterate φ^k maps bijectively the interval $I_{\nu+1}$ onto $I_{\nu+k+1}$, for every $k = 1,2,\ldots,$ and

$$I_{\nu+k+1} \subset (x_{\nu+n}, x_0) \quad \text{for} \quad k = 1,2,\ldots,n-2 .$$

Thus the function $\varphi^{n-1}: I_{\nu+1} \to I_{\nu+n}$ is of class C^1 in $I_{\nu+1}$, and its derivative

$$(\varphi^{n-1})'(x) = \prod_{k=0}^{n-2} \varphi'(\varphi^k(x_i))$$

is positive in $I_{\nu+1}$. This shows that the inverse function

$$\varphi^{-n+1} : I_{\nu+n} \to I_{\nu+1}$$

belongs to the class $C^1(I_{\nu+n})$. By (2), we have

$$\varphi(x) = g(\varphi^{-n+1}(x)) \quad \text{for} \quad x \in I_{\nu+n} , \tag{25}$$

which yields $\varphi \in C^1(I_{\nu+n})$ and $\varphi'(x) > 0$ in $I_{\nu+n}$. But

$$(x_{\nu+n+1}, x_0) = I_{\nu+n} \cup [x_{\nu+1}, x_0) \quad \text{and} \quad [x_{\nu+1}, x_0) \subset (x_{\nu+n}, x_0) ,$$

which ends the proof of (*).

Consider now the assertion:

(**) If φ^{-1} is a C^1-function and $[\varphi^{-1}]'(x) > 0$ in $(x_n, x_{-\nu+1})$, then it has the same properties in $(x_n, x_{-\nu})$, for $\nu = 1,2,\ldots$

The proof of (**) is similar to that of (*); in proving (**) we make use of the relation

$$\varphi^{-1}(x) = g^{-1}(\varphi^{n-1}(x)), \quad x \in (a,b) ,$$

instead of (25).

Since $\varphi : (x_{n-1}, x_0) \to (x_n, x_1)$, and since φ is of class C^1 and has a positive derivative in (x_{n-1}, x_0), it follows that

$$\varphi^{-1} \in C^1(x_n, x_1) \quad \text{and} \quad [\varphi^{-1}]'(x) > 0 \quad \text{in} \quad (x_n, x_1) ,$$

so that, by (**), induction yields $\varphi \in C^1(x_n, b)$.

Similarly, assertion (*), together with $\varphi \in C^1(x_n, x_0)$, implies

$\varphi \in C^1(a,x_0)$. Thus $\varphi \in C^1(a,b)$, which ends the proof of Theorem 2.

4. SOLUTIONS WITH AN ASYMPTOTIC PROPERTY

Now we shall prove a comparison theorem for the solutions of the inequality

$$\psi^2(x) \le g(x) \tag{26}$$

belonging to the class U^p defined below. In this section, we take $a = 0$.

DEFINITION 2. We say that the function $f \in C_0^b$ belongs to the class U^p of functions $(p > 0)$ if and only if there exists a function $q: :[0,b] \to \mathbb{R}$, continuous in $[0,b]$, such that

$$f(x) = q(x)x^p \quad \text{for} \quad x \in [0,b], \quad q(0) > 0 \; .$$

The following lemma will be useful in the sequel:

LEMMA 4. Let (y_m,k) be a double sequence. If

$$\lim_{m\to\infty} y_{m,k} = y_k \quad \text{and} \quad \lim_{k\to\infty} y_k = y_0 \; , \tag{27}$$

then there exist sequences $i_n \to \infty$, $j_n \to \infty$ of positive integers such that

$$\lim_{n\to\infty} y_{i_n,j_n} = y_0 \; .$$

Proof. If infinitely many of the sequences $(y_{m,k})_{k\in N}$ are almost constant, say

$$y_{m,j_n} = y_{j_n} \quad \text{for} \quad m \_ m_n$$

and $j_n \to \infty$, then putting $i_n = \max(m_n,j_n)$ we have

$$\lim_{k\to\infty} y_{i_n,j_n} = y_0 \; .$$

In the other case, denoting by Z the set

$$Z := \{y_{m,k} : m \ge k\} \; ,$$

and by Z^d the derivative of Z, we have $y_0 \in Z^d$. Hence there exists a sequence $z_n \in Z$ that converges to y_0. This implies that there are sequences of indices $i_n \to \infty$, $j_n \to \infty$ such that $z_n = y_{i_n,j_n}$, which was to be proved.

THEOREM 3. Let the function $g \in U^\alpha$ $(\alpha > 0)$ fulfill hypothesis (H).

If the function φ is a continuous solution of the equation

$$(28) \qquad \varphi^2(x) = g(x) \quad \text{for} \quad x \in [0,b] \, ,$$

and φ satisfies the inequality

$$(29) \qquad \varphi(x) \geq f(x) \quad \text{for} \quad x \in [0,b_1] \subset [0,b] \, ,$$

where $f \in U^\beta$, $\beta = \sqrt{\alpha}$, then $\varphi \in U^\beta$.. Moreover, if we write

$$(30) \qquad g(x) = p(x)x^\alpha \, , \quad d := p(0) > 0 \, ,$$

and

$$(31) \qquad \varphi(x) = \eta(x)x^\beta \, ,$$

then

$$(32) \qquad \eta(0) = d^{1/(1+\beta)} \, .$$

Proof. Let φ and f satisfy the assumptions of the theorem. Since $f \in U^\beta$, it follows that $f(x) = q(x)x^\beta$, $q(0) > 0$, so that (29) and (30) yield

$$(33) \qquad \liminf_{x \to 0^+} \eta(x) > 0 \, .$$

Let $x_n \to 0^+$. We shall prove that the sequence $(\eta(x_n))$ is bounded. Assume the contrary. Then there is a sequence (k_n) of positive integers, $k_n \to \infty$, such that the sequence $\eta(x_{k_n}))$ converges to infinity. By (28), (30), and (31), we have

$$\varphi^2(x_{k_n}) = \eta(\varphi(x_{k_n}))[\eta(x_{k_n})]^\beta \, x_{k_n}^\alpha = p(x_{k_n})x_{k_n}^\alpha = g(x_{k_n}) \, ,$$

since $\beta^2 = \alpha$. Therefore

$$p(x_{k_n}) = \eta(\varphi(x_{k_n}))[\eta(x_{k_n})]^\beta \, ,$$

and $(p(x_{k_n}))$ converges to the positive limit d (cf. (30)), whence $(\eta(\varphi(x_{k_n})))$ converges to zero, because $([\eta(x_{k_n})]^\beta$ tends to infinity.

But φ is a continuous function in $[0,b]$ fulfilling equation (28) there. This implies that $\varphi(0) = g(0) = 0$, and therefore that the sequence $(\varphi(x_{k_n}))$ converges to zero. By virtue of (33), we have

$$\lim_{n \to \infty} \eta(\varphi(x_{k_n})) > 0 \, ,$$

a contradiction.

Now we shall prove that (32) holds.

Since, for an arbitrarily given sequence $x_n \to 0$, and for a chosen subsequence (x_{k_n}), the sequence $(\eta(x_{k_n}))$ is bounded, it has a convergent subsequence, say $(\eta(x_{k_{\ell_n}}))$. Denote

$$z_n := x_{k_{\ell_n}}, \quad \lim_{n\to\infty} \eta(z_n) =: s > 0 . \tag{34}$$

From (31), (28), and (30), we get

$$\eta(\varphi(x)) = \varphi^2(x)[\varphi(x)]^{-\beta} = g(x)[\eta(x)]^{-\beta}x^{-\beta^2} = p(x)[\eta(x)]^{-\beta} .$$

By (34) and (30), this yields

$$\lim_{n\to\infty} \eta(\varphi(z_n)) = \lim_{n\to\infty} p(z_n)[\eta(z_n)]^{-\beta} = ds^{-\beta} . \tag{35}$$

Now we shall deal with the double sequence having terms $\eta(\varphi(z_n))$. It will be convenient to use the notations:

$$\delta := -\beta , \quad w_{n,k} = \varphi^k(z_n) .$$

We prove by induction that

$$\lim_{n\to\infty} \eta(w_{n,k}) = d^{(1-\delta^k)/(1-\delta)} \cdot s^{\delta^k} , \quad k = 1,2,\ldots . \tag{36}$$

By (35), relation (36) is true for $k = 1$. Assume it to hold for an integer $k \geq 1$, and consider

$$\eta(w_{n,k+1}) = \eta(\varphi(w_{n,k})) .$$

Since $\varphi^k(0) = 0$ for any k, it follows that

$$\lim_{n\to\infty} w_{n,k} = 0 .$$

Thus, as in (35), we get

$$\begin{aligned}\lim_{n\to\infty} \eta(w_{n,k+1}) &= \lim_{n\to\infty} p(w_{n,k})[\eta(w_{n,k})]^{\delta} = d^{1+\delta(1-\delta^k)/(1-\delta)} \cdot s^{\delta^{k+1}} \\ &= d^{(1-\delta^{k+1})/(1-\delta)} \cdot s^{\delta^{k+1}} ,\end{aligned}$$

and the proof of (36) is complete.

Next we put

$$s = rd^{1/(1-\delta)} , \quad r > 0 .$$

The limit (36) then reads

$$\lim_{n\to\infty} \eta(w_{n,k}) = d^{1/(1-\delta)}\, r^{\delta^k} = d^{1/(1+\beta)} r^{(-1)^k\beta^k} . \tag{37}$$

We are going to prove that $r = 1$.

If $r > 1$, then taking k odd and denoting

$$y_{n,k} := \eta(w_{n,k}) ,$$

from (37) we have

$$y_k := \lim_{n \to \infty} y_{n,k} = d^{1/(1+\beta)} r^{-\beta^k} ,$$

i.e.,

$$\lim_{k \to \infty} y_k = 0 .$$

Lemma 4 thus yields the existence of sequences of integers $i_n \to \infty$ and $j_n \to \infty$ such that

$$\lim_{n \to \infty} y_{i_n, j_n} = \lim_{n \to \infty} \eta(\varphi^{i_n}(z_{j_n})) = 0 .$$

This relation contradicts (33).

If $r \in (0,1)$, then, taking k even, we arrive at the relations

$$\lim_{k \to \infty} y_k = \lim_{k \to \infty} d^{1/(1+\beta)} r^{\beta^k} = 0 ,$$

which by Lemma 4 also lead to a contradiction of (33).

Consequently, $r = 1$ and $s = d^{1/(1+\beta)}$.

Thus we have proved that, given a sequence $x_n \to 0$ and an arbitrary sequence $(\eta(x_{k_n}))$ chosen from the $(\eta(x_n))$, one can find a subsequence

$$\eta(z_n) = \eta(x_{k_{\ell_n}}) \quad \text{of the} \quad (\eta(x_{k_n}))$$

that converges to s. This implies the existence of the limit of $(\eta(x_n))$ and the relation

$$\lim_{n \to \infty} \eta(x_n) = s ,$$

for any $x_n \to 0$, which is equivalent to the relation

$$\lim_{x \to 0^+} \eta(x) = s .$$

We put $\eta(0) := s = d^{1/(1+\beta)}$, and the proof of Theorem 3 is complete.

As a consequence of Theorem 3, we obtain the following:

THEOREM 4. Let the following assumptions be fulfilled:

(i) the function $g \in U^{\alpha}$, $\alpha > 1$, fulfills hypothesis (H);

(ii) the function $\psi \in U^{\beta}$, $\beta = \sqrt{\alpha}$, is a solution of inequality (26);

(iii) there exists a point $x_0 \in (0,b)$ such that $\psi^2(x_0) = g(x_0)$, and the function ψ is strictly increasing in the interval $[\psi(x_0), x_0]$.

If the function ψ is commutable with g, then there exists a solution $\varphi \in U^{\beta}$ of equation (28) such that

$$(38) \qquad \psi(x) \leq \varphi(x) \quad \text{for} \quad x \in [0,b] .$$

The solution φ is unique, and $\psi(x) = \varphi(x)$ for $x \in [\psi(x_0),x_0]$.

Proof. Let φ be the continuous solution of equation (28) fulfilling the condition

$$\varphi(x) = \psi(x) \quad \text{for} \quad x \in [\psi(x_0),x_0] .$$

By virtue of Theorem 1, the function φ satisfies inequality (38). Hence, putting ψ in place of f in (29), from Theorem 3 we get $\varphi \in U^{\beta}$, which ends the proof of Theorem 4.

REMARKS. (i) The assumption of commutability of ψ with g, made in Theorems 1, 2, and 4, is rather strong. Replacing this assumption by the inequality

$$(39) \qquad \psi(g(x)) \leq g(\psi(x)) \quad \text{for} \quad x \in [a,x_0], \quad x_0 < b ,$$

we get local versions of these theorems, which means that we can prove the inequality $\psi(x) \leq \varphi(x)$ only in the interval $[a,x_0]$. Similarly, Theorems 1, 2, and 4 will be true only in a left-hand neighborhood of the point b, if we replace (3) by the inequality (39), restricted this time to $x \in [x_1,b]$, $a < x_1$. Proofs of these versions are almost the same as those of the original theorems (see, e.g., the proof of Theorem 1).

(ii) Results of the present paper were presented at the First International Conference on General Inequalities by Bogdan Choczewski.

REFERENCES

1. G. Brauer, Functional inequalities, Amer. Math. Monthly 68 (1961), 638-642.

2. D. Brydak, On the continuous solutions of a functional inequality in a single variable, Dissertationes Math. (to appear).

3. M. Kuczma, On the functional equation $\varphi^n(x) = g(x)$, Ann. Polon. Math. 11 (1961), 161-175.

4. M. Kuczma, Functional equations in a single variable, Polish Scientific Publishers, Warszawa, 1968.

INTEGRABLE SOLUTIONS OF A LINEAR FUNCTIONAL INEQUALITY

J. Matkowski
Department of Mathematics
Silesian University
40-007 Katowice
ul. Bankowa 14
POLAND

ABSTRACT. In this note we establish conditions on functions f and g under which the functional inequality $|\varphi(x)| \leq g(x)|\varphi[f(x)]|$ has no nontrivial integrable solutions.

1. INTRODUCTION

We shall consider integrable solutions of the linear functional inequality in a single variable

$$|\varphi| \leq g|\varphi\circ f|, \tag{1}$$

where f and g are given, φ is an unknown function, and $g|\varphi\circ f|$ denotes the function $x \to g(x)|\varphi[f(x)]|$. The inequality (1) appears in a natural way in the problem of the uniqueness of a general functional equation of the first order (cf. [2]).

Integrable solutions of the corresponding functional equation

$$\varphi = g\ \varphi\circ f \tag{2}$$

were considered by M. Kuczma [1] and in [2].

2. CONDITIONS

We assume the following:

(*) (X,S,μ) is a σ-finite measure space, $\mu(X) > 0$; $f:X \to X$ is one-to-one; f and f^{-1} are S-measurable; the measures μf^{-1} and μf are absolutely continuous with respect to μ; and

$$\mu(X\setminus\bigcup_{k=0}^{\infty}[f^k(X)\setminus f^{k+1}(X)]) = 0. \tag{3}$$

REMARKS. (i) Let us note that if $X = \langle 0,1\rangle$, μ is Lebesgue measure, and $f:\langle 0,1\rangle \to \langle 0,1\rangle$ satisfies the condition

$$0 < f(x) < x, \qquad x \in \langle 0,1\rangle,$$

then (3) is fulfilled. In particular, it follows from (3) that the set of fixed points of f has measure zero.

(ii) Conditions (*) imply that every two measures in the sequence μf^n, $n = 0, \pm 1, \ldots$, are absolutely continuous with respect to each other. Moreover, if $\mu(X) < \infty$, then $\lim_{n\to\infty} \mu f^n(X) = 0$.

In fact, if $\mu f^n(A) = 0$, then by (*) we have

$$\mu f^{n-1}(A) = \mu f^{-1}(f^n(A)) = 0, \qquad \mu f^{n+1}(A) = \mu f(f^n(A)) = 0,$$

which implies the first statement contained in Remark (ii). To prove the second one, note that $f^{n+1}(X) \subset f^n(X)$, $n = 0,1,\ldots$. Therefore

$$\bigcap_{n=0}^{\infty} f^n(X) \subset (X \setminus \bigcup_{n=0}^{\infty} [f^n(X)\setminus f^{n+1}(X)]),$$

and, consequently, by (3), we have

$$\lim_{n\to\infty} \mu f^n(X) = \mu(\bigcap_{n=0}^{\infty} f^n(X)) = 0.$$

3. SOLUTION

Put

$$A_n = f^n(X)\setminus f^{n+1}(X), \quad n = 0,1,\ldots \tag{4}$$

For the function $g: X \to \mathbb{R}$, we write

$$N_g = \{x \in X : g(x) = 0\}$$

and

$$\kappa_n = \operatorname*{ess\,inf}_{A_n \setminus N_g} g^{-p} \frac{d\mu f}{d\mu}, \quad n = 0,1,\ldots \tag{5}$$

THEOREM. Let (*) be fulfilled and let $g: X \to \mathbb{R}$ be measurable and nonnegative. If the series

$$\sum_{n=0}^{\infty} \prod_{i=0}^{n-1} \kappa_i \tag{6}$$

diverges and $\varphi \in L^p(X,S,\mu)$ satisfies inequality (1) a.e. in X, then $\varphi = 0$ a.e. in X.

<u>Proof</u>. By (*) and (4), we have

$$A_n \in S, \quad f(A_n) = A_{n+1}, \quad f^{-1}(A_{n+1}) = A_n, \quad A_n \cap A_k = \emptyset,$$

$n \neq k$; $n,k = 0,1,\ldots$. Moreover, it follows from (3) that

$$\int_X |\varphi|^p d\mu = \sum_{n=0}^{\infty} \int_{A_n} |\varphi|^p d\mu, \quad \varphi \in L^p(X,S,\mu). \tag{7}$$

Suppose that $\varphi \in L^p(X,S,\mu)$ satisfies (1) a.e. in X. Note that by (1) we have $\varphi = 0$ a.e. in N_g. From (*), (1), and (5), we obtain

$$\int_{A_{n+1}} |\varphi|^p d\mu = \int_{f^{-1}(A_{n+1})} |\varphi\circ f|^p d\mu f = \int_{A_n} |\varphi\circ f|^p \frac{d\mu f}{d\mu}\, d\mu$$

$$\geq \int_{A_n\setminus N_g} \frac{d\mu f}{d\mu}\, g^{-p}\, |\varphi|^p d\mu \geq \kappa_n \int_{A_n} |\varphi|^p d\mu.$$

Having repeated this procedure n times, we arrive at the inequality

$$\int_{A_{n+1}} |\varphi|^p d\mu \geq \Big(\prod_{i=0}^{n} \kappa_i\Big) \int_{A_0} |\varphi|^p d\mu.$$

Hence, using (7), we have

$$\infty > \int_X |\varphi|^p d\mu \geq \Big(\sum_{n=0}^{\infty} \prod_{i=0}^{n-1} \kappa_i\Big) \int_{A_0} |\varphi|^p d\mu.$$

Now the divergence of the series (6) implies $\varphi = 0$ a.e. in A_0. Since (cf. [1])

$$\sum_{n=0}^{\infty} \prod_{i=0}^{n-1} \kappa_i = \sum_{n=0}^{k} \prod_{i=0}^{n-1} \kappa_i + \Big(\prod_{i=0}^{k} \kappa_i\Big) \sum_{n=0}^{\infty} \prod_{i=0}^{n-1} \kappa_{k+i+1},$$

the series $\Sigma_{n=0}^{\infty} \Pi_{i=0}^{n-1} \kappa_{k+i+1}$ diverges for $k = 1,2,\ldots$. Replacing A_0 by A_k and X by $f^k(X)$ in the above argumentation, we obtain $\varphi = 0$ a.e. in A_k for $k = 1,2,\ldots$. In view of (3) and (4), we have $\mu(X\setminus\bigcup_{n=0}^{\infty} A_n) = 0$ and, consequently, $\varphi = 0$ a.e. in X. This completes the proof.

REFERENCES

1. M. Kuczma, On integrable solutions of a functional equation, Bull. Acad. Polon. Sci., Sér Sci. Math. Astronom. Phys. 19 (1971), 593-596.

2. J. Matkowski, Integrable solutions of functional equations, Dissertationes Mathematicae 127 (1975).

Differential and Integral Inequalities

A NOTE ON FIRST-ORDER LINEAR PARTIAL DIFFERENTIAL INEQUALITIES

L. Losonczi
Department of Mathematics
Kossuth Lajos University
4010 Debrecen, pf. 12.
HUNGARY

ABSTRACT. A partial differential identity is applied to obtain information concerning solutions of a given first-order linear partial differential inequality.

1. INTRODUCTION

Let $\mathbb{R}^{n+1}$ be the n+1-dimensional Euclidean space, D be a domain in $\mathbb{R}^{n+1}$, and $h, f_i : D \to \mathbb{R}^1$ $(i = 1,\ldots,n)$ be given functions. Our aim is to investigate the inequality

$$h(t,x)\frac{\partial u}{\partial t}(t,x) + \sum_{i=1}^{n} f_i(t,x)\frac{\partial u}{\partial x_i}(t,x) \geq 0 , \quad (t,x) \in D , \tag{1}$$

assuming that

(i) $h(t,x) > 0$ for $(t,x) \in D$,

(ii) $h, f_i \in C_1(D)$ for $i = 1,\ldots,n$,

where $x = (x_1,\ldots,x_n)$, and $C_1(D)$ is the class of functions from D into $\mathbb{R}^1$ having continuous first-order partial derivatives on D. Similar inequalities with more special coefficients have been studied in [1].

Denote by $\varphi(t,\tau,\xi)$ the solution of the initial-value problem

$$\frac{dx}{dt} = \frac{1}{h(t,x)} f(t,x) , \quad x(\tau) = \xi , \tag{2}$$

where $f = (f_1,\ldots,f_n)$, and where φ, ξ also are n-dimensional vectors. It is known (see [2], pp. 22-32) that for any $(t_0,x_0) \in D$, the function φ exists on $T = I \times U$, where I and U are suitable neighborhoods of t_0 and (t_0,x_0), respectively. Moreover, condition (ii) ensures that $\varphi \in C_1(T), \varphi$ satisfies the equation

$$\frac{\partial \varphi}{\partial \tau}(t,\tau,\xi) + \sum_{i=1}^{n} \frac{1}{h(\tau,\xi)} f_i(\tau,\xi)\frac{\partial \varphi}{\partial \xi_i}(t,\tau,\xi) = 0 , \quad (t,\tau,\xi) \in T , \tag{3}$$

(see [2], p. 40, problem 9), and the determinant of the matrix consisting of the vectors $\partial\varphi/d\xi_i$ is not zero. Restricting the neighborhood U if necessary, we can ensure that the system

$$\varphi(t_0,\tau,\xi) = \eta \tag{4}$$

has a unique solution ξ for $(\tau,\xi) \in U$. Let us denote this solution by

$$\xi = \psi(\tau,\eta) \ . \tag{5}$$

2. THE MAIN RESULT

Our result concerning (1) is the following:

THEOREM. Suppose that (i), (ii) are satisfied. If $u : D \to \mathbb{R}^1$ is partially differentiable on D, then

$$h(t,x)\frac{\partial u}{\partial t}(t,x) + \sum_{i=1}^{n} f_i(t,x)\frac{\partial u}{\partial x_i}(t,x) = h(t,x)\frac{d}{dt}u(t,\psi(t,\eta))\Big|_{\eta=\varphi(t_0,t,x)} \tag{6}$$

holds for $(t,x) \in U$, where U is the neighborhood of $(t_0,x_0) \in D$ mentioned above.

<u>Proof</u>. Calculating the right-hand side of (6), we obtain

$$h(t,x)\frac{\partial u}{\partial t}(t,x) + \sum_{i=1}^{n}\frac{\partial u}{\partial x_i}(t,x)\frac{\partial \psi_i}{\partial \tau}(t,\varphi(t_0,t,x))h(t,x) \ .$$

Thus we have to show that

$$\frac{\partial \psi}{\partial \tau}(t,\varphi(t_0,t,x))h(t,x) = f(t,x) \ . \tag{7}$$

If we differentiate the identity

$$\varphi(t_0,t,\psi(t,\eta)) = \eta$$

with respect to t, we get

$$\frac{\partial \varphi}{\partial \tau}(t_0,t,\psi(t,\eta)) + \sum_{i=1}^{n}\frac{\partial \varphi}{\partial \xi_i}(t_0,t,\psi(t,\eta))\frac{\partial \psi_i}{\partial \tau}(t,\eta) = 0 \ .$$

Substituting here $\eta = \varphi(t_0,t,x)$ and comparing the result with (3), we obtain (7), which completes the proof.

3. DISCUSSION

The identity (6) shows that u satisfies the inequality

$$h(t,x)\frac{\partial u}{\partial t}(t,x) + \sum_{i=1}^{n} f_i(t,x)\frac{\partial u}{\partial x_i}(t,x) \geq 0 \ , \quad (t,x) \in U \ , \tag{8}$$

if and only if $u(t,\psi(t,\eta))$ is an increasing function of t on $t \in J_\eta$ for every fixed $\eta \in \varphi(t_0,U)$, where J_η is an interval depending on η. Let

$$V = \{(\tau,\eta) \mid \eta = \varphi(t_0,\tau,\xi), \ (\tau,\xi) \in U\}$$

be the image of U under the transformation (4). Then (8) is equivalent to

$$(9) \qquad \frac{d}{dt} u(t,\psi(t,\eta)) \geq 0 , \quad (t,\eta) \in V ,$$

or to the inequality

$$u(t + s,\psi(t + s,\eta)) \geq u(t,\psi(t,\eta)) , \quad (t,\eta) \in V ,$$

where $0 \leq s < \delta(t,\eta)$. Using the transformation $\eta = \varphi(t_0,t,x)$, we have

$$(10) \quad u(t + s,\psi(t + s,\varphi(t_0,t,x))) \geq u(t,x) \quad (t,x) \in U \text{ and } 0 \leq s < \delta .$$

Thus we can state our result as follows.

Suppose that the conditions of the theorem are satisfied and $u : D \to \mathbb{R}^1$ is partially differentiable on D. Then the inequality (8) holds if and only if (10) is valid. Now (10) is an algebraic inequality which characterizes the solutions of (8). If we change the inequality sign in (8), then we should change it also in (10); and if we write equality instead of inequality in (8), then we should do so also in (10). In this latter case, s may vary in a full neighborhood of 0. This gives a functional equation for the solution of the partial differential equation corresponding to (8). In a number of cases, U may be taken to be D.

4. EXAMPLE

Finally, we show an example to illustrate our results.

Consider the inequality

$$(11) \qquad \frac{\partial u}{\partial t}(t,x) + (x + e^t) \frac{\partial u}{\partial x}(t,x) \geq 0 , \quad (t,x) \in R^2 .$$

An easy calculation shows that $\varphi(t,\tau,\xi) = e^t(t + \xi e^{-\tau} - \tau)$; and taking $t_0 = 0$, we get $\psi(\tau,\eta) = e^\tau(\tau + \eta)$. The identity (6) has now the form

$$(12) \qquad \frac{\partial u}{\partial t}(t,x) + (x + e^t) \frac{\partial u}{\partial x}(t,x) = \frac{d}{dt} u(t,e^t(t + \eta))\Big|_{\eta=xe^{-t}-t} .$$

It is interesting to remark that (12) can be obtained in the following "formal" way. The solutions of the equation corresponding to (11) are $u(t,x) = \varphi(xe^{-t} - t)$, with arbitrary continuously differentiable φ. Substitute here $\eta = xe^{-t} - t$; then $u(t,e^t(t + \eta)) = \varphi(\eta)$, which can be written as $0 = \frac{d}{dt} u(t,e^t(t + \eta))$, and the right-hand side of this is exactly the expression on the right-hand side of (12) if we set $\eta = xe^{-t} - t$. Using (12), we see that (11) holds if and only if $u(t_1 e^t(t + \eta))$ is an increasing function of t for every real value of η, or u satisfies the inequality

$$u(t + s, e^s(x + se^t)) \geq u(t,x)$$

for $(t,x) \in \mathbb{R}^2$ and $s \geq 0$.

REFERENCES

1. L. Losonczi, A generalization of the Gronwall-Bellman lemma and its applications, J. Math. Anal. Appl. 44 (1973), 701-709.

2. E. A. Coddington and N. Levinson, Theory of ordinary differential equations, McGraw-Hill, New York, Toronto, and London, 1955.

RAYLEIGH'S PRINCIPLE BY EQUIVALENT PROBLEMS

Donald R. Snow
Department of Mathematics
Brigham Young University
Provo, Utah 84602
U.S.A.

ABSTRACT. Rayleigh's Principle provides an inequality which gives upper bounds to an eigenvalue of a differential equation by evaluating a ratio of quadratic functionals with any function from a prescribed class. It also shows that the value of the functional evaluated with the eigenfunction is exactly the eigenvalue. This paper shows how minimizing the functional

$$J[u] = \iint [p\nabla u \cdot \nabla u + (q - \lambda r)u^2]$$

by a modification of Carathéodory's equivalent-problems method yields Rayleigh's Principle for the partial-differential-equation (PDE) eigenvalue problem $\nabla \cdot (p\nabla u) - (q - \lambda r)u = 0$ on D, $u = 0$ on ∂D. The approach leads to a Hamilton-Jacobi equation for a vector variable, and seeking special forms of solution to this leads to a scalar PDE $\nabla \cdot \sigma + \frac{1}{p}\sigma \cdot \sigma = q - \lambda r$ for the vector variable $\sigma(x,y)$. This PDE is an obvious generalization of a Riccati ordinary differential equation, and it can be linearized by the transformation $\sigma = p\,\frac{\nabla w}{w}$. The resulting linear PDE is the original eigenvalue PDE, which has a nonvanishing solution by hypothesis. This guarantees the existence of a "nice" equivalent problem which immediately yields Rayleigh's Principle, both the upper-bound and equality parts. The proof given here is for the two-dimensional case, but the specialization to the one-dimensional case and the generalization to the more-than-two-dimensional cases are immediate.

1. INTRODUCTION

Rayleigh's Principle gives a characterization in terms of "Rayleigh's Ratio" of the eigenvalues and eigenfunctions of a differential equation. In this paper, we treat the two-dimensional eigenvalue problem

$$\nabla \cdot (p\nabla u) - (q - \lambda r)u = 0 \qquad \text{on} \quad D \subset \mathbb{R}^2 \quad ,$$

with $u = 0$ on ∂D. The coefficient functions are assumed to satisfy

$$p(x,y) > 0 \quad , \quad r(x,y) > 0 \quad , \quad q(x,y) \geq 0 \qquad \text{on} \quad D \quad .$$

The eigenvalue problem is to determine λ such that there exists a nontrivial $u(x,y)$ satisfying the equation and boundary condition. We shall assume that such an eigenpair (λ_1,u_1) exists and that $u_1 \neq 0$ on the interior of the domain D. This paper obtains the characterization for the two-dimensional problem only, but the notation will allow immediate generalization to $\mathbb{R}^n$ and specialization to $n = 1$. The case for $N = 1$ was first presented in [7].

The classical proof of Rayleigh's Principle is by a standard variational argument. The approach here will use a modification of Carathéodory's equivalent-problems idea [1] in the Calculus of Variations. For information on this method and other modifications of it, see [2]-[6]. The result here gives both upper bounds to the eigenvalue λ_1 for arbitrary functions and the exact value of λ_1 when the eigenfunction u_1 is used in the ratio.

2. EQUIVALENT-PROBLEMS DEVELOPMENT

We consider the Calculus of Variations problem: Minimize the functional

$$J[u] = \iint_D [p\|\nabla u\|^2 + (q - \lambda r)u^2] \, dxdy$$

subject to $u = 0$ on ∂D. The integrand $p\nabla u \cdot \nabla u + (q - \lambda r)u^2$ will be denoted $L(x,y,u,u_x,u_y)$. Carathéodory's equivalent-problems idea is essentially as follows: Let $S(x,y,u) = (S_1(x,y,u),S_2(x,y,u))$ be any C^2 vector function, unspecified at present. Define

$$L^* = L - \nabla \cdot S(x,y,u(x,y)) \quad ,$$

in which u is assumed to be a function of x and y, and ∇ is the divergence in x and y. We then define the functional

$$J^*[u] = \iint_D L^* \, dxdy = J[u] - \iint_D \nabla \cdot S(x,y,u(x,y)) \, dxdy \quad .$$

By the Divergence Theorem (Green's Theorem in $\mathbb{R}^2$), the divergence integral reduces to a boundary integral

$$\int_{\partial D} S \cdot \underline{n} \, ds \quad ,$$

where $\underline{n}$ is the unit outer normal around ∂D. Since u is known $(u = 0)$ on ∂D, this is a computable constant for any given vector function S. Hence, given S, we have

$$J^*[u] = J[u] + \text{computable constant}$$

for any u satisfying the boundary condition. Thus minimizing J^* is equivalent to minimizing J, since they always differ by the same constant. We call such problems equivalent problems.

The above is true for any such vector function S, so there are infinitely many problems equivalent to the original problem. We shall obtain a particularly nice one, not as Carathéodory does by setting the partial derivatives of L^* equal to zero, but by a simple algebraic approach. Consider the new integrand

$$L^* = p\nabla u \cdot \nabla u + (q - \lambda r)u^2 - \nabla \cdot S - S_u \cdot \nabla u \quad ,$$

where $\nabla \cdot S$ means $S_{1x} + S_{2y}$ and $S_u = (S_{1u}, S_{2u})$. We shall "complete the square" on the dot product in L^*:

$$L^* = p(\nabla u \cdot \Delta u - \frac{1}{p} S_u \cdot \nabla u + \frac{1}{4p^2} S_u \cdot S_u) - \frac{1}{4p} S_u \cdot S_u$$

$$- \nabla \cdot S + (q - \lambda r)u^2$$

$$= p\|\nabla u - \frac{1}{2p} S_u\|^2 - [\nabla \cdot S + \frac{1}{4p} S_u \cdot S_u - (q - \lambda r)u^2] \quad .$$

Now if the vector S is chosen to satisfy the scalar partial differential equation

$$\nabla \cdot S + \frac{1}{4p} S_u \cdot S_u = (q - \lambda r)u^2 \quad , \tag{2.1}$$

then

$$L^* = p\|\nabla u - \frac{1}{2p} S_u\| \geq 0$$

for any function u and $L^* = 0$ if and only if u satisfies

$$\nabla u = \frac{1}{2p} S_u \quad , \tag{2.2}$$

which is a vector equation for the partial derivatives of $u(x,y)$. With such an S, the equivalent functional satisfies $J^*[u] \geq 0$ for all admissible (i.e., satisfying the boundary condition) functions u, and $J^*[u] = 0$ if and only if u satisfies (2.2). Hence a function u satisfying (2.2) is a minimizing function for J^* and therefore also for J.

Equations (2.1) and (2.2) will be called the fundamental equations. Equation (2.1) is in reality the Hamilton-Jacobi partial differential equation for the original functional, but this fact is not needed here. The existence of a solution vector S to (2.1) will be shown later to depend on the original assumption of existence of an eigenpair to the original eigenvalue problem. At present, to show the relationship of the minimizing function u to the eigenvalue problem, we differentiate the Hamilton-Jacobi equation with respect to u to obtain

$$\nabla \cdot S_u + \frac{1}{4p} 2S_u \cdot S_{uu} = 2(q - \lambda r)u \quad .$$

By (2.2), this becomes

$$\begin{aligned} 2(q - \lambda r)u &= \nabla \cdot S_u + S_{uu} \cdot \nabla u \\ &= \nabla \cdot S_u(x,y,u(x,y)) \\ &= \nabla \cdot (2p\nabla u) \quad , \end{aligned}$$

so the minimizing u satisfies

$$\nabla \cdot (p\nabla u) - (q - \lambda r)u = 0 \quad , \tag{2.3}$$

which is the original eigenvalue problem. We have shown here that the Euler-Lagrange equation for the functional J is the eigenvalue partial differential equation. Hence if S and u exist satisfying the fundamental equations, then u satisfies the eigenvalue problem and minimizes J.

3. SOLUTION OF THE FUNDAMENTAL EQUATIONS

We have seen that any solution to the Hamilton-Jacobi equation will produce a "nice" equivalent problem. We shall try to find a solution in the form

$$S(x,y,u) = (\sigma_1(x,y)u^2, \sigma_2(x,y)u^2) = \sigma u^2 \quad .$$

Used in equation (2.1), the quantity u^2 occurs in each term and can be cancelled, leaving the vector σ to satisfy

$$\nabla \cdot \sigma + \frac{\sigma^2}{p} = q - \lambda r \quad , \tag{3.1}$$

where $\sigma^2 = \sigma \cdot \sigma$. This is a scalar partial differential equation for the vector function $\sigma(x,y)$. When the equation is written in terms of components,

$$\sigma_{1x} + \sigma_{2y} + \frac{1}{p}(\sigma_1^2 + \sigma_2^2) = q - \lambda r \quad ,$$

it is an obvious generalization of a Riccati ordinary differential equation, and it reduces to that in the one-dimensional case. To find solutions to this new type of Riccati partial differential equation, we can change variables by the vector transformation $\sigma = p(\nabla w/w)$, which is a generalization of the usual Riccati transformation and which linearizes this new equation. The transformed equation for w is

$$\nabla \cdot (p\nabla w) - (q - \lambda r)w = 0 \quad ,$$

which is the original eigenvalue equation again. Since we assumed (λ_1, u_1) is an eigenpair for this problem and $u_1 \neq 0$ on the interior of D, we are guaranteed that there is such a transformation function w when $\lambda = \lambda_1$. Hence, when $\lambda = \lambda_1$, the function

$$S(x,y,u) = p\,\frac{\nabla u_1}{u_1}\,u^2$$

provides the desired solution to the Hamilton-Jacobi equation (2.1). With this S, equation (2.2) for the minimizing function u becomes

$$\nabla u = \frac{1}{2p}\,S_u = \frac{\nabla u_1}{u_1}\,u \quad ,$$

or

$$\frac{\nabla u}{u} = \frac{\nabla u_1}{u_1} \quad .$$

These relations between the partial derivatives of the minimizing function u and the eigenfunction u_1 imply that $u = cu_1$, where c is an arbitrary constant. Hence the minimizing function is a constant multiple of the eigenfunction. The boundary condition will not determine this constant, since $u = u_1 = 0$ on ∂D.

4. RAYLEIGH'S PRINCIPLE AND CONSEQUENCES

Now consider the equivalent problem generated by the vector function $S(x,y,u) = p(\nabla u_1/u_1)u^2$, which we found above and for which u_1 is the eigenfunction. This is

$$J^*[u] = J[u] - \int_{\partial D} S \cdot \underline{n}\, ds$$

$$= J[u] - \int_{\partial D} \frac{pu^2}{u_1}(\nabla u_1 \cdot \underline{n})\, ds \quad .$$

Since the admissible functions are required to satisfy $u = 0$ on ∂D, the boundary integral drops out and we have

$$0 \leq J^*[u] = J[u] = \iint_D [p\|\nabla u\|^2 + (q - \lambda_1 r)u^2]\, dxdy$$

for all such admissible functions. Note that the parameter λ is now set at the value λ_1, the first eigenvalue, since this is the value for which the function S is known to exist. Since this inequality holds for all $u \in C^2$, $u = 0$ on ∂D, this gives

$$\lambda_1 \iint_D ru^2\, dxdy \leq \iint_D [p\|\nabla u\|^2 + qu^2]dxdy \quad ,$$

or

$$\lambda_1 \leq \frac{\iint_D [p\|\nabla u\|^2 + qu^2]\, dxdy}{\iint_D ru^2\, dxdy} \quad .$$

When the minimizing function $u = cu_1$ is used, the minimum value of J^* occurs, so $0 = J^*[cu_1] = J[cu_1]$, and therefore

$$\lambda_1 = \frac{\iint_D [p\|\nabla(cu_1)\|^2 + q(cu_1)^2]\, dxdy}{\iint_D r(cu_1)^2\, dxdy} \quad .$$

The arbitrary constant c concels, leaving

$$\lambda_1 = \frac{\iint_D [p\|\nabla u_1\|^2 + qu_1^2]\, dxdy}{\iint_D ru_1^2\, dxdy} \quad . \tag{4.1}$$

The ratio in equation (4.1), the upper-bound inequality, and the equality when evaluated at the eigenfunction, are called Rayleigh's Principle; they have been used and generalized extensively. The Principle provides a characterization of the eigenvalue as the minimum value of the Rayleigh ratio and hence gives a means of obtaining upper bounds for the eigenvalue by evaluating the ratio with any C^2-function satisfying zero boundary values. Various eigenvalue-approximation schemes based on this notion provide ways to determine a sequence of admissible functions which decrease the value of the ratio. This characterization of the eigenvalue shows that for the equation coefficients $p, q, r > 0$ the eigenvalue must be positive, and that if p or q is increased in magnitude, or r is decreased, then λ_1 increases. It thus provides comparison theorems and much information concerning the eigenvalue.

5. SUMMARY AND COMMENTS

This paper has given a new and "nonvariational" proof of Rayleigh's Principle for eigenpairs for a second-order partial differential equation. The approach is a modified version of Carathéodory's idea of equivalent problems in the Calculus of Variations. The proof has been presented for the case $n = 2$, but the one-dimensional and more-than-two-dimensional cases are immediate, and the notation used here requires only minor modification in these cases. A new type of a Riccati differential equation for a vector function occurs naturally in this approach; it has been solved using a linearizing transformation.

Further questions now arise, such as the following: What is the general solution to the Riccati equation (3.1), since the linearizing transformation used here is of a special kind? What is the general solution of the Hamilton-Jacobi PDE for the vector function $S(x,y,u)$, since we reduced it to the new Riccati equation by seeking solutions of a particular form? Will other solutions of the Hamilton-Jacobi equation yield related characterizations of the eigenpair (λ_1,u_1)? To what other types of eigenvalue PDE's does this method apply? More general second-order PDE's? Fourth-order PDE's? And finally, can this method be adapted to characterize the succeeding and higher-order eigenpairs (λ_2,u_2), (λ_3,u_3), ..., by considering appropriate orthogonality restrictions on the admissible class of functions as in the classical applications of Rayleigh's Principle?

Dirichlet's Principle can also be proved easily by the foregoing approach.

REFERENCES

1. C. Carathéodory, Variationsrechnung und partielle Differential-gleichungen

erster Ordnung, Teubner, Leipzig, 1935 (English translation, Holden-Day, San Francisco, 1966).

2. H. Rund, The Hamilton-Jacobi theory in the calculus of variations, Van Nostrand, London, 1966 (reprinted with corrections by Krieger, Huntington, N.Y., 1973).

3. D. R. Snow, Carathéodory-Hamilton-Jacobi theory in optimal control, J. Math. Anal. Appl. 17 (1967), 99-118.

4. D. R. Snow, A sufficiency technique in calculus of variations using Carathéodory's equivalent-problems approach, J. Math. Anal. Appl. 51 (1975), 129-140.

5. D. R. Snow, Transversality and natural boundary conditions by equivalent problems in calculus of variations, in Calculus of variations and control theory, edited by David L. Russell, Academic Press, New York, 1976, 391-404.

6. D. R. Snow, Using equivalent problems to solve Bolza's problem of the calculus of variations, Annual Meeting of the American Mathematical Society, San Antonio, Texas, 22-25 January, 1976; abstract in Notices of the American Mathematical Society, 23 (1976), A-169.

7. D. R. Snow, A new proof for Rayleigh's principle for eigenvalue approximations, Annual Meeting of the American Mathematical Society, Washington, D. C., 21-26 January, 1975; abstract in Notices of the American Mathematical Society, 22 (1975), A-198.

SECOND-ORDER CRITERIA FOR PSEUDO-CONVEX FUNCTIONS*

M. Avriel
Technion-Israel Inst. of Technology
Haifa
ISRAEL

S. Schaible
Industrieseminar der Universität Köln
Köln
WEST GERMANY

ABSTRACT. The aim of this paper is to derive necessary conditions and sufficient conditions for (strictly) pseudo-convex functions which are twice continuously differentiable. The criteria are formulated in terms of extended Hessians and in terms of bordered determinants.

1. DEFINITION AND NOTATION

Psuedo-convexity has proved to be a very useful extension of convexity in nonlinear optimization. A C^1-function $f(x)$ defined on an open convex set K in $\mathbb{R}^n$ is called pseudo-convex (strictly pseudo-convex) if

$$(y - x)^t \nabla f(x) \geq 0 \implies f(y) \geq f(x) \quad (f(y) > f(x))$$

for all $x,y \in K$, $y \neq x$.

Denote:

P = family of pseudo-convex C^2-functions;

H = family of C^2-functions such that $\nabla^2 f(x) + r(x) \nabla f(x) \nabla f(x)^t$ is positive semidefinite for some continuous r;

T = family of C^2-functions for which there exists a C^2-function $G(y)$, $G'(y) > 0$, such that $G(f(x))$ is convex.

2. DISCUSSION

It is shown that $T \subset H \subset P$, where all these inclusions are strict. For large subfamilies of C^2-functions, the gap between T and H vanishes. With certain pathological functions excluded from P, a characterization of P in terms of an extended Hessian is given. In addition, similar criteria for strictly pseudo-convex functions are proved.

Furthermore, sufficient conditions for (strictly) pseudo-convex functions are derived that involve bordered determinants.

For quadratic functions, most of the conditions in this paper turn out to be necessary and sufficient for (strict) pseudo-convexity. In particular,

*Prepublication announcement.

we have $T = H = P$, if these families are restricted to quadratic functions.

The results of this paper are published in [1] and [2].

REFERENCES

1. M. Avriel and S. Schaible, Second-order criteria for pseudo-convex functions, Department of Operations Research, Stanford University, June, 1976; forthcoming.

2. S. Schaible, Second-order characterizations of pseudo-convex quadratic functions, Department of Operations Research, Stanford University, Technical Report 75-29, November 1975; forthcoming in J. Optimization Theory Appl. 12, April, 1977.

INFINITE RIEMANN SUMS, THE SIMPLE INTEGRAL, AND THE DOMINATED INTEGRAL

James T. Lewis
Dept. of Mathematics
Univ. of Rhode Island
Kingston, R.I. 02881
U.S.A.

Charles F. Osgood
Naval Research Lab.
Washington, D.C. 20375
U.S.A.

Oved Shisha
Dept. of Mathematics
Univ. of Rhode Island
Kingston, R.I. 02881
U.S.A.

ABSTRACT. Simple integrability of a function f (defined by Haber and Shisha in [2]) is shown to be equivalent to the convergence of the infinite Riemann sum

$$\sum_{k=1}^{\infty} f(\xi_k)(x_k - x_{k-1})$$

to the improper Riemann integral $\int_0^\infty f$ as the gauge of the partition $(x_k)_{k=0}^{\infty}$ of $[0,\infty)$ converges to 0. An analogous result is obtained for dominant integrability (defined by Osgood and Shisha in [5]). Also certain results of Bromwich and Hardy [1] are recovered.

1. INTRODUCTION

In recent papers [2], [3], [5], [6], the concepts of <u>simple integral</u> and <u>dominated integral</u> have been introduced, and their relationship with quadrature formulas for numerical integration have been studied.

In [4], a connection was established between simple integrability of a function f on $[0,\infty)$ and the convergence of series

$$\sum_{k=1}^{\infty} f(\xi_k)(x_k - x_{k-1}),$$

where $0 = x_0 < x_1 < \cdots$ is an "allowable" partition of $[0,\infty)$ and $\xi_1, \xi_2, \cdots$ are corresponding evaluation points.

In Section 2 of the present paper, this connection is studied further and is extended to partitions that are not necessarily "allowable."

In Section 3, analogous results for the dominated integral are developed.

In the special case of equispaced partitions, the convergence of the series

$$\sum_{k=0}^{\infty} hf(kh) \quad \text{as} \quad h \to 0+$$

to the improper Riemann integral $\int_0^\infty f$ has been investigated long ago; in Section 4 we discuss the connection with our results.

2. THE SIMPLE INTEGRAL

DEFINITION 1 ([2], [3]). A complex function f defined on $[0,\infty)$ is "simply integrable" if and only if there is a complex number I with the following property: For each $\varepsilon > 0$, there are positive numbers B and Δ such that if

$$(*) \qquad \begin{cases} 0 = x_0 < x_1 < \cdots < x_n, \quad x_n > B; \quad x_{k-1} \leq \xi_k \leq x_k \\ \text{and} \quad x_k - x_{k-1} < \Delta \quad \text{for} \quad k = 1,\cdots,n , \end{cases}$$

then

$$\left|I - \sum_{k=1}^{n} f(\xi_k)(x_k - x_{k-1})\right| < \varepsilon .$$

It is known that if f is simply integrable, then it is Riemann integrable on each $[0,R]$, $0 < R < \infty$,

$$\int_0^\infty f = \lim_{R\to\infty} \int_0^R f$$

converges, and $I = \int_0^\infty f$.

A sequence $(x_k)_{k=0}^\infty$ is called an "allowable partition" if and only if $0 = x_0 < x_1 < x_2 < \cdots$, $x_k \to \infty$, and

$$0 < \inf_{k\geq 1} (x_k - x_{k-1}) \leq \sup_{k\geq 1} (x_k - x_{k-1}) < \infty .$$

In [4], the following result was obtained.

THEOREM 1. Let $(x_k)_{k=0}^\infty$ be an allowable partition, and let f be a complex function, Riemann integrable on each $[0,R]$, $0 < R < \infty$. A necessary and sufficient condition for f to be simply integrable is that

$$(1) \qquad \sum_{k=1}^{\infty} f(\xi_k)(x_k - x_{k-1})$$

converges whenever $x_{k-1} \leq \xi_k \leq x_k$, $k = 1,2,\cdots$.

We can think of such a series (1) as an "infinite Riemann sum" for $\int_0^\infty f$.

DEFINITION 2. A complex function f on $[0,\infty)$ satisfies the "infinite-Riemann-sum condition on $[0,\infty)$" if and only if there is a complex number I^* with the following property: For each $\varepsilon > 0$ there is $\Delta > 0$ such that if

$$(**) \qquad \begin{cases} 0 = x_0 < x_1 < \cdots , \quad x_k \to \infty; \quad x_k - x_{k-1} < \Delta \\ \text{and} \quad x_{k-1} \leq \xi_k \leq x_k \quad \text{for} \quad k = 1,2,\cdots , \end{cases}$$

then the series

$$\sum_{k=1}^{\infty} f(\xi_k)(x_k - x_{k-1})$$

converges and

$$|I^* - \sum_{k=1}^{\infty} f(\xi_k)(x_k - x_{k-1})| < \varepsilon .$$

We shall show that f is simply integrable if and only if it satisfies the infinite-Riemann-sum condition on $[0,\infty)$, in which case

$$I^* = \int_0^{\infty} f .$$

LEMMA 1. If f satisfies the infinite-Riemann-sum condition on $[0,\infty)$, then, for each $R > 0$, the Riemann integral $\int_0^R f$ exists.

Proof. It is sufficient to prove the lemma for the case that f is real. Given $\varepsilon > 0$, there is a $\Delta > 0$ such that (**) implies

$$|I^* - \sum_{k=1}^{\infty} f(\xi_k)(x_k - x_{k-1})| < \varepsilon/2 .$$

Let

$$0 = y_0 < y_1 < \cdots < y_M = R$$

be a partition of $[0,R]$ with

$$\max_{1 \le k \le M} (y_k - y_{k-1}) < \Delta .$$

Suppose that

$$y_{k-1} \le t_k \le y_k, \quad y_{k-1} \le t_k' \le y_k, \qquad k = 1,\ldots,M .$$

Setting

$$x_k = y_k, \quad k = 1,\ldots,M; \qquad x_k = R + (k - M)(\delta/2), \quad k = M+1, M+2, \cdots ,$$

we have

$$\Big|\sum_{k=1}^{M} f(t_k')(y_k - y_{k-1}) - \sum_{k=1}^{M} f(t_k)(y_k - y_{k-1})\Big|$$

$$= \Big|\Big(\sum_{k=1}^{M} f(t_k')(x_k - x_{k-1}) + \sum_{k=M+1}^{\infty} f(x_k)(x_k - x_{k-1}) - I^*\Big)$$

$$- \Big(\sum_{k=1}^{M} f(t_k)(x_k - x_{k-1}) + \sum_{k=M+1}^{\infty} f(x_k)(x_k - x_{k-1}) - I^*\Big)\Big| < \varepsilon .$$

Hence

$$\sum_{k=1}^{M} O(f,y_{k-1},y_k)(y_k - y_{k-1}) \leq \varepsilon ,$$

where, for $0 \leq u < v$,

$$O(f,u,v) = \sup_{u \leq \alpha < \beta \leq v} |f(\beta) - f(\alpha)| < \infty ,$$

and thus $\int_0^R f$ exists.

THEOREM 2. A function f is simply integrable if and only if it satisfies the infinite-Riemann-sum condition on $[0,\infty)$.

Proof. ($\Leftarrow$) Let $\Delta > 0$ correspond to $\varepsilon = 1$ in Definition 2, let $(x_k)_{k=0}^{\infty}$ be an allowable partition with each

$$x_k - x_{k-1} < \Delta ,$$

and let

$$x_{k-1} \leq \xi_k \leq x_k, \quad k = 1,2,\cdots .$$

Then the series (1) converges, and f is simply integrable by Theorem 1 and Lemma 1.

($\Rightarrow$) Let $\varepsilon > 0$ be given. Let $B > 0$ and $\Delta > 0$ be such that (*) implies

$$\left| \int_0^{\infty} f(x)dx - \sum_{k=1}^{n} f(\xi_k)(x_k - x_{k-1}) \right| < \varepsilon/2 .$$

Let

$$0 = x_0 < x_1 < \cdots , \quad x_k \to \infty; \quad x_k - x_{k-1} < \min\{\Delta,1\}$$
$$\text{and} \quad x_{k-1} \leq \xi_k \leq x_k \quad \text{for } k = 1,2,\cdots .$$

We first show that the series (1) converges. This would follow from the "necessary" part of Theorem 1, except that here $(x_k)_{k=0}^{\infty}$ need not satisfy

$$0 < \inf_{k \geq 1} (x_k - x_{k-1}) .$$

However, an examination of the proof of that part of Theorem 1 (cf. [4, p. 495]) shows that we need only verify that

$$\sum_{k=1}^{\infty} O(f,k-1,k) < \infty$$

implies

$$\sum_{k=1}^{\infty} O(f,x_{k-1},x_k)(x_k - x_{k-1}) < \infty .$$

This implication follows immediately from the inequality

$$\sum_{k=1}^{\infty} O(f,x_{k-1},x_k)(x_k - x_{k-1}) \leq 4 \sum_{k=1}^{\infty} [O(f,k-1,k) + O(f,k,k+1)] .$$

This last inequality, in turn, can be verified as follows: Let k_1 be the least k such that

$$x_k > 1 .$$

Since each

$$x_k - x_{k-1} < 1, \quad \text{we have} \quad x_{k_1} < 2 .$$

Let k satisfy $1 \leq k \leq k_1$, and assume that

$$x_{k-1} \leq t_k' < t_k \leq x_k .$$

Then

$$|f(t_k) - f(t_k')| \leq |f(t_k) - f(1)| + |f(1) - f(t_k')|$$

$$\leq 2[O(f,0,1) + O(f,1,2)] .$$

Hence

$$O(f,x_{k-1},x_k) \leq 2[O(f,0,1) + O(f,1,2)] .$$

So

$$\sum_{k=1}^{k_1} O(f,x_{k-1},x_k)(x_k - x_{k-1}) \leq 4[O(f,0,1) + O(f,1,2)] .$$

Similarly, letting k_2 be the least k such that $x_k > 2$, we can show that

$$\sum_{k=k_1+1}^{k_2} O(f,x_{k-1},x_k)(x_k - x_{k-1}) \leq 4[O(f,1,2) + O(f,2,3)] .$$

Continuing in this way, and adding the corresponding inequalities, we obtain the desired inequality above.

Now if $x_n > B$, we have

$$\left| \int_0^{\infty} f(x)dx - \sum_{k=1}^{n} f(\xi_k)(x_k - x_{k-1}) \right| < \varepsilon/2 .$$

Thus,

$$\left| \int_0^{\infty} f(x)dx - \sum_{k=1}^{\infty} f(\xi_k)(x_k - x_{k-1}) \right| \leq \varepsilon/2 < \varepsilon .$$

Hence f satisfies the infinite-Riemann-sum condition on $[0,\infty)$, and the proof of the theorem is complete.

COROLLARY 1. If f satisfies the infinite-Riemann-sum condition on $[0,\infty)$, then $\int_0^{\infty} f$ converges and equals I^* of Definition 2.

Proof. I^* of Definition 2 is necessarily equal to I of Definition 1, as is easily seen, and the latter is $\int_0^\infty f$.

3. THE DOMINATED INTEGRAL

The concept of the dominated integral of a complex function f on $(0,1]$ was introduced in [5]. It was shown in [6] that, for such an f, its dominated integral exists if and only if the improper Riemann integral $\int_{0+}^1 f$ converges and for every sequence

$$(\Phi_n^*)_{n=1}^\infty$$

of quadrature formulas of a general type,

$$\Phi_n^*(f) \to \int_{0+}^1 f .$$

DEFINITION 3 ([5]). Let f be a complex function on $(0,1]$. A dominated integral of f is a complex number I having the property: For each $\varepsilon > 0$ there exist $\delta \in (0,1)$ and $X \in (0,1)$ such that, if

$$0 < t_n < \cdots < t_0 = 1, \quad t_n < X, \quad \frac{t_j}{t_{j-1}} > 1 - \delta \quad \text{and}$$

$$t_j \le \tau_j \le t_{j-1} \quad \text{for} \quad j = 1,\dots,n ,$$

then

$$\left| I - \sum_{j=1}^n f(\tau_j)(t_{j-1} - t_j) \right| < \varepsilon .$$

Dominant integrability of f means existence of such an I, which implies [5, Theorem 1] that f is Riemann integrable on each $[a,1]$, $0 < a < 1$, and

$$\lim_{a \to 0+} \int_a^1 f = I = \int_{0+}^1 f .$$

DEFINITION 4. A complex function f on $(0,1]$ satisfies the "infinite-Riemann-sum condition on $(0,1]$" if and only if there is a complex number I^* with the following property: For each $\varepsilon > 0$ there exists $\delta \in (0,1)$ such that if

$$0 < \cdots < t_n < \cdots < t_0 = 1, \quad \lim_{j\to\infty} t_j = 0; \quad \frac{t_j}{t_{j-1}} > 1 - \delta \quad \text{and}$$

$$t_j \le \tau_j \le t_{j-1} \quad \text{for} \quad j = 1,2,\dots ,$$

then the series

$$\sum_{j=1}^\infty f(\tau_j)(t_{j-1} - t_j)$$

converges and

$$\left| I^* - \sum_{j=1}^{\infty} f(\tau_j)(t_{j-1} - t_j) \right| < \varepsilon .$$

LEMMA 2. If f satisfies the infinite-Riemann-sum condition on $(0,1]$, then, for each $a \in (0,1)$, the Riemann integral $\int_a^1 f$ exists.

The proof is similar to that of Lemma 1 in [5] and therefore will be omitted.

We say ([5]) that a complex function f satisfies the "Riemann condition for the dominated integral" (RCDI) if and only if

(i) f is defined and bounded on $[a,1]$ for each $a \in (0,1)$;

(ii) for each $\varepsilon > 0$, there exists $\delta \in (0,1)$ such that if

$$0 < t_n < \cdots < t_0 = 1 \quad \text{and} \quad \frac{t_j}{t_{j-1}} > 1 - \delta \quad \text{for} \quad j = 1,\ldots,n ,$$

then

$$OS(f;t_n,\cdots,t_0) = \sum_{j=1}^{n} O(f,t_j,t_{j-1})(t_{j-1} - t_j) < \varepsilon .$$

By Theorem 2 of [5], a complex function f on $(0,1]$ is dominantly integrable if and only if it satisfies RCDI.

THEOREM 3. A function f is dominantly integrable if and only if it satisfies the infinite-Riemann-sum condition on $(0,1]$.

Proof. It is sufficient to prove the theorem in case f is real.

($\Leftarrow$) We shall show that f satisfies RCDI. By Lemma 2, f is bounded on $[a,1]$ for each $a \in (0,1)$. If (ii) fails, then there exists an $\varepsilon_0 > 0$, and for each sequence

$$(\delta^{(k)})_{k=1}^{\infty}$$

with

$$0 < \delta^{(k)} < 1, \quad k = 1,2,\ldots, \quad \text{and} \quad \delta^{(k)} \to 0, \quad \text{say,} \quad \delta^{(k)} \equiv (2k)^{-1} ,$$

there exists, for $k = 1, 2, \ldots,$ a partition

$$0 < t_{n_k}^{(k)} < \cdots < t_0^{(k)} = 1$$

with

$$\min_{1 \leq j \leq n_k} t_j^{(k)}/t_{j-1}^{(k)} > 1 - \delta^{(k)} \quad \text{but} \quad OS(f;t_{n_k}^{(k)},\ldots,t_0^{(k)}) \geq \varepsilon_0 .$$

Hence there exist, for $k = 1,2,\dots;\ j = 1,2,\dots,n_k$, numbers

$$\tau_j^{(k)},\ \hat{\tau}_j^{(k)} \quad \text{with} \quad t_j^{(k)} \le \tau_j^{(k)} \le t_{j-1}^{(k)},\ t_j^{(k)} \le \hat{\tau}_j^{(k)} \le t_{j-1}^{(k)},$$

such that

$$\sum_{j=1}^{n_k} [f(\tau_j^{(k)}) - f(\hat{\tau}_j^{(k)})](t_{j-1}^{(k)} - t_j^{(k)}) \ge \varepsilon_0/2 .$$

For each $k \ge 1$, we extend the partition $t^{(k)}$ to an infinite sequence

$$0 < \cdots < t_{n_k+1}^{(k)} < t_{n_k}^{(k)} < \cdots < t_0^{(k)} = 1$$

with

$$\lim_{j\to\infty} t_j^{(k)} = 0 \text{ , and } \quad t_j^{(k)}/t_{j-1}^{(k)} > 1 - \delta^{(k)}$$

for $j = 1,2,\cdots$. Also, let

$$\tau_j^{(k)} = \hat{\tau}_j^{(k)} = t_j^{(k)} \quad \text{for} \quad j = n_k+1, n_k+2,\cdots,\quad k = 1,2,\cdots .$$

Then

$$\sum_{j=1}^{\infty} f(\tau_j^{(k)})(t_{j-1}^{(k)} - t_j^{(k)}) - \sum_{j=1}^{\infty} f(\hat{\tau}_j^{(k)})(t_{j-1}^{(k)} - t_j^{(k)}) \ge \varepsilon_0/2$$

for all $k \ge 1$, which contradicts the infinite-Riemann-sum condition.

($\Rightarrow$) We can write $f = f^+ - f^-$, where

$$f^+(x) \equiv \max\{f(x),0\} \ge 0 \quad \text{and} \quad f^-(x) \equiv \max\{-f(x),0\} \ge 0 .$$

Since f^+ and f^- are dominantly integrable as f is, and since f satisfies the infinite-Riemann-sum condition on $(0,1]$ if f^+ and f^- do, it suffices to prove the implication assuming $f \ge 0$ on $(0,1]$. Let $\varepsilon > 0$. There exist $\delta \in (0,1)$ and $\chi \in (0,1)$ such that if

$$0 < t_n < \cdots < t_0 = 1,\ t_n < \chi;\ \frac{t_j}{t_{j-1}} > 1 - \delta \quad \text{and}$$

$$t_j \le \tau_j \le t_{j-1} \quad \text{for } j = 1,\cdots,n ,$$

then

$$\left| \int_{0+}^{1} f - \sum_{j=1}^{n} f(\tau_j)(t_{j-1} - t_j) \right| < \varepsilon/2 . \tag{2}$$

Let

$$0 < \cdots < t_j < \cdots < t_0 = 1,\ t_j \to 0;\ \frac{t_j}{t_{j-1}} > 1 - \delta \quad \text{and}$$

$$t_j \le \tau_j \le t_{j-1} \quad \text{for } j = 1,2,\dots .$$

The partial sums of the series

$$\sum_{j=1}^{\infty} f(\tau_j)(t_{j-1} - t_j)$$

whose terms are ≥ 0 are bounded above (by (2)), and hence the series converges. From (2),

$$\left| \int_{0+}^{1} f - \sum_{j=1}^{\infty} f(\tau_j)(t_{j-1} - t_j) \right| \leq \varepsilon/2 < \varepsilon ,$$

and so f satisfies the infinite-Riemann-sum condition on $(0,1]$.

COROLLARY 2. If f satisfies the infinite-Riemann-sum condition on $(0,1]$, then $\int_{0+}^{1} f$ converges and equals I^* of Definition 4.

<u>Proof</u>. I^* of Definition 4 is necessarily equal to I of Definition 3, as is easily seen, and the latter is $\int_{0+}^{1} f$.

4. CONNECTION WITH RESULTS OF BROMWICH AND HARDY

In a 1908 paper [1], Bromwich and Hardy studied conditions on a real function f on $[0,\infty)$ sufficient to guarantee that

$$\lim_{h \to 0+} h \sum_{k=0}^{\infty} f(kh) = \int_{0}^{\infty} f . \tag{3}$$

Two of their results are the following.

(i) If f is nonincreasing on $[0,\infty)$, and $\int_0^\infty f$ converges, then (3) holds.

(ii) If φ is real, nonnegative, continuous, and nonincreasing on $[0,\infty)$, with $\int_0^\infty \varphi < \infty$, and if $F(x)$ is real, continuous, and bounded on $[0,\infty)$, then (3) holds for $f(x) \equiv \varphi(x)F(x)$.

We shall now show that (i) and (ii) can be recovered from our Theorem 2 and Corollary 1, and from the following result [3, p. 9, (b) and Theorem 3]: If f is a complex function, Riemann integrable on $[0,R]$ for each $R > 0$, and if throughout $[0,\infty)$, we have $|f(x)| \leq g(x)$, where g is nonincreasing there and $\int_0^\infty g < \infty$, then f is simply integrable.

Note that since, for $h > 0$,

$$h \sum_{k=0}^{\infty} f(kh)$$

is a special infinite Riemann sum, if a function f satisfies the infinite-Riemann-sum condition on $[0,\infty)$, then (3) follows from the definition of the condition and from Corollary 1. Since by Theorem 2 this condition is equivalent to simple integrability, (i) follows from the quoted result from [3] by taking $g(x) \equiv f(x)$ (≥ 0). Likewise (ii) follows by taking

$$g(x) \equiv \varphi(x)\sup_{0\leq x<\infty}|F(x)| \quad .$$

In [1], it is also asserted that

$$\lim_{c\to 1-} (1 - c) \sum_{k=0}^{\infty} c^k f(c^k) = \int_{0+}^{1} f$$

if f is real, nonnegative, and nonincreasing on $(0,1]$, $\int_{0+}^{1} f < \infty$, and

$$\lim_{x\to 0+} f(x) = \infty \ .$$

This result can be recovered as follows.

(i) Note that

$$(1 - c) \sum_{k=0}^{\infty} c^k f(c^k)$$

is a special infinite Riemann sum of the form considered in Definition 4, obtained by taking $t_k = \tau_{k+1} = c^k$, $k = 0,1,\cdots$.

(ii) f is dominantly integrable by Theorem 3 of [5].

(iii) Use our Theorem 3.

ACKNOWLEDGMENT. The third author gratefully acknowledges National Science Foundation support through Grant No. MCS 76-07448.

REFERENCES

1. T.J. I'a Bromwich and G.H. Hardy, The definition of an infinite integral as the limit of a finite or infinite series, Quarterly J. Pure and Applied Math. 39 (1908), 222-240.

2. S. Haber and O. Shisha, An integral related to numerical integration, Bull. Amer. Math. Soc. 79 (1973), 930-932.

3. S. Haber and O. Shisha, Improper integrals, simple integrals, and numerical quadrature, J. Approximation Theory 11 (1974), 1-15.

4. C.F. Osgood and O. Shisha, On simple integrability and bounded coarse variation, Approximation Theory II, Proc. of a symposium held in Austin, Texas, Jan. 1976; G.G. Lorentz, C.K. Chui, and L.L. Schumaker, eds., Academic Press, N.Y., 1976, 491-501.

5. C.F. Osgood and O. Shisha, The dominated integral, J. Approximation Theory 17 (1976), 150-165.

6. C.F. Osgood and O. Shisha, Numerical quadrature of improper integrals and the dominated integral, to appear in J. Approximation Theory.

7. O. Szász and J. Todd, Convergence of Cauchy-Riemann sums to Cauchy-Riemann integrals, J. Res. Nat. Bureau of Standards 47 (1951), 191-196.

AN EXTREMAL PROBLEM FOR HARMONIC MEASURE*

F. Huckemann
Technische Universität Berlin
Fachbereich Mathematik
D-1000 Berlin 12
WEST GERMANY

ABSTRACT. Upper and lower bounds for the harmonic measure of members of a class of continua in the unit disc are established.

1. INTRODUCTION

If K is a continuum in the unit disc E, the open set $E\setminus K$ has exactly one doubly connected component $E(K)$; $M(K)$ denotes its modulus [$= (2\pi)^{-1} \log r$ when $E(K)$ is conformally equivalent to the annulus $\{w;\ 1 < |w| < r\}$]; ω_K denotes the harmonic measure of K in E. Let there be given two distinct points a, b in $E\setminus\{0\}$ such that $\mathrm{Im}(b/a) \neq 0$. We denote by $\mathcal{K}$ the class of continua K satisfying

$$\text{(i)} \quad \{a,b\} \subset K \subset E\setminus\{0\}, \qquad \text{(ii)} \quad 0 \in E(K).$$

In particular, the Poincaré segment s_{ab} [= segment between a and b of the part of the circle C_{ab} through $a, b, 1/\bar{b}, 1/\bar{a}$ which lies in E] is in $\mathcal{K}$, and it is well known that $K \in \mathcal{K}$ implies $M(K) \leq M(s_{ab})$, equality holding only for $K = s_{ab}$. Let now $0 < M < M(s_{ab})$, and put $\mathcal{K}_M = \{K;\ K \in \mathcal{K},\ M(K) = M\}$. We ask for (best possible) inequalities for $\omega_K(0)$ when $K \in \mathcal{K}_M$.

*Prepublication announcement.

2. QUADRATIC DIFFERENTIALS

It turns out that for small $M(s_{ab}) - M$ the problem is closely related to quadratic differentials of the form

$$\sigma_c = -\frac{(z - c)(1 - \bar{c}z)}{z(z - a)(1 - \bar{a}z)(z - b)(1 - \bar{b}z)} dz^2,$$

where $c \in E\setminus\{0,a,b\}$ is such that the directional field $\sigma_c > 0$ has a regular trajectory structure. The general case $0 < M < M(s_{ab})$ is also related to certain quadratic differentials; it is more complicated, though, and will not be discussed here. Also the limiting case $\mathrm{Im}(b/a) = 0$ is left aside.

We recall that $\sigma_c > 0$ determines a directional field on $\hat{\mathbb{C}}\setminus A$, where $A = \{0,\infty,a,1/\bar{a},b,1/\bar{b}\}$ is the singular set of σ_c in the extended complex plane $\hat{\mathbb{C}}$. Any maximal solution curve of this directional field is a <u>trajectory</u> of $\sigma_c > 0$; either such a trajectory is an analytic Jordan curve, or it may be parametrized by a bijective analytic map γ of the open intervall $\langle 0,1\rangle$ into $\hat{\mathbb{C}}\setminus A$. If, in the latter case, $\gamma(t)$ tends in $\hat{\mathbb{C}}$ to a limit (necessarily contained in A) for $t \to 0$ as well as for $t \to 1$, we call that trajectory a <u>two-ended critical trajectory</u>, and we denote it by $T_{uv}(c)$ when the (possibly coinciding) limits are u and v; $\sigma_c > 0$ has a <u>regular trajectory</u> structure if all trajectories are either Jordan curves or two-ended critical trajectories.

3. THEOREMS

Without loss of generality, we assume $a > 0$ and $b = |b|e^{i\beta}$, where $0 < \beta < \pi$. The following results obtain.

THEOREM 1. Let $0 < M(s_{ab}) - M$ be sufficiently small. Then there are two points $d = d(M)$ and $e = e(M)$, close to but different from zero, with the following properties:

(i) $0 < |d| < \min(a,|b|)$, $0 < \arg d < \beta$,

$0 < |e| < \min(a,|b|)$, $\pi < \arg e < \pi + \beta$;

(ii) $\sigma_d > 0$ has a unique two-ended critical trajectory $T_{ab}(d)$,

$\sigma_e > 0$ has a unique two-ended critical trajectory $T_{ab}(e)$;

(iii) the closure $K(d)$ of $T_{ab}(d)$ and the closure $K(e)$ of $T_{ab}(e)$ are both in $\mathcal{K}_M$.

THEOREM 2. Let $0 < M(s_{ab}) - M$ be sufficiently small, and let $K(d)$ and $K(e)$ be as in Theorem 1. The, with $\omega_2 = \omega_{K(d)}$ and $\omega_1 = \omega_{K(e)}$, for any $K \in \mathcal{K}_M$ we have the double inequality

$$\omega_1(0) \leq \omega_K(0) \leq \omega_2(0),$$

and the double inequality is strict unless $K = K(d)$ or $K = K(e)$.

To prove Theorem 1, the trajectory structure of $\sigma_c > 0$ is determined by consideration of an Abelian differential τ_c on $s_{ab}\setminus\{a,b\}$ satisfying $\tau_c^2 = \sigma_c$. Theorem 2 is proved by mapping $E(K)$, $E(K(d))$, $e(K(e))$ conformally onto the same annulus $R = \{w;\ 1 < |w| < \exp[2\pi M(\mathbf{k})]\}$; the pertinent relations about the positions of the image of 0 (which determine the harmonic measures in question) under these mappings are obtained from well-known properties of certain extremal decompositions of R.

A detailed exposition will be submitted to Commentarii Mathematici Helvetici.

Geometric and Topological Inequalities

E. PICARDUS AB OMNI NAEVO LIBERATUS
(ON THE AXIOMATICS OF VECTOR ADDITION)

Dedicated to A.D. Wallace on his 70th Birthday

J. Aczél
Faculty of Mathematics
University of Waterloo
Waterloo, Ontario
CANADA

ABSTRACT. Some moles are removed from E. Picard's axiomatic treatment of vector addition.

1. INTRODUCTION

The (Latin) title of this paper refers, of course, to Saccheri's book "Euclides ab omni naevo liberatus," in which he tried to free Euclid's axiomatics of Geometry from the "moles" of the postulate of parallels and from other flaws (and in the course of which he inadvertently proved a number of theorems of non-Euclidean Geometry).

In 1928, E. Picard [3] gave an axiomatic treatment of the parallelogram rule of vector addition based on the following postulates:

(1) Vector addition is <u>invariant under</u> (really <u>covariant with</u>) <u>the rotations</u> of the (3-dimensional Euclidean) space; that is, the <u>"resultant"</u> undergoes the same rotation as the two <u>"components."</u>

(2) Vector addition is <u>commutative and associative</u>.

(3) <u>For two vectors of the same direction, vector addition reduces to "algebraic addition"</u>; that is, the resultant points in that same direction and its length is the sum of the lengths of the components.

2. THREE MOLES

While Picard's treatment is itself an improvement of that given in 1811 by Poisson [4], there remained on it a few moles, some of which were cut out in 1966 and 1969 by Aczél [1], [2].

In the first place, Picard also assumed tacitly that

(4) the resultant of two vectors of equal length depends <u>continuously on the measure of their angle</u>,

and that

(5) it also depends <u>continuously on their length</u>.

While both assumptions (4) and (5) could be weakened simultaneously, one purpose of this note is to eliminate (5) completely, while retaining (4) in its present form, partly because this will serve to eliminate other "moles" from Picard's deliberations.

First we state two consequences of (1) which were observed, explicitly or implicitly, also by E. Picard in [3]:

(6) The resultant of two vectors of equal length lies in the plane spanned by them; more exactly, it lies in the bisector of their angle.

(7) The resultant of two vectors of equal length and opposite directions is the zero vector.

As usual, the zero vector is a vector of zero length and arbitrary direction.

A second mole in Picard's 1928 presentation [3] is the following: He derived (cf. (1), (19), and (20), below) the result that the length of the resultant of two unit vectors, enclosing an angle of measure 2φ, is $2\cos a\varphi$. From (7), $2\cos a\frac{\pi}{2} = 0$ follows; but from this he infers that $a = 1$, while $a = 2k + 1$ (k a nonnegative integer) is also a solution.

More importantly, the third mole is the implication that (6) means that the resultant of two vectors of equal length points in the direction of the bisector of their angle of lesser measure, which by no means follows from (1). However, we shall prove this inequality, which will also help us eliminate (5).

3. ANALYSIS

The following observation will be important. By (3), the zero vector is a unit element of the set of vectors under addition; and, by (7), vectors of equal length and opposite directions are inverses. In view of (2), we therefore have the following:

(8) The set of vectors forms an Abelian group under addition.

Since, in groups,

(9) inverses are unique,

in the formula (20) below, the expression

$$2\cos(2k + 1)\varphi$$

for the length of the resultant of two unit vectors with angle of measure 2φ, any choice $k \neq 0$ is impossible because then two unit vectors of the angle of measure $\frac{\pi}{2k+1} \neq \pi$ would have the zero vector as resultant, i.e., be inverses, which contradicts (7) and (9).

Now we come to the main purpose of this note, which is to show that <u>the resultant of two vectors of equal length points in the direction of the bisector of their angle of lesser measure</u>.

Indeed, this is so by (3) if their angle of lesser measure is 0. If there existed two vectors of equal length and of angle measure $2\varphi < \pi$ such that their resultant pointed in the direction of the bisector of their angle of greater measure, then, by the continuity (4), there would exist two vectors of the same lengths and of angle measure $2\varphi_0 \in]0,2\varphi[\subset]0,\pi[$ whose resultant would be the zero vector. But this would again contradict (9). Thus we have indeed proved that

(10) <u>the resultant of two vectors of equal length points in the direction of the bisector of their angle of lesser measure</u>.

(The proposition (7) takes care of the case where there is no angle of lesser measure.)

The statement (10) is a <u>nonnegativity</u> statement which is very useful in Picard's further deliberations. We denote, as usual, by $\|P\|$ the length of the vector P, and by $(P,Q)\measuredangle$ the (lesser) angle measure of P and Q. We take, as Picard does, two vectors P_1 and P_2 of the same length $x = \|P_1\| = \|P_2\|$, with a fixed angle. Their resultant lies, by what we have just proved, in the bisector of their angle of lesser measure. We denote its length by $f(x)$. As a length (absolute value), it is nonnegative:

$$f(x) \geq 0 \ . \tag{11}$$

Taking two further vectors of equal length $y = \|Q_1\| = \|Q_2\|$, Q_1 in the direction of P_1 and Q_2 in the direction of P_2, we have, by (2), (3), and (10), and by the definition of f,

$$f(x+y) = \|(P_1 + Q_1) + (P_2 + Q_2)\| = \|(P_1 + P_2) + (Q_1 + Q_2)\|$$

$$= \|P_1 + P_2\| + \|Q_1 + Q_2\| = f(x) + f(y)$$

for all nonnegative x,y. This is Cauchy's functional equation. As is known (see, e.g., Picard [3], Aczél [1], [2]), if we also have (11) then, <u>without any continuity hypothesis (5)</u>,

$$f(x) = cx \tag{12}$$

follows. Here c is some nonnegative constant, depending of course on the angle of the two vectors which we have kept fixed until now. Equation (12) means that

(13) <u>the length of the resultant of two vectors of equal length is proportional to their length</u>.

From (12), c <u>is</u> evidently the <u>length of the resultant of two unit vectors</u>. For convenience, we denote its dependence on the <u>half</u> angle measure φ of the two unit vectors by

$$c = 2g(\varphi) \ . \tag{14}$$

We take, again with Picard [3] (cf. Aczél [1], [2]), for unit vectors P_1, P_2, Q_1, Q_2, with angle measures $(P_1,P_2)\measuredangle = (Q_1,Q_2)\measuredangle = 2\psi$, $(P_1,Q_1)\measuredangle = 2(\varphi+\psi)$, $(P_2,Q_2)\measuredangle = 2(\varphi - \psi)$. Then, by (10), $(P_1 + P_2, Q_1 + Q_2)\measuredangle = 2\varphi$ and [cf. (14)]

$$\|P_1 + P_2\| = \|Q_1 + Q_2\| = 2g(\psi) \tag{15}$$

and $\|P_1 + Q_1\| = 2g(\varphi + \psi)$, $\|P_2 + Q_2\| = 2g(\varphi - \psi)$. Also, P_1,Q_1 and P_2,Q_2 have a common bisector and, by (2), (3), (13) and (14),

$$2g(\varphi+\psi) + 2g(\varphi-\psi) = \|(P_1+Q_1)+(P_2+Q_2)\| = \|(P_1+P_2)+(Q_1+Q_2)\| = 2g(\psi)2g(\varphi).$$

So we have, for g, d'Alembert's functional equation

$$g(\varphi+\psi)+g(\varphi-\psi) = 2g(\varphi)g(\psi) \quad \text{for all} \quad 0 \le \psi \le \varphi \le \pi/2 \ , \tag{16}$$

and therefore (cf. Picard [3], Aczél [1], [2], where the domain is different but the proof applies also in this case), under the continuity assumption (4), all solutions are given by

$$g(\varphi) = 0 \ , \tag{17}$$

$$g(\varphi) = \cosh c\varphi \ , \tag{18}$$

$$g(\varphi) = \cos c\varphi \ , \qquad 0 \le \varphi \le \pi/2 \ , \tag{19}$$

where c is a constant which can be taken as nonnegative since both cosh and cos are even functions. From (3) and (7), we have [cf. (14)]

$$g(0) = 1 \quad \text{and} \quad g(\pi/2) = 1 \ ,$$

which excludes (17) and (18) and reduces (19) to

$$g(\varphi) = \cos(2k + 1)\varphi, \qquad k \text{ being a nonnegative integer.} \tag{20}$$

As mentioned above, however, (7) and (9) imply $k = 0$, and we have

$$g(\varphi) = \cos \varphi \ .$$

Therefore, by (12) and (14),

$$2x \cos \varphi$$

is the length of the resultant of two vectors having equal length x and enclosing an angle of measure 2φ. The direction of the resultant lies, by (10), in the bisector of their angle of lesser measure. Thus we have proved the paralellogram rule for vectors of equal length.

(The above argument shows also that another way of avoiding (10) does not work. If E is the unit vector of the bisector of the angle of lesser measure of two vectors of equal length x, then the resultant of these two vectors could be defined as $f(x)E$, and in this case $f(x)$ could be negative. However, then [even with (5)] c in (12) and thus g in (14) could be negative, while, as seen in (15), $2g(\psi)$ is the length [absolute value] of a vector and thus nonnegative.)

The extension of the parellelogram rule to vectors of unequal length goes by a purely geometric reasoning correctly contained in Picard [3] or Aczél [1], where also the illustrations to those arguments, and to the ones reproduced here, can be found.

4. A FOURTH MOLE

Only a fourth, small mole has still to be removed: Picard's proof of the paralellogram rule does not apply to the case where the two vectors P and Q are of opposite direction. This, however, is easy: Let, for instance, P be the longer vector. Then $P = R - Q$, where R is of the same direction as P. By (2) and by the definition of the inverse, $P + Q = (R - Q) + Q = R$, in accordance with the (degenerated case of) the paralellogram rule.

This research has been supported in part by the National Research Council of Canada Grant Nr. A-2972.

REFERENCES

1. J. Aczél, Lectures on functional equations and their applications, Academic Press, New York and London, 1966.

2. J. Aczél, On applications and theory of functional equations, Birkhäuser, Basel,and Academic Press, New York, 1969.

3. E. Picard, Lecons sur quelques équations fonctionnelles avec des applications à divers problèmes d'analyse et de physique mathématique, Gauthier-Villars, Paris, 1928.

4. S. D. Poisson, Traité de mécanique, Bachelier, Paris, 1811.

LES FONCTIONS DU TRIANGLE POUR LES ESPACES NORMÉS ALÉATOIRES

D. H. Mouchtari
Faculté des Mathématiques
Université de Kazan,
Kazan, 420 000
U.R.S.S.

A. N. Šerstnev
Faculté des Mathématiques
Université de Kazan,
Kazan, 420 000
U.R.S.S.

RÉSUMÉ. On étudie des propriétés algébriques telles que la distributivité à gauche ou à droite des fonctions du triangle pour les espaces normés aléatoires.

1. INTRODUCTION

Dans la théorie des espaces munis d'une métrique aléatoire, il y a quelques problèmes qui concernent la fonction du triangle et qui n'existent pas dans la théorie traditionnelle des espaces métriques. Pour les espaces normés aléatoires, ces problèmes sont plus spécifiques à cause du lien entre des propriétés de la norme aléatoire et les opérations algébriques.

Rappelons quelques définitions (voir p. ex. [1]).

Soit B l'ensemble dont les éléments sont les fonctions sur $\mathbb{R}$ de type $\xi = 1 - F$, ou F est la fonction de distribution d'une variable aléatoire non négative. B est muni de l'ordre naturel: $\xi \leq \eta$ signifie $\xi(x) \leq \eta(x)$ $(x \in \mathbb{R})$. La fonction $\Delta \in B$ caractérisée par la propriété: $\Delta(x) = 0$ si $x > 0$, est l'élément minimal de B.

Dans B on introduit la multiplication par les nombres réels $a \geq 0$:

$$0 \circ \xi = \Delta, \quad a \circ \xi(x) = \xi(\tfrac{x}{a}).$$

Une fonction du triangle (f.t.) est une loi de composition qui définit dans B la structure d'un semi groupe commutatif ordonné d'élément neutre Δ. Disons qu'une f.t. μ est distributive à droite (resp. à gauche) si pour tous $\xi, \eta \in B$, $a > 0$:

$$\mu(a \circ \xi, a \circ \eta) = a \circ \mu(\xi,\eta)$$

$$(\text{ou resp.} \quad (a + b) \circ \xi = \mu(a \circ \xi, b \circ \xi)).$$

Les f.t. les plus utilisées sont les f.t. de Menger et la f.t. de Wald (voir p. ex. [2]). Ces f.t. sont distributives à droite. La plus importante est la f.t. de Menger μ^0 suivante:

$$\mu^0(\xi,\eta)(x) = \inf_{0 \leq t \leq 1} \max\{\xi(tx), \eta((1 - t)x)\}. \tag{1}$$

Il est très facile de démontrer que μ^0 est distributive à gauche.

Un espace vectoriel L sur le champ $\Lambda(= \mathbb{C}$ ou $\mathbb{R})$ est dit espace normé aléatoire (e.n.a) s'il existe une application $\|\cdot\| : L \to B$ telle que pour tous $\varphi,\psi \in B$, $a \in \mathbb{R}$:

(I) $\|\varphi\| = \Delta$ entraine $\varphi = \theta$ (θ = zero de l'espace L),

(II) $\|a\varphi\| = |a| \circ \|\varphi\|$,

(III) $\|\varphi + \psi\| < \mu(\|\varphi\|, \|\psi\|)$, ou μ est une f.t.

Soit $M(L)$ la classe de toutes les f.t. μ telles que (III) a lieu pour μ et tous $\varphi, \psi \in L$. Pour l'e.n.a. L nous ne distinguons pas les f.t. $\mu', \mu'' \in M(L)$ qui sont égales sur l'emsemble $B_L \times B_L$, où $B_L = \{\|\varphi\| | \varphi \in L\}$. Disons que $\mu' < \mu''(\mu', \mu'' \in M(L))$ si $\mu'(\xi,\eta) < \mu''(\xi,\eta)$ pour tous $\xi, \eta \in B_L$.

2. THÉORÈMES

THÉORÈME 1. Soit μ une f.t. distributive à droite telle que:

$$(a + b) \circ \xi \leq \mu(a \circ \xi, b \circ \xi), \ \xi \circ B, \ a, b > 0. \tag{2}$$

Alors il existe un e.n.a. $(L,\|\cdot\|)$ tel que $B_L = B$ et μ est l'élément le plue petit de l'ensemble $M(L)$.

Ce théorème est pratiquement démontré dans [1], bien qu'il soit formulé sous une forme plus faible. Le résultat admet une réciproque.

THÉORÈME 2. Soit L un e.n.a., et soit μ l'élément le plus petit de $M(L)$. Alors μ est distributive à droite sur B_L.

Démonstration. Il est facile de voir que si $\mu \in M(L)$ et

$$\mu_a(\xi,\eta) \overset{\text{déf.}}{=} a \circ \mu(\frac{1}{a} \circ \xi, \frac{1}{a} \circ \eta), \; a > 0, \; \xi, \; \eta \in B,$$

μ_a est dans $M(L)$ pour tout $a > 0$. Le théorème résulte immédiatement des relations:

$$a \circ \mu(\xi,\eta) = \mu_a(a \circ \xi, a \circ \eta) \geq \mu(a \circ \xi, a \circ \eta)$$

$$= a \circ \mu_{1/a}(\xi,\eta) \geq a \circ \mu(\xi,\eta),$$

où $a > 0$ et ξ, η sont des éléments arbitraires de B_L.

Soit

$$\mu_*(\xi,\eta)(\cdot) \text{déf.} = \inf_{a>0} \mu_a(\xi,\eta)(\cdot), \; (\xi,\eta \in B).$$

Il est connu ([1], Proposition 12) que l'égalité $\mu = \mu_*$ est nécessaire et suffisante pour que μ soit distributive à droite. En général, il est naturel d'essayer d'améliorer l'inégalité du triangle (III) en remplacant μ par μ_*. Mais, μ_* peut ne pas être une f.t. (μ_* a toutes les propriétés de f.t. sauf peut être l'associativité).

Dans [1] on a trouvé des conditions suffisantes pour que μ soit une f.t.

THÉORÈME 3 [1]. Soit μ une f.t. et

(a) $\{\mu_a(\cdot, \cdot)\}$, $a > 0$, filtrant à gauche,

(b) par rapport à la métrique de Lévy sur B, $\eta_n \downarrow \eta$ entraine $\mu(\xi,\eta_n) \downarrow \mu(\xi,\eta)$ pour tout $\xi \in B$.

Alors μ_* est une f.t., et $\mu_* \in M(L)$ chaque fois que $\mu \in M(L)$.

3. EXEMPLE

L'exemple suivant montre que la condition (a) du Théorème 3 n'est pas nécessaire pour l'associativité de μ_*:

Soit

$$f(x) = \begin{cases} 2x & \text{pour } x < 1, \\ x + 1 & \text{pour } x > 1. \end{cases}$$

La loi de composition

$$(x,y) = f^{-1}(f(x) + f(y))$$

définit un semi groupe ordonné commutatif sur $[0,\infty)$ d'élément neutre 0.

Posons

$$\nu(\xi,\eta)(x) = \inf_{(x',x'')=x} \max\{\xi(x'), \eta(x'')\}.$$

On vérifie facilement que ν est une f.t. et que $\nu_* = \mu^0$. Mais, si

$$\xi(x) = \begin{cases} 1 & \text{pour } x < 0, \\ \frac{1}{1+x} & \text{pour } x \geq 0, \end{cases}$$

pour tous $a > b > 0$, nous avons

$$\nu_a(\xi,\xi)(\tfrac{3}{b}) < \nu_b(\xi,\xi)(\tfrac{3}{b}),$$

$$\nu_a(\xi,\xi)(\tfrac{3}{a}) > \nu_b(\xi,\xi)(\tfrac{3}{a}).$$

C'est-à-dire, les éléments $\nu_a(\xi,\xi)$ et $\nu_b(\xi,\xi)$ sont incomparables. Cela signifie que (a) n'a pas lieu.

4. UNICITÉ

Nous avons déjà remarqué que μ^0 est distributive à gauche.

THÉORÈME 4. μ^0 est l'unique f.t. distributive à gauche.

<u>Démonstration</u>. Le théorème est la conséquence des lemmes suivants:

LEMME 1. Si une f.t. μ a la propriété (2), alors $\mu^0 \leq \mu$.

<u>Démonstration</u>. Pour $\xi, \eta \in B$ et $x > 0$ fixés, posons:

$$x' = \inf\{y \mid \xi(y) \leq \mu^0(\xi,\eta)(x)\},$$

$$x'' = x - x'.$$

Il est facile de voir que $\eta(x'') \geq \mu(\xi,\eta)(x)$. Posons:

$$\zeta(x) = \begin{cases} \mu^0(\xi,\eta)(x) & \text{pour } 0 < x \leq x', \\ 0 & \text{pour } x > x'. \end{cases}$$

Nous avons : $\zeta \leq \xi$, $\frac{x''}{x'} \circ \zeta \leq \eta$,

et donc

$$\mu(\xi,\eta)(x) \geq \mu(\zeta, \frac{x''}{x'} \circ \zeta)(x) \geq (\frac{x''}{x'} + 1) \circ \zeta(x) = \mu^0(\xi,\eta)(x).$$

LEMME 2. Soit pour tous $\xi \in B$, $a, b > 0$,

$$(a + b) \circ \xi \geq \mu(a \circ \xi, b \circ \xi).$$

Alors $\mu < \mu^0$.

<u>Démonstration</u>. Soient $\xi, \eta \in B$, $x > 0$ arbitraires. Pour chaque $\varepsilon > 0$ on peut choisir x', x'' tels que:

$$\max\{\xi(x'), \eta(x'')\} < \mu^0(\xi,\eta)(x) + \varepsilon.$$

Posons:

$$\zeta(x) = \max\{\xi(x), \frac{x'}{x''} \circ \eta(x)\}, \ x \in \mathbb{R}.$$

Alors $\zeta \geq \xi$, $\frac{x''}{x'} \circ \zeta > \eta$.

Ainsi:

$$\mu(\xi,\eta)(x) \leq \mu(\zeta, \frac{x''}{x'} \circ \zeta)$$

$$\leq (1 + \frac{x''}{x'}) \circ \zeta(x)$$

$$= \max\{\xi(x'), \eta(x'')\} < \mu^0(\xi,\eta)(x) + \varepsilon.$$

BIBLIOGRAPHIE

1. D. H. Mouchtari, et A. N. Šerstnev, Sur les meilleures inégalités du triangle pour les espaces normés aléatoires (en russe), Učen. zap. KGU 125 (1965/66), livre 2, 102-113.

2. A. N. Šerstnev, Sur la généralisation probabiliste des espaces métriques, Učen. zap. KGU 124 (1964), livre 2, 3-11.

QUADRATIC FUNCTIONALS SATISFYING A SUBSIDIARY INEQUALITY

Jürg Rätz
Department of Mathematics
University of Bern
CH-3012 Bern
SWITZERLAND

ABSTRACT. A mapping $q : X \to K$ defined on a normed K-vector space is called a quadratic functional if $q(x + y) + q(x - y) = 2q(x) + 2q(y)$ $(x,y \in X)$ holds. The main question dealt with in this paper is the role of the subsidiary inequality condition:

(Q') There exists a real constant $c \geq 0$ such that $|q(x) - q(y)| \leq c\|x\|^2$ for all $x,y \in X$ with $\|x\| = \|y\|$.

It is shown in Theorem 1 that (Q') together with another boundedness condition implies continuity of q, while (Q') alone is too weak to guarantee q to be continuous (Theorem 2). An occurrence of a condition (Q) of type (Q') in the theory of ordinary differential equations is mentioned at the beginning of Section 3, in which some additional logical connections are discussed. Remark 1 in Section 2 indicates essential differences between the behavior of quadratic functionals and that of additive mappings.

1. PRELIMINARIES AND NOTATION

In this paper, K denotes the field $\mathbb{R}$ of real numbers or the field $\mathbb{C}$ of complex numbers, and $(X,\|\cdot\|)$ denotes a normed K-vector space. It is advantageous to exclude the trivial space $X = \{0\}$ from our considerations.

If $(X,\|\cdot\|)$ is a normed vector space over $\mathbb{C}$, we obtain its so-called real restriction $(\tilde{X},\|\cdot\|)$ by putting $\tilde{X} = X$, leaving addition and norm unchanged, and restricting multiplication $\cdot : \mathbb{C} \times X \to X$ to $\mathbb{R} \times X$. In the special case where $(X,\langle\cdot,\cdot\rangle)$ is a complex inner-product space, i.e., where $\langle\cdot,\cdot\rangle$ is a $\mathbb{C}$-valued positive definite hermitian sesquilinear functional on $X \times X$, it is easily verified that

$$\langle x,y\rangle = \mathrm{Re}\langle x,y\rangle + i \cdot \mathrm{Re}\langle x,iy\rangle \quad (x,y \in X), \tag{1}$$

that $\mathrm{Re}\langle\cdot,\cdot\rangle$ is a real-valued positive definite symmetric bilinear functional on $\tilde{X} \times \tilde{X}$, and that $\langle\cdot,\cdot\rangle$ and $\operatorname{Re}\langle\cdot,\cdot\rangle$ induce the same norm $\|\cdot\|$ on X. Thus $(\tilde{X}, \operatorname{Re}\langle\cdot,\cdot\rangle)$ is a real inner-product space, and of course it is compatible with our former terminology to call it the real restriction of $(X, \langle\cdot,\cdot\rangle)$.

For a simultaneous treatment of the real and the complex case, two conventions are very useful:

(i) The mapping $\lambda \mapsto \overline{\lambda}$ from K into K is the usual conjugation if $K = \mathbb{C}$, and is the identical mapping if $K = \mathbb{R}$. Hence, in the real case, hermitian sesquilinear is equivalent to symmetric bilinear.

(ii) Every real normed space (real inner-product space) is defined to be its own real restriction.

Hence it is true in either case that a normed space and its real restriction have the same additive, metric, uniform, and topological structures and therefore the same continuous functions defined on them. Two more aspects of this close connectedness are expressed by Lemmas 1 and 2, below; their straightforward proofs are omitted.

LEMMA 1. If $B = \{e_\ell : \ell \in L\}$ is a Hamel base of the complex inner-product space X, then $B' := \{e_\ell, ie_\ell : \ell \in L\}$ is a Hamel base of the real restriction $\tilde{X}$ of X. If B is orthonormal, so is B'.

DEFINITION 1. If X is a normed K-vector space, a mapping $q : X \to K$ is called a <u>quadratic functional</u> on X (cf. [6]) if

$$q(x + y) + q(x - y) = 2q(x) + 2q(y) \qquad (x,y \in X) \tag{2}$$

holds.

J. Aczél [1] and M. Hosszú [4] determined the general solution of the functional equation (2) under very general algebraic assumptions.

LEMMA 2. If X is a normed $\mathbb{C}$-vector space, $\tilde{X}$ its real restriction, and $q : X \to \mathbb{C}$ is a quadratic functional, then:

(i) $\operatorname{Re} q$ and $\operatorname{Im} q$ are quadratic functionals from $\tilde{X}$ into $\mathbb{R}$.

(ii) q is continuous on X if and only if $\operatorname{Re} q$ and $\operatorname{Im} q$ are continuous on $\tilde{X}$.

DEFINITION 2. If $(X, \langle\cdot,\cdot\rangle$ is an inner-product space over K, and $V : X \to X$ is an additive mapping, i.e.,

$$V(x + y) = Vx + Vy \qquad (x,y \in X), \tag{3}$$

then the mapping $q_V : X \to K$ defined by

(4) $$q_V(x) := \langle Vx,x\rangle \qquad (x \in X)$$

is called the <u>quadratic</u> <u>functional</u> <u>associated</u> <u>with</u> V. (It is immediate that q_V satisfies (2).)

The terminology ambiguity of "ball" and "sphere" in the literature motivates the following:

DEFINITION 3. For a normed K-vector space $(X,\|\cdot\|)$, a point x of X, and a real number $r > 0$, we call

(i) $T_r(x) := \{y \in X : \|y - x\| \leq r\}$ the r-<u>ball</u>,

(ii) $S_r(x) := \{y \in X : \|y - x\| = r\}$ the r-<u>sphere</u>,

around x.

2. THE SUBSIDIARY INEQUALITY (Q')

The main properties of quadratic functionals q on normed spaces we are interested in are these:

(Q') There exists a real constant $c \geq 0$ such that $|q(x) - q(y)| \leq c\|x\|^2$ for all $x,y \in X$ with $\|x\| = \|y\|$.

(C) q is continuous on X.

In the following Lemma, we state some miscellaneous elementary facts on q.

LEMMA 3. If $q : X \to K$ is a quadratic functional on a normed K-vector space X, then:

(i) (C) implies (Q').

(ii) (Q') implies boundedness of q on each individual sphere $S_r(0)$.

(iii) The homogeneity set

(5) $$H(q) := \{\lambda \in K : q(\lambda x) = |\lambda|^2 q(x) \quad \text{for every } x \in X\}$$

of q contains all rational numbers. It forms a subfield of K if and only if it is closed under addition.

<u>Proof</u>. (i) For $K = \mathbb{R}$, a slight modification of [6], p. 60, Theorem 2, shows that there exists a real constant $c \geq 0$ such that

(6) $$|q(x) \leq c\|x\|^2 \quad \text{for every } x \in X.$$

In the complex case, this argument, applied to Re q and to Im q, and Lemma 2 lead again to (6), and (6) obviously implies (Q').

(ii) Let $r > 0$ be arbitrary, $y_0 \in X$ be fixed such that $\|y_0\| = r$, and x be any vector of $S_r(0)$. Then $|q(x)| - |q(y_0)| \leq |q(x) - q(y_0)| \leq$

cr^2; therefore $|q(x)| \leq cr^2 + |q(y_0)|$.

(iii) The first part is proved as in [6], p. 58. It is easily seen that $\lambda,\mu \in H(q)$ implies $\lambda\mu \in H(q)$, that $\lambda \in H(q) \setminus \{0\}$ implies $1/\lambda \in H(q)$, and that $\lambda \in H(q)$ implies $(-\lambda) \in H(q)$. Now the second assertion follows from a well-known subfield criterion.

REMARK 1. We shall see later (Theorem 2 (i), (ii)) that (Q') does not imply (C) and that, a fortiori, boundedness of q on each sphere $S_r(0)$ is strictly weaker than (C). This expresses an essential contrast between quadratic functionals and additive mappings $f : \mathbb{R}^n \to \mathbb{R}$. If, for $n \geq 2$, f is bounded on one single sphere $S_r(x)$, then it is continuous on $\mathbb{R}^n$. The reason is that the midpoint convex hull of $S_r(x)$ is $T_r(x)$, and thus that it has positive inner Lebesgue measure ([3], p. 159, Theorem 1). A second contrast can be found in the homogeneity behavior: The homogeneity set $H_f := \{\lambda \in K : f(\lambda x) = \lambda f(x)$ for every $x \in X\}$ of an additive mapping f always is a subfield of the scalar field ([8], p. 67, Lemma 1). For quadratic functionals, this need not be the case (Lemma 5 (iii), (iv)).

The question is natural under what additional conditions a quadratic functional satisfying (Q') is continuous. One such condition is presented in this result:

THEOREM 1. Let X be a normed K-vector space, $q : X \to K$ a quadratic functional satisfying (Q'), and A a bounded subset of X such that the set $B := \{\|x\| : x \in A\}$ has positive inner Lebesgue measure and q is bounded on A. Then q is continuous on X.

<u>Proof</u>. It is no loss of generality to assume that $K = \mathbb{R}$: For $K = \mathbb{C}$, we consider Re q and Im q and make use of Lemma 2.

By the definition of inner Lebesgue measure, B contains a subset B' of positive Lebesgue measure. For $A' := \{x \in A : \|x\| \in B'\}$, we then obtain $B' = \{\|x\| : x \in A'\}$. Since $A' \subset A$, q is bounded on A', say

(7) $$|q(x)| \leq b \quad \text{for every } x \in A'.$$

Fix a vector $y \in X$ such that $\|y\| = 1$. Since B' is bounded, there exists a nonnegative real number d with the property

(8) $$0 \leq \beta \leq d \quad \text{for every } \beta \in B'.$$

Let β be any element of B'. There exists $x_\beta \in A'$ with $\|x_\beta\| = \beta$. On the other hand, $\|\beta y\| = |\beta| = \beta$, and (Q') and (8) imply $|q(\beta y) - q(x_\beta)| \leq c\beta^2 \leq cd^2$; i.e., by (7),

$$(9) \qquad |q(\beta y)| \le cd^2 + |q(x_\beta)| \le cd^2 + b \quad \text{for every} \quad \beta \in B'.$$

Next we consider the mapping $g : \mathbb{R} \to \mathbb{R}$ defined by $g(\lambda) = q(\lambda y)$ $(\lambda \in \mathbb{R})$. It is clear that g satisfies (2), and (9) shows that g is bounded on the set B' of positive Lebesgue measure. From a result of S. Kurepa ([6], p. 57, Theorem 1), we conclude that $g(\lambda) = \lambda^2 g(1)$ $(\lambda \in \mathbb{R})$; i.e.,

$$(10) \qquad q(\lambda y) = \lambda^2 q(y) \quad \text{for every} \quad \lambda \in \mathbb{R}.$$

Now choose an arbitrary element x of $T_1(0)$. With respect to $\| \|x\| y\| = \|x\| \le 1$, (Q') implies $|q(x) - q(\|x\|y)| \le c\|x\|^2 \le c$, and hence, by (10), $|q(x)| \le c + |q(\|x\|y)| = c + \|x\|^2 |q(y)| \le c + |q(y)|$. So q is bounded on $T_1(0)$ and therefore continuous on X ([6], p. 60, Theorem 2, which, as a matter of fact, holds for any normed $\mathbb{R}$-vector space).

REMARK 2. The hypothesis on A in Theorem 1 is satisfied in each of the following important cases:

(i) B contains an open interval.

(ii) A is a segment $[x,y] := \{\lambda x + (1 - \lambda)y : \lambda \in \mathbb{R},\ 0 \le \lambda \le 1\}$ with $\|x\| \ne \|y\|$. Then $[x,y]$ is connected and $\|\cdot\|$ is continuous; therefore, B contains an open interval.

(iii) X is strictly convex and A is any segment $[x,y]$ with $x \ne y$: If $\|x\| \ne \|y\|$, then (ii) applies; if $\|x\| = \|y\|$, then $\|y\| > \|\frac{1}{2}(x + y)\|$, and we may consider $[\frac{1}{2}(x + y),y]$ instead of $[x,y]$.

(iv) $X = \mathbb{R}^n$, and A is a bounded subset of $\mathbb{R}^n$ of positive Lebesgue measure. It follows from a standard procedure in measure theory (cf., e.g., [7], pp. 277-278, 380-384) that B has positive inner Lebesgue measure.

REMARK 3. It should be noticed that in Remark 2 (iv) measurability of A $(A \subset \mathbb{R}^n,\ n \ge 2)$ does not imply measurability of B: Let e be a unit vector of $\mathbb{R}^n$, C a non-Lebesgue measurable subset of the unit interval $[0,1]$, and $A := \{\lambda e : \lambda \in C\} \cup T_1(2e)$. Then A has positive n-dimensional Lebesgue measure, but $B = C \cup [1,3]$ is not Lebesgue measurable in $\mathbb{R}$. Thus it is adequate in Theorem 1 to require merely that B has positive inner Lebesgue measure.

3. QUADRATIC FUNCTIONALS ON INNER-PRODUCT SPACES

In his paper [10], H. M. Riemann considers the following conditions for functions $V : \mathbb{R}^n \to \mathbb{R}^n$. These are formulated here for arbitrary inner-product spaces $(X,\langle\cdot,\cdot\rangle)$:

(L) There exists a real constant $c \geq 0$ such that $\|Vx - Vy\| \leq c\|x - y\|$ for all $x,y \in X$ (Lipschitz condition).

(Q) There exists a real constant $c \geq 0$ such that $|\langle V(x + z) - Vz,x\rangle - \langle V(y + z) - Vz,y\rangle| \leq c\|x\|^2$ for all $x,y,z \in X$ with $\|x\| = \|y\|$.

(Z) There exists a real constant $c \geq 0$ such that $\|V(x+y) + V(x-y) - 2Vx\| \leq c\|y\|$ for all $x,y \in X$ (Zygmund condition).

It is shown in [10] that (L) $\Rightarrow$ (Q) $\Rightarrow$ (Z) and that (Z) $\Rightarrow$ (Q) for $X = \mathbb{R}^1$. While (L) implies continuity of V, (Z) does not, and the question is whether (Q) does. We are going to show that the answer to this question is negative (cf. the Corollary at the end of the paper), and that an additive mapping serves as a counterexample. From now on, $V : X \to X$ always denotes an additive mapping. Condition (Q) then specializes to the following condition (Q'_V) for the quadratic functional q_V associated with V (cf. (4)):

(Q'_V) q_V satisfies (Q').

For brevity, we introduce two more conditions concerning V and q_V:

(I) V is continuous on X.

(II) q_V is continuous on X.

The remainder of this paper is devoted to the study of the logical connections between our conditions (L), (I), (II), (Q'_V), and (Z).

LEMMA 4. If X is an inner-product space over K and $V : X \to X$ is additive, then:

(i) (L) $\Longleftrightarrow$ (I) $\Rightarrow$ (II) $\Rightarrow$ (Q'_V) $\Rightarrow$ (Z).

(ii) If $\dim \tilde{X} = 1$, i.e., $X = \mathbb{R}^1$, then (L) $\Longleftrightarrow$ (I) $\Longleftrightarrow$ (II) $\not\Longleftrightarrow$ $(Q'_V) \Longleftrightarrow$ (Z).

(iii) If $\dim \tilde{X} \geq 2$, then (Z) $\not\Rightarrow$ (Q'_V).

(iv) If $K = \mathbb{R}$ and X is noncomplete, then (II) $\not\Rightarrow$ (I).

(v) The homogeneity set $H(q_V)$ defined in (5) is a subfield of K.

<u>Proof</u>. (i) (L) $\Rightarrow$ (I) $\Rightarrow$ (II) $\Rightarrow$ (Q'_V) $\Rightarrow$ (Z) is clear from (4), Lemma 3 (i), and the fact that (Z) holds with $c = 0$.

(I) $\Rightarrow$ (L): As a continuous additive mapping, V is $\mathbb{R}$-linear, and (L) follows from a well-known continuity criterion for linear mappings.

(ii) (II) $\Rightarrow$ (I): (II) implies that $x \mapsto x \cdot Vx$ is bounded on the interval $[1/2,1]$, i.e., that V is bounded there. Now (I) follows from [2], p. 34, Theorem 1.

(Z) $\Rightarrow$ (Q'_V): Since $\|x\| = \|y\|$ is equivalent here to $x = \pm y$, q_V satisfies (Q') with $c = 0$. Hence (Q'_V) is trivially true.

$(Q'_V) \not\Rightarrow$ (II): Any discontinuous additive mapping $V : \mathbb{R} \to \mathbb{R}$ serves as a counterexample.

(iii) Let $\{e_1,e_2\}$ be an orthonormal subset and $\{e_\ell : \ell \in L\}$ a Hamel base of the real restriction $\tilde{X}$ of X containing e_1 and e_2. Let $f : \mathbb{R} \to \mathbb{R}$ be a discontinuous additive function and $V : \tilde{X} \to \tilde{X}$ be defined by $Vx = f(\xi_1)e_1$ for $x = \sum_{\ell\in L} \xi_\ell e_\ell$. Then V is additive and trivially satisfies (Z). Furthermore, it turns out that $q_V(e_2) = 0$. For $1/2 \leq \beta \leq 1$, we define $x_\beta := \beta e_1 + \sqrt{1-\beta^2}\, e_2$ and conclude that $\|x_\beta\| = 1$ and $q_V(x_\beta) = f(\beta)\cdot\beta$. (Q'_V) together with $q_V(e_2) = 0$ would imply that $\beta \mapsto f(\beta)\cdot\beta$ is bounded on $[1/2,1]$ and finally that f is bounded there, a contradiction to its discontinuity ([2], p. 34, Theorem 1). Hence (Q'_V) does not hold.

(iv) See [9], where more information about the logical connection between (I) and (II) can be found.

(v) Let $\lambda,\mu \in H(q_V)$. If $\lambda\mu = 0$, then $\lambda + \mu \in \{\lambda,\mu\} \subset H(q_V)$. In the remaining case $\mu \neq 0$, $\lambda \neq 0$, we proceed as follows. Let $x \in X$ be arbitrary. Then $q_V[(\lambda+\mu)x] = \langle V(\lambda x),\lambda x\rangle + \langle V(\lambda x),\mu x\rangle + \langle V(\mu x),\lambda x\rangle + \langle V(\mu x),\mu x\rangle = q_V(\lambda x) + \bar{\mu}\langle V(\lambda x),x\rangle + \bar{\lambda}\langle V(\mu x),x\rangle + q_V(\mu x) = q_V(\lambda x) + (\bar{\mu}/\bar{\lambda})q_V(\lambda x) + (\bar{\lambda}/\bar{\mu})q_V(\mu x) + q_V(\mu x) = [\bar{\lambda}\lambda + \bar{\mu}\lambda + \bar{\lambda}\mu + \bar{\mu}\mu]\cdot q_V(x) = |\lambda+\mu|^2 q_V(x)$. Since x is an arbitrary element of X, we infer that $(\lambda+\mu) \in H(q_V)$, and the assertion follows from Lemma 3 (iii).

The next step, which will culminate in showing that $(Q'_V) \not\Rightarrow$ (II) also in spaces other than $\mathbb{R}^1$ (see Lemma 4 (ii) and Theorem 2), consists in bringing two classes of quadratic functionals into contact, namely the ones associated with additive mappings (Definition 2) and those introduced in Lemma 5.

LEMMA 5. Let X be an inner product space over K, $\|\cdot\|$ the norm derived from the inner product, and $f : \mathbb{R} \to \mathbb{R}$ an additive function. Then the following statements hold:

(i) $q_f := f \circ \|\cdot\|^2$ is a quadratic functional satisfying (Q').

(ii) If f is discontinuous, then so is q_f.

(iii) If the homogeneity field

(11) $$H_f := \{\lambda \in \mathbb{R} : f(\lambda\alpha) = \lambda f(\alpha) \quad \text{for every} \quad \alpha \in \mathbb{R}\}$$

of f ([8], p. 67, Lemma 1) contains a positive element which is not the square of an element of H_f, then the homogeneity set $H(q_f)$ of q_f (see (5)) is not a subfield of K.

(iv) Not every quadratic functional of type q_f is associated in the sense of (4) with an additive mapping $V : X \to X$.

Proof. (i) $\|\cdot\|^2$ satisfies the so-called parallelogram law $\|x + y\|^2 + \|x - y\|^2 = 2\|x\|^2 + 2\|y\|^2$ $(x,y \in X)$, and therefore q_f is a solution of (2). $\|x\| = \|y\|$ implies $q_f(x) = q_f(y)$; i.e., (Q') holds with $c = 0$.

(ii) Assume that q_f is continuous on X, i.e., on $\tilde{X}$. Then there exists $c \geq 0$ such that $|q_f(x)| \leq c$ for every $x \in T_1(0)$ ([6], p. 60, Theorem 2). For an arbitrary $\beta \in [0,1]$, we find $x \in T_1(0)$ such that $\beta = \|x\|^2$. Then $|f(\beta)| = |f(\|x\|^2)| = |q_f(x)| \leq c$; thus f is bounded on $[0,1]$, in contradiction to our hypothesis on f ([2], p. 34, Theorem 1).

(iii) Let $\lambda \in H_f$, $\lambda > 0$, such that $\sqrt{\lambda} \notin H_f$. Then $q_f(\sqrt{\lambda}\, x) = f(\|\sqrt{\lambda}\, x\|^2) = f(\lambda\|x\|^2) = \lambda f(\|x\|^2) = (\sqrt{\lambda})^2 q_f(x)$ for every $x \in X$; hence $\sqrt{\lambda} \in H(q_f)$. Clearly also $1 \in H(q_f)$. Since $\sqrt{\lambda} \notin H_f$, there exists $\alpha \in \mathbb{R}$ with the property $f(\sqrt{\lambda}\,\alpha) \neq \sqrt{\lambda}\, f(\alpha)$. Certainly, $\alpha \neq 0$, and with respect to oddness of f, it is no loss of generality to assume $\alpha > 0$. Now we choose $x \in X$ such that $\alpha = \|x\|^2$. Therefore $f(\sqrt{\lambda}\ \|x\|^2) \neq \sqrt{\lambda}\, f(\|x\|^2)$, and furthermore $q_f[(\sqrt{\lambda} + 1)x] = f(\|(\sqrt{\lambda} + 1)x\|^2) = f[(\sqrt{\lambda} + 1)^2\|x\|^2] = f(\lambda\|x\|^2) + 2f(\sqrt{\lambda}\ \|x\|^2) + f(\|x\|^2) \neq f(\lambda\|x\|^2) + 2\sqrt{\lambda} f(\|x\|^2) + f(\|x\|^2) = (\sqrt{\lambda} + 1)^2 q_f(x)$; i.e., $(\sqrt{\lambda} + 1) \notin H(q_f)$, which shows that $H(q_f)$ is not a subfield of K.

(iv) There exist additive functions $f : \mathbb{R} \to \mathbb{R}$ with the field of rational numbers as their homogeneity field ([8], p. 68, Theorem 3). Then, by (iii), $H(q_f)$ is not a subfield of K, and now the assertion follows from Lemma 4(v).

REMARK 4. Lemma 5 (i) produces quadratic functionals q_f satisfying (Q'), and Lemma 5 (iii), (iv) show that a careful choice of f has to be made if we wish to construct a q_f which is associated with an additive mapping $V : X \to X$. From the view of Lemma 5 (iv), derivations, namely functions $f : \mathbb{R} \to \mathbb{R}$ such that

(12) $$f(\alpha + \beta) = f(\alpha) + f(\beta), \quad f(\alpha\beta) = \alpha f(\beta) + \beta f(\alpha) \qquad (\alpha,\beta \in \mathbb{R})$$

might be good candidates, and they really are. The author thanks Professor S. Kurepa for a helpful hint in this direction.

THEOREM 2. If X is an inner-product space over K with an orthonormal Hamel base, then there exists an additive mapping $V : X \to X$ such that:

(i) q_V satisfies (Q'); i.e., (Q'_V) holds.

(ii) q_V is not continuous on X; i.e., (II) does not hold.

(iii) V is not continuous on X; i.e., (I) does not hold.

Proof. We form the real restriction $\tilde{X}$ of X. By Lemma 1 and convention (ii) in Section 1, $\tilde{X}$ has an orthonormal Hamel base $\{e_\ell : \ell \in L\}$, no matter whether $K = \mathbb{R}$ or $K = \mathbb{C}$. Let $f : \mathbb{R} \to \mathbb{R}$ be a discontinuous derivation (for the existence of f, consult the extension theory of derivations in fields, e.g., in [5], p. 167 ff. or [11], p. 121 ff.). For $x \in \tilde{X}$, $x = \sum \xi_\ell e_\ell$, we define $Vx := \sum f(\xi_\ell)e_\ell$; notice that these sums contain only a finite number of terms different from the zero vector. We then obtain, by virtue of (12),

$$q_V(x) = \langle Vx,x\rangle = \sum \xi_\ell f(\xi_\ell) = \sum \frac{1}{2} f(\xi_\ell^2) = \frac{1}{2} f\left(\sum \xi_\ell^2\right) = \frac{1}{2} f(\|x\|^2),$$

and q_V is of the type considered in Lemma 5. Statements (i) and (ii) follow from Lemma 5 (i) and (ii), while statement (iii) is a consequence of (ii) and Lemma 4 (i).

COROLLARY. For mappings $V : \mathbb{R}^n \to \mathbb{R}^n$, (Q) does not imply continuity of V.

Proof. The euclidean space $\mathbb{R}^n$ has an orthonormal Hamel base. The additive mapping V supplied by Theorem 2 satisfies (Q) but is not continuous.

REFERENCES

1. J. Aczél, The general solution of two functional equations by reduction to functions additive in two variables and with the aid of Hamel bases, Glasnik Mat.-Fiz. Astr. 20 (1965), 65-73.

2. J. Aczél, Lectures on functional equations and their applications, Academic Press, New York and London, 1966.

3. R. Ger and M. Kuczma, On the boundedness and continuity of convex functions and additive functions, Aeq. Math. 4, (1970), 157-162.

4. M. Hosszú, A remark on the square norm., Aeq. Math. 2 (1969), 190-193.

5. N. Jacobson, Lectures in abstract algebra, Volume III, Van Nostrand, Princeton, New Jersey, 1964.

6. S. Kurepa, On the quadratic functional, Publ. Inst. Math. Beograd 13, (1959), 57-72.

7. I. P. Natanson, Theorie der Funktionen einer reellen Veränderlichen. 2. Auflage, Akademie-Verlag, Berlin, 1961.

8. J. Rätz, On the homogeneity of additive mappings, Aeq. Math. 14 (1976), 67-71.

9. J. Rätz, Additive mappings on inner product spaces. To appear.

10. H. M. Riemann, Ordinary differential equations and quasiconformal mappings. To appear in Inventiones Mathematicae.

11. O. Zariski and P. Samuel, Commutative Algebra, Volume I, Van Nostrand, Princeton, New Jersey, 1963.

ON OPERATORS IN HILBERT SPACE DEPENDING ANALYTICALLY ON A PARAMETER

I. Fenyö
Polytechnical University of Budapest
Budapest XI
HUNGARY

ABSTRACT. A simple proof is given of the fact that if $A(t)$ is a linear, completely continuous operator defined in a separable Hilbert space for every value t of a connected domain D of the complex plane, and if $A(t)$ depends analytically on t, then the set of values t for which the operator $E + A(t)$ (E is the identity operator) is singular either has no finite limit point or is the whole connected domain D.

1. INTRODUCTION

The fundamental problem in spectral theory is to determine, for an operator S in a given space, the values of t for which the operator $E + tS$ has an inverse, where E denotes the identity operator. In other words, in spectral theory we look for numbers t for which $E + A(t)$ is regular, with $A(t) = tS$. A natural generalization of this question is to consider an operator $A(t)$ depending not necessarily linearly on the parameter t, asking for which values of t the operator $E + A(t)$ is regular.

This generalization of the fundamental problem occurred for the first time in a paper of J. D. Tamarkin [2], who considered integral equations with kernels depending on a parameter. Applications of such integral equations are known, and a technique for the numerical calculation of eigenvalues of their kernels has been worked out by R. E. Kalaba and M. K. Scott [1].

We formulate a theorem concerning the set of numbers for which $E + A(t)$ has no inverse. The proof is based on two classical ideas: The first is an approximation theorem for operators in Hilbert space, the second an idea of E. Schmidt to reduce a given integral equation to an integral equation with degenerate kernel.

2. DEFINITIONS

Let H be a separable Hilbert space and $C(H)$ the vector space of the linear, completely continuous mappings from H into H. D denotes a connected domain in the complex plane $\mathbf{C}$. We say that $A(\cdot): D \to C(H)$ is <u>analytic</u> in D if, for every $t \in D$, $A(t)$ can be represented by a power series,

$$A(t) = \sum_{j=0}^{\infty} a_j t^j ,$$

where $a_j \in C(H)$ $(j = 0,1,2,3,\ldots)$, converging to $A(t)$ with respect to the operator norm.

We define t as a <u>regular point</u> of $A(\cdot)$ if $[E + A(t)]^{-1}$ exists and is an element of $C(H)$. Every point of D which is not a regular point is called a <u>singular point</u> of $A(t)$.

3. THEOREM AND PROOF

We now prove the following result.

THEOREM. If $A(\cdot): D \to C(H)$ is analytic in D, then either every t is a singular point of $A(\cdot)$, or the singular points have no finite limit point in D.

<u>Proof</u>. We have only to show that if a finite limit point of the singular points exists and is in D, then every point of D is singular.

Suppose t_0 is a finite limit point of the singular points of $A(\cdot)$. For an arbitrary given $\varepsilon: 0<\varepsilon<1$, consider $A_n(t) := \Sigma_{j=0}^{n} a_j t^j$ such that

$$\|A(t_0) - A_n(t_0)\| < \frac{\varepsilon}{2} .$$

As a_j $(j = 0,1,2,\ldots)$ is a completely continuous operator, it is the limit of finite-dimensional linear operators; that is, there exists a finite-dimensional linear operator $\mathring{a}_j: H \to H$ such that $\|a_j - \mathring{a}_j\| < \varepsilon/2\rho$ $(j = 0,1,2,\ldots,n)$, where $\rho = \Sigma_{j=0}^{n} |t_0|^j$. If we put

$$\mathring{A}_n(t) := \sum_{j=0}^{n} \mathring{a}_j t^j ,$$

then

$$\|A_n(t_0) - \mathring{A}_n(t_0)\| < \frac{\varepsilon}{2\rho} \sum_{j=0}^{n} |t_0|^j = \frac{\varepsilon}{2} .$$

By the triangle inequality, we therefore have

$$\|A(t_0) - \mathring{A}_n(t_0)\| \leq \|A(t_0) - A_n(t_0)\| + \|A_n(t_0) - \mathring{A}_n(t_0)\| < \varepsilon < 1 .$$

But the function $\|A(t) - \mathring{A}_n(t)\|$ is continuous; therefore, there exists a neighborhood $U(t_0)$ of t_0 in which

$$\|A(t) - \mathring{A}_n(t)\| < \varepsilon < 1$$

holds. Then

$$[E + A(t) - \mathring{A}_n(t)]^{-1} = B(t) \in C(H)$$

for every $t : t \in U(t_0)$. This means that, for a given arbitrary $z \in H$, the equation

$$x + [A(t) - \mathring{A}_n(t)]x = z$$

implies

$$x = B(t)z \quad . \tag{1}$$

Let us now consider the equation

$$x + A(t)x = y \qquad (t \in U(t_0)) \tag{2}$$

for a given $y \in H$. Then we have

$$x + [A(t) - \mathring{A}_n(t)]x + \mathring{A}_n(t)x = y \qquad (t \in U(t_0)) \ ;$$

and if we put

$$z := x + [A(t) - \mathring{A}_n(t)]x \ ,$$

we get

$$x = B(t)z \qquad (t \in U(t_0)) \tag{3}$$

With this substitution, we transform (2) into

$$z + \mathring{A}_n(t)B(t)z = y \qquad (t \in U(t_0)) \ . \tag{4}$$

It is easy to see that (4) is equivalent to (2) because every step leading from (2) to (4) is reversible.

$\mathring{A}_n(t)$ is a linear combination of finite-dimensional operators; therefore it is itself a finite-dimensional operator. As the set of finite-dimensional operators is an ideal in the algebra $C(H)$ of complete continuous operators, $\mathring{A}_n(t)B(t)$ is of finite dimension (for every $t : t \in U(t_0)$).

Our aim is, on account of this fact, to transform (4) into a system of algebraic equations. In order to do this, let us denote by $R(\cdot)$ the range of an operator. Let $X_j = R(\mathring{a}_j)$ $(j = 0,1,2,\ldots)$, which is a finite-dimensional linear subspace of H, and let X be the minimal linear subspace of H containing the set $\bigcup_{j=0}^{n} X_j$. X is a finite-dimensional subspace; an orthonormal basis for it is $\{x_1,x_2,\ldots,x_N\}$, and for every $x \in H$ the relation

$$\mathring{a}_j\, x = \sum_{k=1}^{N} \alpha_{jk}(x)x_k \qquad (j = 0,1,2,\dots,n)$$

is valid, where the α_{jk} are linear bounded functionals on H. In a similar way, let $Y_j = R(\mathring{a}_j^*)$, where the star denotes the adjoint operator with respect to $(.,.)$. Now we form the linear subspace Y from the subspaces Y_j just as we formed X from the subspaces X_j. Let a basis for this space be denoted by $\{y_1,y_2,\dots,y_N\}$. (As $\dim R(\mathring{a}_j) = \dim R(\mathring{a}_j^*)$ for $j = 0,1,2,\dots,n$, the relation $\dim X = \dim Y$ holds.)

By definition of $\alpha_{jk}(x)$,

$$\alpha_{jk}(x) = (x_k,\mathring{a}_j\, x) = (\mathring{a}_j^*\, x_k,x) \qquad (j = 0,1,\dots,n, \quad k = 1,2,\dots,N) \;;$$

and as $\mathring{a}_j^*\, x_k \in R(\mathring{a}_j)$, we can write

$$\mathring{a}_j^*\, x_k = \sum_{r=1}^{N} c_{jkr}\, y_r \qquad (j = 0,1,2,\dots,n, \quad k = 1,2,3,\dots,N) \;,$$

where the constants c_{jkr} are uniquely determined. We therefore have

$$\alpha_{jk}(x) = \sum_{r=1}^{N} c_{jkr}(y_r,x) \qquad (j = 0,1,2,\dots,n, \quad k = 1,2,3,\dots,N) \;.$$

The explicit expression for the finite-dimensional operator in (4) is the following:

$$\text{(5)} \qquad \begin{aligned} \mathring{A}_n(t)B(t)z &= \sum_{j=1}^{n} \sum_{k=1}^{N} \sum_{r=1}^{N} c_{jkr}\, t^j (y_r,x)x_k \\ &= \sum_{k=1}^{N} \sum_{r=1}^{N} \beta_{kr}(t)(y_r,x)x_k \;, \end{aligned}$$

where

$$\beta_{kr}(t) = \sum_{j=0}^{n} c_{jkr}\, t^j \qquad (k,r = 1,2,\dots,N) \;.$$

From (3) and (5), it follows that

$$\text{(6)} \qquad \begin{aligned} \mathring{A}(t)B(t)z &= \sum_{k=1}^{N} \sum_{r=1}^{N} \beta_{kr}(t)(y_r,B(t)z)x_k \\ &= \sum_{k=1}^{N} \sum_{r=1}^{N} \beta_{kr}(t)(B^*(t)y_r,z)x_k \\ &= \sum_{k=1}^{N} \sum_{r=1}^{N} \beta_{kr}(t)(v_r(t),z)x_k \;, \end{aligned}$$

where

$$v_r(t) := B^*(t)y_r \in H \qquad (t \in U(t_0), \quad r = 1,2,\dots,N) \;.$$

Since $B^{-1}(t) = E + A(t) - \mathring{A}_n(t)$ exists, $B^*(t)$ is also invertible; therefore the elements $v_r(t)$ are linearly independent for every $t \in U(t_0)$, since the elements y_r $(r = 1,2,\ldots,N)$ have this property.

Now let us define

$$p_r := (v_r(t),z) \qquad (r = 1,2,\ldots,N) .$$

By (6), the equation (4) can be written in the form

$$z + \sum_{k=1}^{N} \sum_{r=1}^{N} \beta_{kr}(t)p_r x_k = y . \tag{7}$$

Taking the scalar product with $v_s(t)$ $(s = 1,2,\ldots,N)$, we obtain the following system of linear equations, in which the p_r numbers are considered as unknowns:

$$p_s + \sum_{k=1}^{N} \sum_{r=1}^{N} \beta_{kr}(t)(v_s(t),x_k)p_r = (v_s(t),y) .$$

Thus, by introducing the notations

$$\sum_{r=1}^{N} \beta_{kr}(t)(v_s(t),x_k) = c_{sr}(t) \qquad (s,r = 1,2,\ldots,N) ,$$

$$(v_s(t),y) = g_s(t) ,$$

we have

$$p_s + \sum_{r=1}^{N} c_{sr}(t)p_r = g_s(t) \qquad (s = 1,2,\ldots,N) . \tag{8}$$

It is easy to see that any solution z of (4) yields a solution of (8):

$$p_s = (v_s(t),z) \qquad (s = 1,2,\ldots,N) .$$

Conversely, any solution of (8) provides a solution of (4), namely, by (7), the following:

$$z = y - \sum_{k=1}^{N} \sum_{r=1}^{N} \beta_{kr}(t)p_r x_k .$$

Thus (4) is solvable for every y if and only if (8) is. As we remarked, the elements $v_k(t)$ are linearly independent $(t \in U(t_0))$; therefore, the numbers $g_s(t)$ can be taken as any N numbers by the proper choice of y. It follows that the system of equations (8) has a solution for all $y \in H$ if and only if t is not a root of

$$D(t) = \begin{vmatrix} 1+C_{11}(t) & C_{12}(t) & \cdots & C_{1N}(t) \\ C_{21}(t) & 1+C_{22}(t) & \cdots & C_{2N}(t) \\ \cdots & \cdots & \cdots & \cdots \\ C_{N1}(t) & C_{N2}(t) & \cdots & 1+C_{NN}(t) \end{vmatrix} \tag{9}$$

for every $t \in U(t_0)$.

By definition of $B(t)$, we have

$$B(t) = \sum_{k=0}^{\infty} (-1)^k [A(t) - \mathring{A}_n(t)]^k . \tag{10}$$

Every member of this series is analytic, and because $\|[A(t) - \mathring{A}_n(t)]^k\| < \varepsilon^k$ for every $t \in U(t_0)$, it is uniformly convergent in $U(t_0)$. Hence $B(t)$ is analytic in $U(t_0)$. The same holds for $B^*(t)$, and so $v_r(t)$ is also analytic in $U(t_0)$. This implies the analyticity of $(v_s(t), z_k)$ in the domain $U(t_0)$ for $s,k = 1,2,\ldots,N$.

As the $\beta_{kr}(t)$ are polynomials, the coefficients $C_{sr}(t)$ are analytic, and for this reason $D(t)$ is also analytic in $U(t_0)$. If now t is a singular point of $A(t)$, then $D(t) = 0$, because in the contrary case (8) would have a unique solution for every $y \in H$. From this it follows that (4) and also (2) would have a unique solution for every $y \in H$, and t would not be a singular point.

As t_0 is a limit point of singular points, $D(t)$ vanishes on a set with limit point t_0. But, as $D(t)$ is analytic, it therefore vanishes identically in the whole subdomain $U(t_0)$ of D; that is, every point of $U(t_0)$ is a singular point of $A(t)$. Using the principle of analytic continuation, we can extend this result to all points of D, since D is connected.

REFERENCES

1. R. E. Kalaba and M. R. Scott, An initial-value method for integral operators: V-kernels depending upon a parameter, submitted for publication in Appl. Math. Comput.

2. G. D. Tamarkin, On Fredholm's integral equations whose kernels are analytic in a parameter, Ann. of Math. (2) 28 (1926 - 1927), 127-152.

THE COUNTERSPHERICAL REPRESENTATION OF A MINIMAL SURFACE

E. F. Beckenbach
Department of Mathematics
University of California
Los Angeles, California 90024
U.S.A.

ABSTRACT. The "counterspherical representation" of a surface in Euclidean 3-space is introduced in this paper. For meromorphic minimal surfaces in isothermal representation, the familiar spherical representation and the new counterspherical representation together give a physical interpretation of the "visibility function" that appears when the Nevanlinna theory of meromorphic functions of a complex variable is extended to these surfaces. In particular, the spherical and counterspherical representations can be used to show that the fundamental theorem of algebra for rational and logarithmico-rational minimal surfaces can be regarded not merely as an analytic theorem but also as a topological result.

1. INTRODUCTION: MINIMAL SURFACES AND INEQUALITIES

In the calculus of variations, a minimal surface is defined to be a surface for which the first variation of the area integral vanishes.

In differential geometry, the defining condition given above for S to be a minimal surface translates into the vanishing of the mean curvature K' of S:

$$K' = 0 \; . \tag{1}$$

Since the mean curvature of S at a point P of S is the sum of the principal curvatures of S at P, and since the Gaussian curvature K of S at P is the product of these two curvatures, it follows that every minimal surface is in particular a surface of nonpositive Gaussian curvature:

$$K \leq 0 \; . \tag{2}$$

A surface S in 3-dimensional Euclidean space E^3 is said to be given in terms of <u>isothermal parameters</u> u, v, or to be given in <u>isothermal representation</u>, if and only if the representation

$$S : x_j = x_j(u,v) \; , \qquad j = 1,2,3 \; ,$$

or simply

$$S : x = x(u,v) , \tag{3}$$

where $x(u,v)$ denotes the vector function

$$(x_1(u,v), x_2(u,v), x_3(u,v)) ,$$

for (u,v) in some domain D of definition, is such that

$$E = G = \lambda(u,v) , \quad F = 0, \tag{4}$$

in which the scalar products

$$E = \frac{\partial x}{\partial u} \cdot \frac{\partial x}{\partial u} , \quad F = \frac{\partial x}{\partial u} \cdot \frac{\partial x}{\partial v} , \quad G = \frac{\partial x}{\partial v} \cdot \frac{\partial x}{\partial v} \tag{5}$$

are the coefficients of the first fundamental quadratic form of S. Such an isothermal representation is conformal, or angle-preserving, except at points where $\lambda(u,v) = 0$.

Minimal surfaces in isothermal representation can be seen to furnish a generalization of the theory of analytic functions of a complex variable $w = u + iv$ through the following:

THEOREM OF WEIERSTRASS. A necessary and sufficient condition that a surface S given in isothermal representation by (3) be minimal is that the coordinate functions of S be harmonic.

Minimal surfaces in isothermal representation are doubly related to potential theory through subharmonic functions (functions that are dominated by harmonic functions in much the same way that convex functions are dominated by linear functions):

(i) A necessary and sufficient condition that a continuous surface S given by (3) be a minimal surface in isothermal representation is that, for each point $a = (a_1, a_2, a_3) \in E^3$, the logarithm of the distance function,

$$\log\{[x(u,v) - a] \cdot [x(u,v) - a]\}^{1/2} , \tag{6}$$

be subharmonic [10].

In fact, for a minimal surface S in isothermal representation, the Laplacian of (6) is given, at points where $x(u,v) \neq a$, by

$$\Delta \log\{[x(u,v)-a] \cdot [x(u,v)-a]\}^{1/2} = \frac{2\{[x(u,v)-a] \cdot X(u,v)\}^2}{\{[x(u,v)-a] \cdot [x(u,v)-a]\}^2} \lambda(u,v) , \tag{7}$$

in which $X(u,v)$ denotes the directed unit normal vector of S. Notice that the right-hand member of (7) clearly is nonnegative, whence the function (6) is subharmonic.

Henceforth, we shall ordinarily write, for example, (7) simply as

$$\Delta \log[(x - a) \cdot (x - a)]^{1/2} = \frac{2[(x - a) \cdot X]^2}{[(x - a) \cdot (x - a)]^2} \lambda . \tag{8}$$

(ii) A necessary and sufficient condition that a surface S given in isothermal representation by (3) be of nonpositive Gaussian curvature is that the logarithm of the area-deformation function,

$$\log \lambda(u,v) = \log E = \log G,$$

be subharmonic [11].

In fact, when a surface is given in isothermal representation, the formula [12, p. 140] for its Gaussian curvature K reduces, at points where $\lambda \neq 0$, to

$$K = -\frac{1}{2\lambda} \Delta \log \lambda , \tag{9}$$

so that $\Delta \log \lambda \geq 0$, or $\log \lambda$ is subharmonic, if and only if $K \leq 0$.

Since, as we have pointed out, every minimal surface is in particular a surface of nonpositive Gaussian curvature, $\log \lambda(u,v)$ is subharmonic for every minimal surface in isothermal representation.

The applicability of the theory of inequalities to minimal surfaces in isothermal representation results from (i) and (ii), above, in the following way:

The <u>principle of the maximum</u> in complex-variable theory depends on the fact that for an analytic function $f(w)$, $\log|f(w)|$ is subharmonic (actually harmonic except where $f(w) = 0$). The principle applies to all nonnegative functions having subharmonic logarithms. In general, then, the inequalities of geometric function theory that involve $|f(w)|$ have analogues for minimal surfaces, while those concerned with $|f'(w)|$ extend not only to minimal surfaces but more inclusively to all surfaces of nonpositive Gaussian curvature.

Thus, for example, the Lemma of Schwarz has an analogue for minimal surfaces [10],[3], and the isoperemetric inequality extends to (and characterizes) surfaces of nonpositive Gaussian curvature [11].

2. MEROMORPHIC MINIMAL SURFACES

If, for some (u_0,v_0), the minimal surface S is given in isothermal representation by (3) in the annulus

(10) $$R_1 < r < R_2 ,$$

where $u - u_0 = r \cos \theta$, $v - v_0 = r \sin \theta$, then, by the Theorem of Weierstrass given above, the vector function $x(u,v)$ is harmonic in (10), and accordingly $x(u,v)$ can be represented [14] there by

(11) $$x(u,v) = c \log r + \sum_{k=-\infty}^{\infty} r^k (a_k \cos k\theta + b_k \sin k\theta) .$$

The constant vector coefficient b_0 is arbitrary; for simplicity, we take $b_0 = 0$. The other constant vector coefficients c, a_k, b_k are uniquely determined by $x(u,v)$.

For the present theory, we extend the (u,v)-plane, and also Euclidean 3-space, by postulating a single ideal _point at_ ∞ for each.

If $R_1 = 0$ in (10), we say that S has an _isolated singularity_ at (u_0,v_0). Similarly, if $R_2 = \infty$ in (10) we say that S has an isolated singularity at ∞; by means of an inversion, we treat an isolated singularity at ∞ as an isolated singularity at the origin [7].

As in complex-variable theory, we say that the isolated singularity of S at (u_0,v_0) is _essential_ if

(12) $$a_k \cdot a_k + b_k \cdot b_k \neq 0$$

for an infinitude of negative indices k.

If the isolated singularity is not essential, and if the lowest index $k = t$ for which (12) holds is negative, then we say that S has a _pole of order_ $|t|$ at (u_0,v_0). Otherwise, since, as we shall presently see, S cannot have an ∞-point that is merely logarithmic, we say that the singularity is _removable_.

If the isolated singularity is removable, then we adjoin to S the point a_0 corresponding to (u_0,v_0), if indeed this correspondence was not already given in the definition of S. Then the vector function (3) gives an isothermal map of the disc $r < R_2$ onto the (extended) surface, which we again denote by S, and we say that S is _regular_ at (u_0,v_0).

If S is regular at (u_0,v_0), then either $x(u,v) \equiv a_0$, S reduces to a point, and we say that S is a _constant minimal surface_; or there is a lowest positive index $k = t$ for which (12) holds, and we say that S has an a_0_-point of order_ t at (u_0,v_0). In particular, if S has an a_0-point of order t at (u_0,v_0), and a_0 is the null vector, $a_0 = 0$, then we say that S has a _zero of order_ t at (u_0,v_0).

The harmonic vector function $x(u,v)$ given by (11) also satisfies the conditions (4), since S is in isothermal representation. In terms of the coefficients in (11), the conditions (4) are equivalent [8], [9], [2] to

$$(13)\quad \begin{cases} 2ka_k \cdot c + \sum_{j=-\infty}^{\infty} j(k-j)(a_j \cdot a_{k-j} - b_j \cdot b_{k-j}) = 0\,, & k = \pm 1, \pm 2, \cdots\,, \\ c \cdot c - 2 \sum_{j=1}^{\infty} j^2 (a_j \cdot a_{-j} - b_j \cdot b_{-j}) = 0\,, & \\ kb_k \cdot c + \sum_{j=-\infty}^{\infty} j(k-j) a_j \cdot b_{k-j} = 0\,, & k = 0, \pm 1, \pm 2, \cdots\,. \end{cases}$$

The equations (13) are more tractable than they might at first glance appear to be. In the plane case $x_3 \equiv 0$, they reduce simply to

$$b_{1,k} = \pm a_{2,k}\,, \quad b_{2,k} = \mp a_{1,k}\,, \quad c_1 = c_2 = 0\,, \quad k = \pm 1, \pm 2, \cdots\,.$$

If (12) does not hold for any $k < 0$, then it follows from the second equation in (13) that $c \cdot c = 0$ and therefore that:

A minimal surface given in isothermal representation by (3) cannot have an isolated ∞-point that is merely logarithmic.

If S has a pole of order $-t > 0$ or an a_0-point of order $t > 0$, then for $k = 2t$ the first and third equations in (13) reduce respectively to

$$t^2(a_t \cdot a_t - b_t \cdot b_t) = 0 \quad \text{and} \quad t^2(a_t \cdot b_t) = 0\,,$$

whence

$$(14)\qquad a_t \cdot a_t = b_t \cdot b_t \neq 0 \quad \text{and} \quad a_t \cdot b_t = 0\,.$$

If S has a pole of order $-t > 0$ at (u_0, v_0), then from (11) and (14) we obtain

$$(15)\qquad x(u,v) \cdot x(u,v) = r^{2t}\, a_t \cdot a_t + o(r^{2t})$$

as $r \to 0$. Similarly, if S has an a-point of order $t > 0$ at (u_0, v_0), then

$$(16)\qquad [x(u,v) - a] \cdot [x(u,v) - a] = r^{2t}\, a_t \cdot a_t + o(r^{2t})\,.$$

By (14), (15), and (16) we thus see that if S does not reduce to a point, then not only the poles but also the finite a-points of S are isolated.

By the same <u>method of coefficients</u>, it can be shown that if S is a nonconstant minimal surface, and S has no singularities other than poles, then the zeros and infinities of $\lambda(u,v)$ are isolated, and further that S has a continuous directed normal vector function $X(u,v)$ even at the zeros and infinities of $\lambda(u,v)$.

If, except for poles, S is a regular minimal surface given in isothermal representation by (3) for (u,v) in a domain D, then we say that S is a <u>meromorphic minimal surface for</u> (u,v) <u>in</u> D; if D is the entire finite plane, then we simply say that S is a <u>meromorphic minimal surface</u>.

3. NEVANLINNA THEORY

The classical Nevanlinna theory [15] of meromorphic functions of a complex variable is concerned with the distribution of values of these functions. His first fundamental theorem is the statement that, for a given nonconstant meromorphic function $f(w)$, the sum of the "enumerative function" and the "proximity function" is essentially the same for all points in the closed complex plane, and his second fundamental theorem shows that the enumerative function ordinarily dominates.

Is it reasonable to expect a similar theory to hold for meromorphic minimal surfaces?

We have seen that the poles and the finite a-points of a nonconstant meromorphic minimal surface are isolated, so that they are enumerable. This fact is promising.

But a 2-dimensional minimal surface in E^3 cannot contain or even come close to most points of E^3. For example, this is the case if the surface is a plane. So something new must be added.

Analytically, that "something" comes from the properties (i) and (ii) of minimal surfaces discussed in Section 1 of this paper.

Namely, the proof of Nevanlinna's first fundamental theorem involves an application of Green's theorem to $\log|f(w)|$, and the proof of his second fundamental theorem involves an application of Green's theorem to $\log|f'(w)|$. The area integrals of

$$\Delta \log|f(w)| \quad \text{and} \quad \Delta \log|f'(w)| ,$$

which might be expected to appear in an application of Green's theorem to

$$\log|f(w)| \quad \text{and} \quad \log|f'(w)| ,$$

are missing because we have

$$\Delta \log|f(w)| = 0 \quad \text{and} \quad \Delta \log|f'(w)| = 0$$

except at the zeros and poles of $f(w)$ and $f'(w)$.

The corresponding Laplacians for minimal surfaces, given in (8) and (9), ordinarily do not vanish. Accordingly, there are additional terms in the equations that appear in the corresponding theorems for minimal surfaces. The fact that these terms are nonnegative adds to the significance of their presence in the equations.

In the classical Nevanlinna theory [15], for a given meromorphic function $\zeta = f(w)$, and for a in the closed complex ζ-plane and $0 < r < \infty$, the proximity function $m(r,a;f)$ is defined in terms of the $\log^+$ function of distances of points in the ζ-plane.

In the elegant Ahlfors version [1] of the Nevanlinna theory, which we shall follow, the closed complex ζ-plane is projected stereographically onto a sphere of unit diameter, and the spherical proximity function $m^\circ(r,a;f)$ is defined in terms of the logarithm of chordal distances of corresponding points on the sphere.

Similarly, for a given meromorphic minimal surface S given by (3), the closed 3-dimensional Euclidean space containing S is projected stereographically onto the 3-dimensional hypersphere of diameter 1 having center at $(x_1,x_2,x_3,x_4) = (0,0,0,1/2)$ in 4-dimensional Euclidean space. Distances between points in the (x_1,x_2,x_3)-space are replaced with the chordal distances between corresponding points on the hypersphere; and, for a in the closed 3-dimensional space and $0 < r < \infty$, the hyperspherical proximity function $m^\circ(r,a;S)$ is accordingly defined by

$$m^\circ(r,a;S) = \frac{1}{2\pi}\int_0^{2\pi} \log \frac{[1+x(re^{i\theta})\cdot x(re^{i\theta})]^{1/2}(1+a\cdot a)^{1/2}}{\{[x(re^{i\theta})-a]\cdot[x(re^{i\theta})-a]\}^{1/2}}\, d\theta\ , \quad a \neq \infty,$$

$$m^\circ(r,\infty;S) = \frac{1}{2\pi}\int_0^{2\pi} \log[1+x(re^{i\theta})\cdot x(re^{i\theta})]^{1/2}\, d\theta\ ,$$

where $x(re^{i\theta})$ denotes $x(r\cos\theta, r\sin\theta)$.

Proceeding again in analogy with the complex-variable case, for a in the closed 3-dimensional Euclidean space and $0 \leq \rho < \infty$, let $n(\rho,a;S)$ denote the number of a-points of S in

$$|w| = (u^2+v^2)^{1/2} \leq \rho\ .$$

Then the enumerative function $N(r,a;S)$ is defined, for a finite or infinite and $0 < r < \infty$, by

$$N(r,a;S) = \int_0^r \frac{n(\rho,a;S)-n(0,a;S)}{\rho}\, d\rho + n(0,a;S)\log r\ . \tag{17}$$

In the extended theory, for any finite a in 3-dimensional Euclidean space and $0 \leq \rho < \infty$, a new function $h(\rho,a;S)$ is defined by

$$h(\rho,a;S) = \frac{1}{2\pi}\int_{|w|\leq\rho} \Delta \log[(x - a) \cdot (x - a)]^{1/2}\, dA_w ,$$

whence, by (8),

$$h(\rho,a;S) = \frac{1}{\pi}\iint_{|w|\leq\rho} \frac{[(x - a) \cdot X]^2}{[(x - a) \cdot (x - a)]^2} \lambda\, dA_w . \tag{18}$$

In particular, we have

$$h(0,a;S) = 0 ; \tag{19}$$

and since $\lim_{a\to\infty} h(\rho,a;S) = 0$, we define $h(\rho,\infty;S)$ by

$$h(\rho,\infty;S) = 0 . \tag{20}$$

By (18) and (20), we have $h(\rho,a;S) \geq 0$, with equality for $\rho > 0$ if and only if either (i) $a = \infty$ or (ii) S is a plane surface and a lies on S.

Since

$$\frac{(x - a) \cdot X}{[(x - a) \cdot (x - a)]^{1/2}} = \cos \varphi ,$$

where φ is the angle between the vector $x - a$ and the directed unit normal vector X, (18) can be written as

$$h(\rho,a;S) = \frac{1}{\pi}\iint_{|w|\leq\rho} \frac{\cos^2\varphi}{(x - a) \cdot (x - a)} \lambda\, dA_w .$$

For an element of area of measure dA_w in the $w = u + iv$ plane, the corresponding element of S has measure $\lambda\, dA_w$, the component of the latter element perpendicular to the "line of sight" vector $x - a$ has measure $|\cos \varphi|\lambda\, dA_w$, and the central projection from the "nerve center" a of this last element on the unit "eyeball" sphere $\mathfrak{S}_a$ with center a has measure

$$dA_{\mathfrak{S}_a} = \frac{|\cos \varphi|}{(x - a) \cdot (x - a)} \lambda\, dA_w .$$

Hence

$$h(\rho,a;S) = \frac{1}{4\pi}\iint_{|w|\leq\rho} 4|\cos \varphi|\, dA_{\mathfrak{S}_a} . \tag{21}$$

Since the surface area of $\mathfrak{S}_a$ is 4π, (21) shows that $h(\rho,a;S)$ **is** the ratio of the measure of the image of the map of $|w| \leq \rho$ on the eyeball surface $\mathcal{S}_a$, with variable weighting factor $4|\cos\varphi|$, to the surface area of $\mathcal{S}_a$. (The weighting factor $4|\cos\varphi|$ in the ratio is independent of the radius of the eyeball. An attractive alternative radius is $1/2$, since in this case the surface area of the eyeball is simply π.)

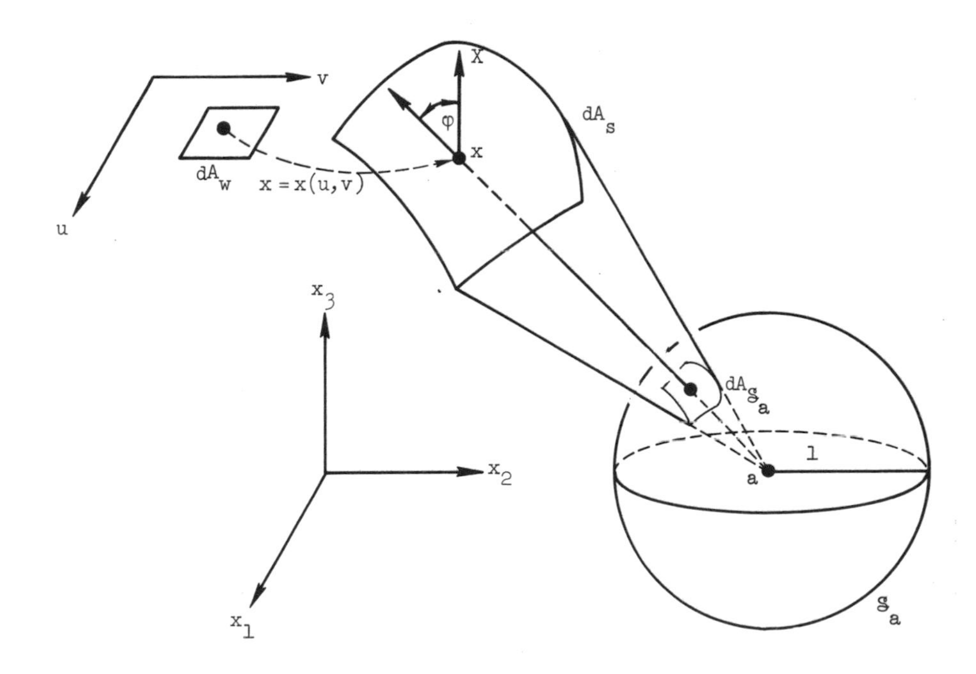

Thus $h(\rho,a;S)$ was introduced [2], [9] as furnishing a weighted measure of the "visibility" of the map of $|w| \leq \rho$ on S, as viewed from a.

In the extended theory, to the hyperspherical proximity function $m^\circ(r,a;S)$ and the enumerative function $N(r,a;S)$ we adjoin the <u>visibility function</u> $H(r,a;S)$, defined, for a finite or infinite and $0 < r < \infty$, by

$$H(r,a;S) = \int_0^r \frac{h(\rho,a;S)}{\rho}\, d\rho \; . \tag{22}$$

In particular, by (20) we have $H(r,\infty;S) = 0$.

Now the <u>hyperspherical affinity function</u> $\mathfrak{A}^\circ(r,a;S)$ is defined, for a finite or infinite and $0 < r < \infty$, by

$$\mathfrak{A}^\circ(r,a;S) = m^\circ(r,a;S) + N(r,a;S) + H(r,a;S) + C(a;S) \; ,$$

with the constant $C(a;S)$ chosen so that

$$\lim_{r\to 0} \mathfrak{A}^\circ(r,a;S) = 0 ;$$

the hyperspherical characteristic function $T^\circ(r;S)$ is defined by

$$T^\circ(r;S) = \mathfrak{A}^\circ(r;\infty;S) ;$$

and the hyperspherical form of the extended first fundamental theorem is the statement that, for each finite a,

$$\mathfrak{A}^\circ(r,a;S) = T^\circ(r;S) . \tag{23}$$

For plane maps, with $x_3(u,v) \equiv 0$, and for a restricted to the plane, $m^\circ(r,a;S)$ reduces to the spherical proximity function $m^\circ(r,a;f)$, $H(r,a;S)$ vanishes identically, and (23) reduces to the standard spherical form of the first fundamental theorem for meromorphic functions of a complex variable.

For the extension of the second fundamental theorem, we need two additional definitions:

First, following the classical theory, we let

$$n_1(\rho;S) = n(\rho,0;S_u) - n(\rho,\infty,S_u) + 2n(\rho,\infty;S) ,$$

where the surface S_u is defined by

$$S_u : x = \frac{\partial x(u,v)}{\partial u} .$$

The function $n_1(\rho;S)$ registers the multiple points of S in the disc $|w| \leq \rho$, in such a manner that a finite or infinite k-fold point of S contributes $k - 1$ to $n_1(\rho;S)$. We then define the second enumerative function $N_1(r;S)$ by

$$N_1(r;S) = \int_0^r \frac{n_1(\rho;S) - n_1(0;S)}{\rho} d\rho + n_1(0;S) \log r .$$

Next we let

$$h_1(\rho;S) = \frac{1}{4\pi}\iint_{|w|\leq\rho} (-K\lambda)\, dA_w . \tag{24}$$

The well-known geometric interpretation of $h_1(\rho;S)$ will be discussed in Sections 5 and 6. We now define the second visibility function $H_1(r;S)$ by

$$H_1(r;S) = \int_0^r \frac{h_1(\rho;S)}{\rho} d\rho .$$

In terms of these new functions, and of others introduced earlier, the hyperspherical form of the extended second fundamental theorem [6], [4] can be stated as follows:

Let S be a nonconstant meromorphic minimal surface, and let $a_1, a_2, \ldots, a_q$ be q points, $q > 2$. Let k be a given number, $k \geq 0$. Then

$$(25) \quad \sum_{j=1}^{q} m^\circ(r,a_j;S) \leq 2T^\circ(r;S) - N_1(r;S) - 2H_1(r;S) + O[\log r] + O[\log T^\circ(r;S)],$$

for r outside an open set Δ_k such that $\int_{\Delta_k} r^k dr < \infty$.

4. RATIONAL MINIMAL SURFACES

By (19) and (20), the definition (22) of $H(r,a;S)$ is analogous to the definition (17) of $N(r,a;S)$, with $h(\rho,a;S)$ in place of $n(\rho,a;S)$.

As a general principle, in the extended theory the sum $N(r,a;S) + H(r,a;S)$ plays the same role as $N(r,a;f)$ in the classical theory, and the sum $n(\rho,a;S) + h(\rho,a;S)$ plays the same role as $n(\rho,a;f)$. This principle will be illustrated in the present section.

Let the minimal surface S, given in isothermal representation by (3), be meromorphic in the <u>closed</u> $w = u + iv$ plane. Then, since the poles of S are isolated, S can have at most a finite number of poles in all. The surface is then a <u>logarithmico-rational minimal surface</u>, or a <u>rational minimal surface</u>, according as there are, or are not, logarithmic terms in any of the expansions (11) at the poles of S.

The coordinate functions of a rational minimal surface can be represented by ratios of polynomials in u,v; and the coordinate functions of a logarithmico-rational minimal surface can be represented by such ratios plus sums of logarithmic terms.

The sum of the orders of the poles of a rational or logarithmico-rational minimal surface S is called the <u>order</u>, or <u>degree</u>, of S.

For example, the minimal surface S of Enneper is given in isothermal representation by

$$x_1(u,v) = 3u - u^3 + 3uv^2 ,$$
$$x_2(u,v) = -3v + v^3 - 3u^2v ,$$
$$x_3(u,v) = 3u^2 - 3v^2 .$$

This surface is a polynomial, or entire rational, minimal surface of degree 3. In the closed $w = u + iv$ plane, S has just one zero; this is at $w = 0$ and is of order 1. The surface has a pole of order 3 at $w = \infty$.

For another example, the functions

$$x_1(u,v) = \frac{(1 + u^2 + v^2)u}{u^2 + v^2},$$

$$x_2(u,v) = \frac{(1 + u^2 + v^2)v}{u^2 + v^2},$$

$$x_3(u,v) = \log(u^2 + v^2)$$

are the coordinate functions of a logarithmico-rational minimal surface (actually a catenoid) in isothermal representation. The surface is of degree 2, with poles of order 1 at $w = 0$ and at $w = \infty$; it has no zero.

We note in passing that for a nonconstant meromorphic minimal surface S given in isothermal representation by (3), $T^\circ(r;S)$ satisfies

$$\liminf_{r \to \infty} \frac{T^\circ(r;S)}{\log r} = c < \infty$$

if and only if c is a positive integer and S is a rational or logarithmico-rational minimal surface of degree c [7].

For a rational or logarithmico-rational minimal surface S of degree $m > 0$, and for any finite or infinite a, let $n_a \geq 0$ be the order of the a-point of S at $w = \infty$ (of course, $n_a = 0$ for all points a except one), and let

$$\text{(26)} \qquad \begin{aligned} n(a;S) &= n_a + \lim_{\rho\to\infty} n(\rho,a;S) , \\ h(a;S) &= \lim_{\rho\to\infty} h(\rho,a;S) . \end{aligned}$$

Then $n(a;S)$ is the total number of a-points, counting multiplicities, of S in the closed w-plane, and $h(a;S)$ is the weighted measure, in the sense discussed above in Section 2, of the visibility of the entire surface S when viewed from a.

In particular, for $a = \infty$ we have $n(\infty;S) = m$, by the definition of the degree of S; and by (20) we have $h(\infty;S) = 0$. Therefore, $n(\infty;S) + h(\infty;S) = m$.

More generally, we have the following [7]:

THEOREM. If S is a rational or logarithmico-rational minimal surface of degree $m > 0$ in the extended E^3-space, then for each a, including $a = \infty$, in this space,

$$\text{(27)} \qquad n(a;S) + h(a;S) = m .$$

Since m is a positive integer, since $n(a;S)$ is a nonnegative integer, and since $h(a;S) \geq 0$, with equality if and only if either $a = \infty$ or S is a plane surface with a on S, it therefore follows from (27) that:

(i) For each a, both $n(a;S)$ and $h(a;S)$ are nonnegative integers.

(ii) $n(a;S) \leq m$, with equality if and only if either $\mathbf{a} = \infty$ or S is a plane surface with a on S.

(iii) $h(a;S) \leq m$, with equality for almost all points of E^3.

(iv) The theorem includes the fundamental theorem of algebra as a special case.

For a rational or logarithmico-rational minimal surface, $h(a;S)$ is discontinuous, as a function of a, between S and the complement of S in E^3, and it is discontinuous on S at the multiple points of S. Nevertheless, as we have just seen, for such a surface $h(a;S)$ takes on only integer values! This fact insistently suggests that there surely must be an alternative "visual" interpretation of the integral in (18) such that:

(i) the visual image of S on the unit eyeball sphere $\mathfrak{S}_a$ covers $\mathfrak{S}_a$ exactly $m - n(a;S)$ times, and

(ii) under the visual-image transformation, we have

$$dA_{\mathfrak{S}_a} = \frac{4[(x-a) \cdot X]^2}{[(x-a) \cdot (x-a)]^2} \lambda \, dA_w \ . \tag{28}$$

The devious way in which this is true will be discussed in what follows.

5. SPHERICAL AND COUNTERSPHERICAL REPRESENTATIONS

For a surface S given by (3), the surface

$$x = X(u,v) = \frac{x_u \times x_v}{(EG - F^2)^{1/2}} , \tag{29}$$

where $X(u,v)$ is the directed unit normal vector function for S, and where subscripts denote differentiation, lies on the unit sphere $\mathfrak{S}_0$ with center at the origin, since

$$X \cdot X = 1 \ . \tag{30}$$

The surface (29), introduced by Gauss, is called the <u>spherical representation</u> of S.

We shall use known properties of the spherical representation (29) of S, and also properties of the surface defined by

$$\mathbf{x} = Y(u,v) = X(u,v) - 2 \frac{x(u,v) \cdot X(u,v)}{x(u,v) \cdot x(u,v)} x(u,v) \ . \tag{31}$$

Notice that

$$(32)\qquad Y \cdot Y = X \cdot X - 4\,\frac{x \cdot X}{x \cdot x}\,(x \cdot X) + 4\,\frac{(x \cdot X)^2}{(x \cdot x)^2}\,(x \cdot x) = X \cdot X = 1\ ,$$

so that the surface (31), like the surface (29), lies on the sphere S_0. We shall call the surface (31) the <u>counterspherical representation</u> of S.

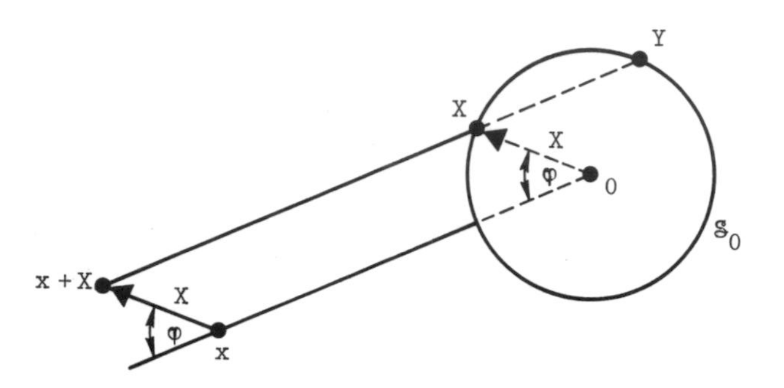

Notice also that for each (u,v), the points X, Y, and x + X are all on the line $P = X + \alpha x$, α real, and that this line is parallel to the line $P = \alpha x$ through the origin and the point x. The figure shows the plane containing all these points. When φ is obtuse, Y lies between X and x + X.

Since X is normal to S, we have

$$(33)\qquad X \cdot dx = 0$$

for every direction dx on S. Using (33), we find from (31) that

$$(34)\qquad dY = dX - 2\,\frac{x \cdot X}{x \cdot x}\,dx - 2\,\frac{x \cdot dX}{x \cdot x}\,x + 4\,\frac{(x \cdot X)(x \cdot dx)}{(x \cdot x)^2}\,x\ .$$

In evaluating $dY \cdot dY$ from (34), we find that all scalar products involving the last two terms in (34) cancel directly, leaving

$$(35)\qquad dY \cdot dY = 4\,\frac{(x \cdot X)^2}{(x \cdot x)^2}\,(dx \cdot dx) - 4\,\frac{x \cdot X}{x \cdot x}\,(dx \cdot dX) + (dX \cdot dX)\ .$$

Since the first, second, and third fundamental quadratic forms of S, $dx \cdot dx$, $-dx \cdot dX$, and $dX \cdot dX$, are interrelated [12, p. 142] through the identity

$$(36)\qquad K(dx \cdot dx) + K'(dx \cdot dX) + (dX \cdot dX) = 0\ ,$$

we can write (35) in the form

$$(37)\qquad dY \cdot dY = \left[4\,\frac{(x\cdot X)^2}{(x\cdot x)^2} - K\right](dx\cdot dx) - \left[4\,\frac{x\cdot X}{x\cdot x} + K'\right](dx\cdot dX)\ .$$

Along with the symbols E, F, G for the coefficients of the first fundamental quadratic form of S, given in (5), for the corresponding coefficients of the second and third fundamental quadratic forms of S we use the symbols e, f, g and $\mathcal{E}, \mathfrak{F}, \mathcal{G}$, respectively.

We shall denote the coefficients of the first fundamental quadratic form $dY \cdot dY$ of the counterspherical representation of S by

$$\mathcal{E}' = Y_u \cdot Y_u,\quad \mathfrak{F}' = Y_u \cdot Y_v,\quad \mathcal{G}' = Y_v \cdot Y_v\ .$$

It should be noted that, unlike the first, second, and third fundamental quadratic forms of S, the form $dY \cdot dY$ is not independent of the position of S in space; indeed, $dY \cdot dY$ owes its usefulness, in part, to this fact.

Now (35) can be written equivalently as

$$\begin{cases} \mathcal{E}' = 4\,\dfrac{(x\cdot X)^2}{(x\cdot x)^2}\,E + 4\,\dfrac{x\cdot X}{x\cdot x}\,e + \mathcal{E}\ , \\[2ex] \mathfrak{F}' = 4\,\dfrac{(x\cdot X)^2}{(x\cdot x)^2}\,F + 4\,\dfrac{x\cdot X}{x\cdot x}\,f + \mathfrak{F}\ , \\[2ex] \mathcal{G}' = 4\,\dfrac{(x\cdot X)^2}{(x\cdot x)^2}\,G + 4\,\dfrac{x\cdot X}{x\cdot x}\,g + \mathcal{G}\ , \end{cases}$$

and (37) can be written equivalently as

$$(38)\qquad \begin{cases} \mathcal{E}' = \left[4\,\dfrac{(x\cdot X)^2}{(x\cdot x)^2} - K\right]E + \left[4\,\dfrac{x\cdot X}{x\cdot x} + K'\right]e\ , \\[2ex] \mathfrak{F}' = \left[4\,\dfrac{(x\cdot X)^2}{(x\cdot x)^2} - K\right]F + \left[4\,\dfrac{x\cdot X}{x\cdot x} + K'\right]f\ , \\[2ex] \mathcal{G}' = \left[4\,\dfrac{(x\cdot X)^2}{(x\cdot x)^2} - K\right]G + \left[4\,\dfrac{x\cdot X}{x\cdot x} + K'\right]g\ . \end{cases}$$

From (38), we obtain

$$\mathcal{E}'\mathcal{G}' - \mathfrak{F}'^2 = \left[4\,\frac{(x\cdot X)^2}{(x\cdot x)^2} - K\right]^2(EG - F^2) + \left[4\,\frac{x\cdot X}{x\cdot x} + K'\right]^2(eg - f^2)$$
$$+ \left[4\,\frac{(x\cdot X)^2}{(x\cdot x)^2} - K\right]\left[4\,\frac{x\cdot X}{x\cdot x} + K'\right](Eg - 2Ff + Ge)\ .$$

Therefore, since [12, p. 109]

$$(39)\qquad eg - f^2 = K(EG - F^2)$$

and

$$(40)\qquad Eg - 2Ff + Ge = K'(EG - F^2)\ ,$$

we have

$$\mathcal{E}'\mathcal{G}' - \mathcal{F}'^2 = \left\{\left[4\,\frac{(x\cdot X)^2}{(x\cdot x)^2} - K\right]^2 + \left[4\,\frac{x\cdot X}{x\cdot x} + K'\right]^2 K \right.$$
$$\left. + \left[4\,\frac{(x\cdot X)^2}{(x\cdot x)^2} - K\right]\left[4\,\frac{x\cdot X}{x\cdot x} + K'\right]K'\right\}(EG - F^2)\ . \tag{41}$$

Since we are here concerned primarily with minimal surfaces S in isothermal representation, we shall henceforth restrict our study of spherical and counterspherical representations to this case.

If S is a minimal surface in isothermal representation, then (1) and (4) hold, and (36) yields

$$\mathcal{E} = \mathcal{G} = -K\lambda\ , \quad \mathcal{F} = 0\ , \tag{42}$$

so that, as is well known, the spherical representation of S also is in isothermal representation.

From (42), we get

$$\mathcal{E}\mathcal{G} - \mathcal{F}^2 = K^2\lambda^2\ ,$$

whence, by (2)

$$(\mathcal{E}\mathcal{G} - \mathcal{F}^2)^{1/2} = -K\lambda\ . \tag{43}$$

By (1) and (4), for a mimimal surface S in isothermal representation, (41) reduces to

$$\mathcal{E}'\mathcal{G}' - \mathcal{F}'^2 = \left\{\left[4\,\frac{(x\cdot X)^2}{(x\cdot x)^2} - K\right]^2 + \left[4\,\frac{x\cdot X}{x\cdot x}\right]^2 K\right\}\lambda^2 = \left[4\,\frac{(x\cdot X)^2}{(x\cdot x)^2} + K\right]^2\lambda^2\ ,$$

whence [recall (2)],

$$\pm(\mathcal{E}'\mathcal{G}' - \mathcal{F}'^2)^{1/2} = \left[4\,\frac{(x\cdot X)^2}{(x\cdot x)^2} + K\right]\lambda \tag{44}$$

according as

$$-K \le 4\,\frac{(x\cdot X)^2}{(x\cdot x)^2} \quad \text{or} \quad -K \ge 4\,\frac{(x\cdot X)^2}{(x\cdot x)^2}\ . \tag{45}$$

Both the directed normal to the spherical representation (29) of S at the point $X(u,v)$, and the directed normal to the counterspherical representation (31) of S at the point $Y(u,v)$, in the respective directions of the unit vectors

$$\frac{X_u \times X_v}{(\mathcal{E}\mathcal{G} - \mathcal{F}^2)^{1/2}} \quad \text{and} \quad \frac{Y_u \times Y_v}{(\mathcal{E}'\mathcal{G}' - \mathcal{F}'^2)^{1/2}}\ ,$$

are so directed that the u-curve, the v-curve, and the normal, in that order, have the same mutual orientation as the x_1-, x_2-, and x_3-axes.

Further, since by (30) and (32) the spherical representation of S and the counterspherical representation of S lie on the unit sphere $\mathcal{S}_0$, the foregoing directed normals are parallel to the respective normals to $\mathfrak{S}_0$ in the directions of the unit vectors $X(u,v)$ and $Y(u,v)$ -- the latter vectors of course being directed outward from $\mathfrak{S}_0$.

Therefore the directed normal to the spherical representation of S and the directed normal to the counterspherical representation of S are directed outward or inward from $\mathfrak{S}_0$ according, respectively, as

$$(46) \qquad \frac{X_u \times X_v}{(\mathcal{E}\mathcal{G} - \mathcal{F}^2)^{1/2}} = \pm X \quad \text{and} \quad \frac{Y_u \times Y_v}{(\mathcal{E}'\mathcal{G}' - \mathcal{F}'^2)^{1/2}} = \pm Y \ .$$

To determine the signs in the equations (46), it is convenient to take scalar products with X and Y, respectively, and then to solve the resulting determinantal equations

$$(47) \qquad (X_u\ X_v\ X) = \pm(\mathcal{E}\mathcal{G} - \mathcal{F}^2)^{1/2}, \quad (Y_u\ Y_v\ Y) = \pm(\mathcal{E}'\mathcal{G}' - \mathcal{F}'^2)^{1/2} \ .$$

For the first equation in (47), since

$$(48) \qquad a \times a = 0 \quad \text{and} \quad a \times b = -b \times a$$

for all vectors a,b, and since [12, p. 137]

$$(49) \qquad X_u = -\frac{e}{\lambda} x_u - \frac{f}{\lambda} x_v \ , \quad X_v = -\frac{f}{\lambda} x_u - \frac{g}{\lambda} x_v \ ,$$

and $x_u \times x_v = \lambda X$, by (4) and (39) we have

$$X_u \times X_v = \frac{eg - f^2}{\lambda^2}(x_u \times x_v) = K\lambda X \ ,$$

and therefore, by (30),

$$(50) \qquad (X_u\ X_v\ X) = K\lambda(X \cdot X) = K\lambda = -(\mathcal{E}\mathcal{G} - \mathcal{F}^2)^{1/2} \ .$$

Because the sign of the right-hand member of (50) is negative, we have the known result that the directed unit normal vector of the spherical representation of S is directed <u>inward</u> from $\mathfrak{S}_0$.

For the determination of the sign in the second equation in (47), from (34) we have

$$(51) \qquad Y_u = X_u - 2\frac{x \cdot X}{x \cdot x} x_u - 2\frac{x \cdot X_u}{x \cdot x} x + 4\frac{(x \cdot X)(x \cdot x_u)}{(x \cdot x)^2} x \ ,$$

$$(52) \qquad Y_v = X_v - 2\frac{x \cdot X}{x \cdot x} x_v - 2\frac{x \cdot X_v}{x \cdot x} x + 4\frac{(x \cdot X)(x \cdot x_v)}{(x \cdot x)^2} x \ .$$

Noting from (1), (4), and (40) that $e + g = 0$, from (48) and (49) we get

$$(53) \qquad (X_u \times x_v) + (x_u \times X_v) = 0 \ .$$

Using (48), (53), and the determinantal values

$$(X_u\ X_v\ X) = K\lambda\ , \quad (x_u\ x_v\ X) = \lambda\ ,$$
$$(X\ X_u\ x) = -X_v \cdot x\ , \quad (x\ X_v\ X) = -X_u \cdot x\ ,$$
$$(X\ x_u\ x) = x_v \cdot x\ , \quad (x\ x_v\ X) = x \cdot x_u\ ,$$

in expanding $(Y_u\ Y_v\ Y)$ from (31), (51), and (52), we find that some terms cancel directly, leaving

$$(54)\qquad \begin{aligned}(Y_u\ Y_v\ Y) = {} & \left[4\,\frac{(x \cdot X)^2}{(x \cdot x)^2} + K\right]\lambda - \frac{8(x \cdot X)^2}{(x \cdot x)^3}\left[(x \cdot x_u)^2 + (x \cdot x_v)^2 + \lambda(x \cdot X)^2\right] \\ & + \frac{2}{x \cdot x}\left[(x \cdot X_u)^2 + (x \cdot X_v)^2 - K\lambda(x \cdot X)^2\right].\end{aligned}$$

By the Pythagorean Theorem, we have

$$\frac{(x \cdot x_u)^2}{\lambda} + \frac{(x \cdot x_v)^2}{\lambda} + (x \cdot X)^2 = x \cdot x$$

and

$$\frac{(x \cdot X_u)^2}{-K\lambda} + \frac{(x \cdot X_v)^2}{-K\lambda} + (x \cdot X)^2 = x \cdot x\ ,$$

and therefore, by (44) and (54),

$$(55)\qquad \begin{aligned}(Y_u\ Y_v\ Y) & = \left[4\,\frac{(x \cdot X)^2}{(x \cdot x)^2} + K\right]\lambda - 2\left[4\,\frac{(x \cdot X)^2}{(x \cdot x)^2} + K\right]\lambda \\ & = -\left[4\,\frac{(x \cdot X)^2}{(x \cdot x)^2} + K\right]\lambda = \mp(\mathcal{E}'\mathcal{G}' - \mathcal{F}'^2)^{1/2},\end{aligned}$$

according as the first or second inequality in (45) holds.

Hence the directed unit normal vector of the counterspherical representation of S is directed <u>inward</u> from $\mathcal{S}_0$ if the <u>first</u> (strict) inequality in (45) holds, and it is directed <u>outward</u> if the <u>second</u> (strict) inequality in (45) holds.

It follows that if one of the two inequalities in (45) holds in one (u,v)-region, and the other inequality holds in an adjacent region, then the counterspherical representation folds back on itself along the common boundary of the maps of the two regions.

6. THE EYEBALL REVISITED

For an arbitrary finite point a of E^3, the analysis of Section 5 can be applied, by means of the translation of axes defined by $z = x - a$, to obtain representations of a given surface S on the unit sphere $\mathcal{S}$ with center at a.

The figure illustrates the plane section containing the pertinent points in the case in which φ is obtuse. The point

$$X_a = a + X$$

is the <u>spherical representation of</u> x <u>on</u> $\mathcal{S}_a$, and the point

$$Y_a = a + X - 2\,\frac{(x - a) \cdot X}{(x - a) \cdot (x - a)}\,(x - a)$$

is the <u>counterspherical representation of</u> x <u>on</u> $\mathcal{S}_a$.

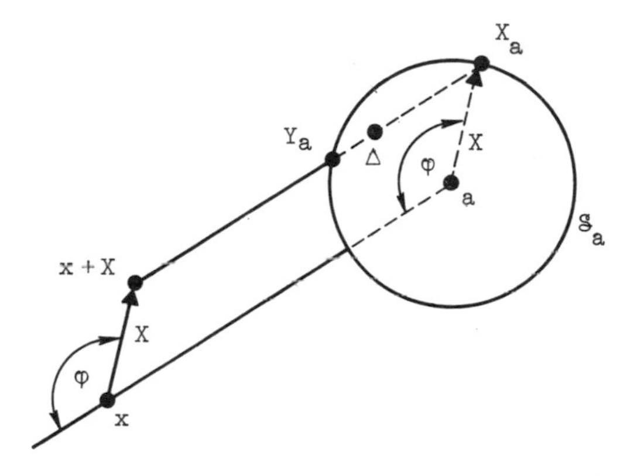

In this context, the spherical and counterspherical representations of S defined in Section 5 would more precisely be called the respective representations of S <u>on</u> $\mathcal{S}_0$.

Of course, the spherical representations of S on $\mathcal{S}_a$ and $\mathcal{S}_b$ are congruent for all a, b.

If S is a minimal surface in isothermal representation, then the spherical representation of S on $\mathcal{S}_a$ is isothermal and never folds back on itself; it reduces to a point if and only if S is a plane surface. The counterspherical representation of S on $\mathcal{S}_a$ is isothermal if and only if S is a plane surface.

We orient the sphere $\mathcal{S}_a$ with its normals directed inward.

From (43), (44), (50), and (55), by the translation we have

$$dA_{\mathcal{S}_{a,s}} + (-1)^{\delta(w)}\, dA_{\mathcal{S}_{a,c}} = 4\,\frac{[(x - a) \cdot X]^2}{[(x - a) \cdot (x - a)]^2}\,\lambda\, dA_w\,, \tag{56}$$

where

$$dA_{\mathfrak{S}_{a,s}} = (\mathcal{E}\mathcal{G} - \mathcal{F}^2)^{1/2}\, dA_w = -K\lambda\, dA_w$$

and

$$dA_{\mathfrak{S}_{a,c}} = (\mathcal{E}'\mathcal{G}' - \mathcal{F}'^2)^{1/2}\, dA_w$$

are the respective elements of area on the spherical representation and on the counterspherical representation of S on $\mathfrak{S}_a$, and where

$$\delta(w) = 0 \quad \text{or} \quad 1$$

according as the normal to the counterspherical representation of S on $\mathfrak{S}_a$ is directed inward or outward from $\mathfrak{S}_a$ at the point corresponding to $w = u + iv$.

It is in the form (56) that the anticipated identity (28) of condition (ii) on page 289 is realized.

Equation (56), along with equations (18) and (26), shows that each of the functions $h(\rho,a;S)$ and $h(a;S)$ is the sum of a spherical component (actually independent of a) and a counterspherical component:

$$(57)\qquad \begin{aligned} h(\rho,a;S) &= h_1(\rho,a;S) + h_2(\rho,a;S)\ , \\ h(a;S) &= h_1(a;S) + h_2(a;S)\ , \end{aligned}$$

where [cf. (24)]

$$(58)\qquad h_1(\rho,a;S) = \frac{1}{4\pi}\iint_{|w|\le\rho} dA_{\mathfrak{S}_{a,s}} = \frac{1}{4\pi}\iint_{|w|\le\rho} (\mathcal{E}\mathcal{G} - \mathcal{F}^2)^{1/2}\, dA_w = \frac{1}{4\pi}\iint_{|w|\le\rho} (-K\lambda)\, dA_w\ ,$$

$$(59)\qquad h_2(\rho,a;S) = \frac{1}{4\pi}\iint_{|w|\le\rho} (-1)^{\delta(w)} dA_{\mathfrak{S}_{a,c}} = \frac{1}{4\pi}\iint_{|w|\le\rho} (-1)^{\delta(w)} (\mathcal{E}'\mathcal{G}' - \mathcal{F}'^2)^{1/2}\, dA_w\ ,$$

$$(60)\qquad h_1(a;S) = \lim_{\rho\to\infty} h_1(\rho,a;S)\ ,$$

$$(61)\qquad h_2(a,S) = \lim_{\rho\to\infty} h_2(\rho,a;S)\ .$$

By (58), $h_1(\rho,a;S)$ is the ratio of the area of the map of $|w| \le \rho$ on the spherical representation of S on $\mathfrak{S}_a$, to the surface area of $\mathfrak{S}_a$; and by (60), $h_1(a;S)$ is this ratio for the entire surface S.

By (59), $h_2(\rho,a;S)$ is the ratio of the "algebraic" area (that is, with account taken of the fact that the surface might fold back on itself and the area be considered as being negative when the normal is directed outward) of $|w| \le \rho$ on the counterspherical representation of S on $\mathfrak{S}_a$, to the surface area of $\mathfrak{S}_a$; and by (61), $h_2(a;S)$ is this ratio for the entire surface S.

If a compact oriented surface without boundary is mapped on an oriented sphere, then the map must cover the sphere "algebraically" an integral number of times, called the <u>order</u>, or <u>degree</u>, of the map [13, p. 124].

The spherical and counterspherical representations of a rational or logarithmico-rational minimal surface S on $\mathfrak{S}_a$ are of the foregoing sort. By the interpretation of (60) and (61) given above, $h_1(a;S)$ and $h_2(a;S)$ are the degrees of the spherical and counterspherical representations of S on $\mathfrak{S}_a$, respectively; that is, $h_1(a;S)$ and $h_2(a;S)$ denote the number of times the spherical and counterspherical representations of S on $\mathfrak{S}_a$ cover $\mathfrak{S}_a$ positively. By (57), this explains why $h(a;S)$ must be an integer [actually, a whole number, since $h(a;S) \geq 0$ by (18) and (26)].

By way of explanation, we note that as the point a approaches the point x of S nontangentially, less and less of the part of S near x is mapped positively on more and more of one or more sheets (depending on the multiplicity of x as a point of S) of the counterspherical representation of S on $\mathfrak{S}_a$, until in the limit, $a = x$, $h_2(a;S)$ decreases and $n(a;S)$ increases discontinuously by the appropriate counting number; $h_1(a;S)$, the number of times the spherical representation of S on $\mathfrak{S}_a$ covers $\mathfrak{S}_a$, remains constant.

The fact that the whole number $h(a;S) = h_1(a;S) + h_2(a;S)$ is $m - n(a;S)$, thus fulfilling condition (i) on page 289, follows from the analytically established [7] generalization of the fundamental theorem of algebra. Defining $h_1(\infty;S)$ and $h_2(\infty;S)$ so that $h_1(\infty;S) + h_2(\infty;S) = 0$, we can now state that result topologically as follows:

THEOREM. If S is a rational or logarithmico-rational minimal surface of degree $m > 0$ in the extended E^3 space, then for each a, including $a = \infty$, in this space,

$$n(a;S) + h_1(a;S) + h_2(a;S) = m ;$$

that is, the sum of the orders of the a-points, the degree of the spherical representation of S on $\mathfrak{S}_a$, and the degree of the counterspherical representation of S on $\mathfrak{S}_a$, is m.

Some problems suggested by the material in this paper are listed [5] in the Questions and Problems section at the end of the present volume.

The following "optical" terms and allusions are included by way of justifying the use of the term "visibility function"; they should not be taken too seriously. The author has found them to be usefully suggestive. Fortunately, they are not "colorful."

For the point x of S, "light" from the "source" $x + X$ travels a linear path, parallel to the "line of sight" from x to the "nerve center" a, striking the "eyeball" $\mathfrak{S}_a$ at X_a, where it forms the "corneal image" of x. Thence it travels along the same line to Y_a, where it forms the "retinal image" of x.

Lengths at x are magnified by an amount $(-K)^{1/2}$ in the corneal image at X_a, and by an "optical law" the "corneal lens" at X_a "focuses" the image at the point Δ at distance $[-K(x-a)\cdot(x-a)]^{1/2}$ along the ray from X_a through Y_a. There is an "image reversal" at Δ, so that beyond Δ the normal to the image is reversed in direction: if Δ is inside the eyeball, then the inward normal to the corneal image at X_a is reversed in the retinal image at Y_a, again to point inward; but if Δ is outside the eyeball, then the normal at X_a is not reversed in the retinal image at Y_a, and accordingly the normal at Y_a points outward from $\mathfrak{S}_a$.

When a comes into coincidence with x, x becomes a tactile "mote" -- or multiple mote, in accordance with the multiplicity of x as a point of S -- in the eye.

The transparent "optic nerve" to the nerve center a fills the eyeball, allowing a to "see" S as a composite of <u>two</u> images, the corneal image and the retinal image. Thus does Nature endow our Cyclops with "steoroscopic vision"!

REFERENCES

1. L. V. Ahlfors, Beiträge zur Theorie der meromorphen Funktionen, Den syvende skandinaviske matematikerkongress i Oslo, 19-22 August, 1929, pp. 84-91, A.W. Brøggers Boktrykkeri A/S, Oslo, 1930.

2. E. F. Beckenbach, An introduction to the theory of meromorphic minimal surfaces, Proceedings of symposia in pure mathematics, vol. 11, Entire functions and related parts of analysis, American Mathematical Society, Providence, R.I., 1968.

3. E. F Beckenbach, Convexity, Hardy's theorem, and the lemma of Schwarz, Inequalities, vol. 3, Academic Press, Inc., New York and London, 1972.

4. E. F. Beckenbach, Defect relations for meromorphic minimal surfaces: An introduction, Lecture notes in mathematics, vol. 419, Topics in analysis, Colloquium on mathematical analysis, Jyväskylä, 1970, pp. 18-41, Springer-Verlag, Berlin, Heidelberg, New York, 1974.

5. E. F. Beckenbach, Some problems in the theory of surfaces, Proceedings of the first international conference on general inequalities, Oberwolfach, 1976, pp.315-318, Birkhäuser Verlag, Basel, Stuttgart, 1977.

6. E. F. Beckenbach and T. A. Cootz, The second fundamental theorem for minimal surfaces, Bull. Amer. Math. Soc. 76 (1970), 711-716.

7. E. F. Beckenbach, F. H. Eng, and R. E. Tafel, Global properties of rational and logarithmico-rational minimal surfaces, Pacific J. Math. 50 (1974), 355-381.

8. E. F. Beckenbach and J. W. Hahn, Triples of conjugate harmonic functions and minimal surfaces, Duke Math. J. 2 (1936), 698-704.

9. E. F. Beckenbach and G. A. Hutchison, Meromorphic minimal surfaces, Bull. Amer. Math. Soc. 68 (1962), 519-522; Pacific J. Math 28 (1969), 17-47.

10. E. F. Beckenbach and T. Radó, Subharmonic functions and minimal surfaces, Trans. Amer. Math. Soc. 35 (1933), 648-661.

11. E. F. Beckenbach and T. Radó, Subharmonic functions and surfaces of negative curvature, Trans. Amer. Math. Soc. 35 (1933), 662-674.

12. W. C. Graustein, Differential geometry, Macmillan Company, New York, 1935.

13. S. Lefschetz, Introduction to topology, Princeton University Press, Princeton, N.J., 1949.

14. W. F. Osgood, Lehrbuch der Funktionentheorie, vol. 1, G. G. Teubner, Leipzig, 1928.

15. R. Nevanlinna, Zur Theorie der meromorphen Funktionen, Acta Math. 46 (1925), 1-99.

Remarks and Problems

REMARKS AND PROBLEMS CONCERNING PRÉKOPA'S INEQUALITY

Georg Aumann

(P) Let f, g be measureable and nonnegative on $\mathbb{R}^1$, and let

$$\rho(x) := \sup\{f(y)g(z) : y, z \in \mathbb{R}^1 \text{ and } y + z = 2x\}, \qquad x \in \mathbb{R}^1.$$

Then [1]

$$\int_{-\infty}^{+\infty} \rho(x)dx \geq \left(\int_{-\infty}^{+\infty} f^2(x)dx\right)^{1/2} \left(\int_{-\infty}^{+\infty} g^2(x)dx\right)^{1/2} .$$

(P) can be used to prove the following theorem on logarithmic concave probability density functions (p.d.f.) in $\mathbb{R}^2$:

(M) If $w : \mathbb{R}^2 \to \mathbb{R}_0^+$ is a logarithmic concave p.d.f., i.e., if w is summable with

$$\int w = 1 \quad \text{and} \quad w\left(\frac{x_1 + x_2}{2}\right)^2 \geq w(x_1)w(x_2)$$

("Tendency to the middle") for all $x_1, x_2 \in \mathbb{R}^2$, then the marginal density μ in each direction is also logarithmic concave. [If $x = (\xi_1, \xi_2)$ is a cartesian coordinate representation in $\mathbb{R}^2$, then

$$\xi_1 \to \mu(\xi_1) := \int_{-\infty}^{+\infty} w(\xi_1, \xi_2)d\xi_2$$

is the marginal density of w (on $\mathbb{R}^1$) in the direction of the ξ_2-axis.]

PROBLEMS. I. Simplify the proof of (P) by weakening the assumptions on f and g, assuming that f and g are (additionally) logarithmic concave.

II. Establish the generalization of (P) to $\mathbb{R}^n$, $n \geq 2$.

III. Look for an inverse of (M) by characterizing probability densities all of whose marginal densities are logarithmic concave.

REFERENCE

[1] A. Prékopa, Logarithmic concave measures with application to stochastic programming, Acta Sci. Math. (Szeged) 32 (1971), 301-316.

A SIMPLE DIFFERENTIAL PROOF OF THE INEQUALITY BETWEEN THE ARITHMETIC AND GEOMETRIC MEANS

O. Shisha

We prove that if $x_1, x_2, \ldots, x_n$ are positive numbers not all equal, then

$$(1) \qquad [(x_1 + \cdots + x_n)/n]^n - x_1 \cdots x_n > 0 \quad .$$

We assume $n > 1$ and that the statement holds for $n - 1$. We may also assume

$$x_n > a = (x_1 + \cdots + x_{n-1})/(n - 1) \quad .$$

Consider

$$f(x) \equiv [(x_1 + \cdots + x_{n-1} + x)/n]^n - x_1 \cdots x_{n-1}x \quad .$$

Then

$$f(a) = [a^{n-1} - x_1 \cdots x_{n-1}]a \geq 0 \quad ;$$

and if $x > a$, then

$$f'(x) = [(x_1 + \cdots + x_{n-1} + x)/n]^{n-1} - x_1 \cdots x_{n-1} > f(a)/a \geq 0 \quad .$$

Hence $f(x_n) > 0$, which is (1).

This proof appears to be somewhat simpler than the differential proof of Liouville [1].

References

[1] J. Liouville, Sur la moyenne arithmétique et la moyenne géométrique de plusieurs quantités positives, J. Math. Pures Appl. 4 (1839), 493-494.

A CONVEXITY PROBLEM IN COMPLEX-VARIABLE THEORY

E. F. Beckenbach

It was observed by Professor Kairies [2] at this converence that the function f defined by $f(x) = 1$ for $1 \leq x < 2$ and by $f(x + 1) = xf(x)$ for $x \geq 2$, which is related to the gamma function in that $f(n) = (n - 1)!$ for each positive integer n, is a convex function of x. Namely, the continuous function is constant, linear, quadratic, and so on, in successive unit intervals, with positive jumps in the derivative at the integers.

A similar function occurs in complex-variable theory. The left-hand member of Jensen's formula for an analytic function f, with $f(0) \neq 0$,

$$\frac{1}{2\pi} \int_0^{2\pi} \log|f(Re^{i\theta})|\,d\theta = \log|f(0)| + \log \sum_j \frac{R}{|z_j|} ,$$

is the logarithm of the geometric mean, or mean of order zero, of $|f(Re^{i\theta})|$, and therefore

$$\mathfrak{M}_0(|f| \; ; R) = |f(0)| \prod_j \frac{R}{|z_j|} .$$

This continuous function of R is constant in the interval $[0,|z_1|]$, linear in $[|z_1|,|z_2|]$, quadratic in $[|z_2|,|z_3|]$, and so on, with positive jumps in the derivative at the radii $|z_j|$, with $|z_j| \leq |z_k|$ for $j < k$, of circles on which zeros of f lie. Thus $\mathfrak{M}_0(|f| \; ; R)$ is a convex function of R.

It has been shown in [1] that the mean of order t, $\mathfrak{M}_t(|f| \; ; R)$, is a convex function of R for all analytic functions f for the values $t = 2, 1, \frac{2}{3}, \ldots,$ that is, for $t = \frac{2}{k}$, $k = 1, 2, 3, \ldots$. The proposed problem is to determine all values of t for which this is true.

REFERENCES

1. E. F. Beckenbach, W. Gustin, and H. Shniad, On the mean modulus of an analytic function, Bull. Amer. Math. Soc. 55 (1949), pp. 184-190.

2. H.-H. Kairies, Convexity in the theory of the gamma function, Proceedings of the First International Conference on General Inequalities, Oberwolfach, 1976, pp. 49-62, Birkhäuser Verlag, Basel, Stuttgart, 1977.

REMARKS ON OSTROWSKI'S INEQUALITY

F. Huckemann

We consider the inequality

$$(1) \qquad \binom{n}{k} p^k (1-p)^{n-k} \leq \exp[-2n(p-\bar{p})^2], \quad \text{where } \bar{p} = \frac{k}{n},$$

for $p \in (0,1)$. The question whether in (1), $2n$ may be replaced by αn with $\alpha > 2$, was proposed at this conference by Ostrowski and already answered negatively by Redheffer [1]. Probabilistic considerations lead to the same result.

Let S_n be the number of successes in n Bernoulli trials with probability p for success at an individual trial. Then

$$P[S_n = k] = \binom{n}{k} p^k (1-p)^{n-k},$$

and $(S_n - np)/\sqrt{np(1-p)}$ is asymptotically normally distributed. In particular, the following limit theorem holds:

If $k = k(n)$ is such that

$$\lim_{n \to \infty} \frac{k(n) - np}{n^{2/3}} = 0,$$

then

$$(2) \qquad \binom{n}{k} p^k (1-p)^{n-k} \sim \frac{1}{\sqrt{2\pi np(1-p)}} \exp\left[-\frac{(k-np)^2}{2np(1-p)}\right]$$

as $n \to \infty$.

With $\bar{p} = \frac{k(n)}{n}$, (2) becomes

$$(3) \qquad \binom{n}{k} p^k (1-p)^{n-k} \sim \frac{1}{\sqrt{2\pi np(1-p)}} \exp\left[-\frac{n}{2p(1-p)}(p-\bar{p})^2\right],$$

showing that for $p = 1/2$ the constant 2 as factor of n in (1) cannot be improved.

Formula (3), however, suggests that (1) may possibly be improved by using a different function of p in the exponent.

REFERENCE

1. Alexander M. Ostrowski and Raymond M. Redheffer, Inequalities related to the normal law, Proceedings of the first international conference on general inequalities, Oberwolfach, 1976, pp. 125-129, Birkhäuser, Verlag, Basel, Stuttgart, 1977.

A DETERMINANT INEQUALITY

Raymond M. Redheffer

Form an n-by-n determinant $D(n)$ in three steps, as follows:

(i) Every element of the first column is 1, every second element of the second column is 1, every third element of the third column is 1, and so on.

(ii) The first element of each column is 1.

(iii) All other elements are 0.

For example,

$$D(5) = \begin{vmatrix} 1 & 1 & 1 & 1 & 1 \\ 1 & 1 & 0 & 0 & 0 \\ 1 & 0 & 1 & 0 & 0 \\ 1 & 1 & 0 & 1 & 0 \\ 1 & 0 & 0 & 0 & 1 \end{vmatrix}.$$

You are asked to prove or disprove the following inequality:

$$|D(n)| \leq (n^{1/2+\varepsilon}) \quad \text{as} \quad n \to \infty .$$

This holds for all $\varepsilon > 0$ if, and only if, the Riemann Hypothesis is true. For details, see the Proceedings of the Oberwolfach Conference on Optimization, 1976.

ON FAVARD'S THEOREM FOR ORTHOGONAL POLYNOMIALS

Kurt Endl

The following problem was posed by G. Alexits:

What can be said about systems $\{f_n(x)\}_0^\infty$ of real functions satisfying a recurrence relation

$$f_{n+1}(x) = (\psi(x) - \alpha_n)f_n(x) - \beta_n f_{n-1}(x)$$

$$(\alpha_n,\beta_n) \in \mathbb{R}_n \quad (n\geq 0); \quad \beta_n > 0 \quad (n\geq 1); \quad f_0(x) \equiv 1, \ f_{-1}(x) \equiv 0),$$

where $\psi(x)$ is an arbitrary function?

For $\psi(x) \equiv x$, by Favard ([3],[4]), $\{f_n(x)\}_0^\infty$ is a system of orthogonal polynomials. In the general case, there is also a statement of orthogonality [1]. The answer is a special case of the following result.

THEOREM 1. Suppose X is a commutative algebra over $\mathbb{R}$, without diviosrs of zero, with unit element w. Then the following statements for a system $\{f_n\}_0^\infty \subset X$ are equivalent.

I. The system satisfies a recurrence relation

$$f_{n+1} = (\psi - \alpha_n)f_n - \beta_n f_{n-1}$$

$$(\alpha_n,\beta_n \in \mathbb{R}_n \quad (n\geq 0); \quad \beta_n > 0 \quad (n\geq 1); \quad f_0 = e, \quad f_{-1} = 0) .$$

II. The system is "orthogonal"; i.e., there exists a linear functional μ on X with

$$\mu(f_m \cdot f_n) = 0, \quad m, n \geq 0, \ m \neq n \ ,$$

$$\mu(f_n^2) > 0, \quad n \geq 0 \ .$$

An application can be made to the question: Under what conditions does a system $\{f_n\}_0^\infty$ satisfying I have involutory forms? Here we say that $\{f_n\}_0^\infty$ has <u>involutory forms</u> if signs $\{c_n\}_0^\infty$ can be found such that the coefficient matrix of

$$\{c_n f_n\}_0^\infty = \{c_n \cdot \sum_{\nu=0}^{n} c^{(n)}\psi^\nu\}$$

is involutory ([2]):

THEOREM 2. Suppose $\{f_n\}_0^\infty$ is orthogonal in the sense of Theorem 1,

$\{\psi^n\}_0^\infty$ is a linearly independent set, and

$$\alpha_0 + \cdots + \alpha_{n-1} \neq 0 \qquad (n \geq 1) \quad .$$

If $\{f_n\}_0^\infty$ has involutory forms, then the moments

$$\mu_n = \mu(\psi^n)$$

satisfy

$$(\alpha_n - \alpha_{n-1})\mu_n = (\beta_n - \beta_{n-1} + \beta_1)\mu_{n-1} \qquad (n \geq 1) \quad .$$

In the case of the Laguerre polynomials, this leads to the functional equation of the Γ-Function:

$$\mu_n = n\mu_{n-1} \quad .$$

REFERENCES

1. K. Endl, Eine Bemerkung zum Satz von Favard über orthogonale Polynomsysteme (to appear in the Acta Math. Acad. Sci. Hung.).

2. K. Endl, Über involutörische Matrizen, die durch dreigliedrige Rekursionsformeln erzeugt werden (to appear in the Mitt. Math. Seminar Giessen).

3. J. Favard, Sur les polynômes de Tchebycheff, C. R. Acad. Sci. Paris 200 (1935), 2052-2053.

4. G. Freud, Orthogonale Polynome, Birkhäuser Verlag, Basel, Stuttgart, 1969.

A PROBLEM IN PARABOLIC DIFFERENTIAL INEQUALITIES

Raymond M. Redheffer and Wolfgang Walter

Let us consider the parabolic operator

$$Pu := u_t - u_{xx} - f(t,x,u,u_x) \quad \text{in} \quad G: \; 0 < t \leq T, \; 0 < x < d \quad ,$$

and let $\Gamma = \bar{G} - G$ (parabolic boundary of G).

The <u>comparison principle</u> is the statement that

$$Pu \leq Pv \;\text{ in }\; G \;,\; u \leq v \;\text{ on }\; \Gamma \Rightarrow u \leq v \;\text{ in }\; G \;.$$

The principle holds, for example, if f is of class C^1 and if u and u_x (or v and v_x) are bounded.

We have constructed counterexamples for the comparison principle in the case

$$f(t,x,u,u_x) = (u + a(t,x))|u_x|^{\alpha} \quad , \quad \text{where} \quad \alpha > 2$$

(a is continuous in $\bar{G}$, u and v are continuous in $\bar{G}$, and u_x and v_x are continuous in G).

The following conjecture is posed:

If $f(t,x,z,p)$ satisfies an inequality

$$|f(t,x,z,p)| \leq K(|z|)(1 + p^2) \quad ,$$

where $K(s)$ is a continuous, positive function, then the comparison principle holds true (for functions u, v such that $u, v \in C^0(\bar{G})$, $u_x, v_x \in C^0(G)$).

A PROBLEM ORIGINATING IN TRANSMISSION-LINE THEORY

Raymond M. Redheffer

Consider an obstacle with:

left-hand reflection coefficient	a
left-hand transmission coefficient	$b \neq 0$
right-hand reflection coefficient	c
right-hand transmission coefficient	$d \neq 0$

At first, a, b, c, d are complex numbers. If this obstacle is followed at the right by a termination having the complex reflection z, multiple reflection gives

$$f(z) = a + bz(1 - cz)^{-1}d$$

for the overall reflection at the left. We assume that $1 - cz$ is nonsingular, which, for the moment, means that $cz \neq 1$.

Turning the obstacle end-for-end gives the left-hand reflection

$$g(z) = c + dz(1 - az)^{-1}b$$

when $1 - az$ is nonsingular. We set $\|z\| = \sup|z\xi|$ for $|\xi| = 1$. Then physical considerations suggest that

$$\text{(1)} \qquad \|f(z)\| \leq 1 \ \text{ for } \ \|z\| \leq 1 \Longleftrightarrow \|g(z)\| \leq 1 \ \text{ for } \ \|z\| \leq 1 \ ,$$

where, of course, z is in the domain of f or g, respectively. This is, in fact, true for complex numbers, for n-by-n matrices, and even when a, b, c, d, z are linear operators on a Hilbert space $\mathcal{H}$.

The norm on $\mathcal{H} \times \mathcal{H}$ is given from that on $\mathcal{H}$ by the Pythagorean formula, and we use the sup norm for operators on $\mathcal{H} \times \mathcal{H}$ just as for operators on $\mathcal{H}$. Then it is not hard to show that both conditions (1) follow from

$$\text{(2)} \qquad \left\| \begin{matrix} \lambda b & a \\ c & d/\lambda \end{matrix} \right\| \leq 1 \ \text{ for some } \ \lambda > 0 \ .$$

It is also true that if the two conditions (1) hold, then (2) holds. A simple proof of this in the general case of operators on a Hilbert space is desired. The only proof known to the author is much harder than the proofs of the other statements.

AN INTEGRAL INEQUALITY CONNECTED WITH COMPLETENESS

Raymond M. Redheffer

Let C be the class of complex-valued functions f such that

$$\int_{-\pi}^{\pi} e^{i\lambda t} f(t)dt = 0, \qquad f \in L^p[-\pi,\pi],$$

where λ is a complex constant and $1 \leq p \leq \infty$. It is required to find the best constant $\alpha = \alpha(p,\lambda)$ such that

$$\|e^{-i\lambda x} \int_{-\pi}^{x} e^{i\lambda t} f(t)dt\|_p \leq \alpha\|f\|_p .$$

The value for $p = 2$ is known, and for some other cases, but not in general. This problem is interesting because of its connection with completeness of complex exponentials.

Let $\{\lambda_n\}$ and $\{\nu_n\}$ be sequences of complex numbers, ordered so that $|\lambda_n|$ increases with n $(n = 1,2,3,\ldots)$, and let

$$d_n = |\lambda_n - \nu_n|\alpha_n, \qquad \alpha_n = \alpha(p,\lambda_n).$$

Let $E(\lambda)$ be the Paley-Wiener excess $L^p[-\pi,\pi]$ of $\{\exp i\lambda_n x\}$, and $E(\nu)$ of $\{\exp i\nu_n x\}$. Then the condition

$$\sum_{1}^{\infty} \frac{(1 + d_1)(1 + d_2) \cdots (1 + d_n)}{n^2} < \infty$$

implies $|E(\lambda) - E(\nu)| \leq 1$. Clearly, the convergence or divergence of the above series is affected by minute changes in the value of α_n.

It should be observed, however, that some error is tolerable. For example, if α_n is approximated by β_n, where

$$\sum_1^\infty |\lambda_n - \nu_n| \, |\alpha_n - \beta_n| < \infty ,$$

then no error is introduced in the above criterion. Because of this slight latitude, the problem is perhaps not completely inaccessible.

SOME PROBLEMS IN THE THEORY OF SURFACES

E. F. Beckenbach

The material in the author's paper [1] in the earlier pages of this volume suggests several problems and problem areas, including the following:

1. According to the fundamental theorem of the theory of surfaces (Reference [12, p. 139] of [1]), the first and second fundamental differential quadratic forms of a surface S uniquely determine S to within its position in space. Similarly (Reference [12, p. 144] of [1]), the second and third fundamental differential quadratic forms of S uniquely determine S to within its position in space.

Unlike the three fundamental differential quadratic forms of S, the differential quadratic form $dY_a \cdot dY_a$, introduced in [1], where S is considered as being "seen" by the "nerve center" a through the spherical representation X_a and the counterspherical representation Y_a on the sphere $\mathfrak{S}_a$ with unit radius and center at a, is not independent of the position of S in space.

It would be interesting to establish, both for minimal surfaces and for surfaces in general, the extent to which X_a and Y_a determine S, and what the "optical illusions" of such "vision" are; and generally to investigate the properties of $dY_a \cdot dY_a$ and its relationships with S and its three fundamental differential quadratic forms.

2. In [1], the extended fundamental theorem of algebra for minimal surfaces, which earlier had been formulated analytically (Reference [7] of [1]), was expressed in terms of topological entities in the statement that, for a rational or logarithmico-rational minimal surface S of degree m, and for an arbitrary finite or infinite point a in the extended enclosing 3-dimensional Euclidean space E^3, the sum of the orders of the a-points on S, the degree of the spherical representation of S (on $\mathfrak{S}_a$), and the degree of the counterspherical representation of S on $\mathfrak{S}_a$, is equal to m:

$$n(a;S) + h_1(a;S) + h_2(a;S) = m \quad . \tag{1}$$

An independent, topological proof (cf. [3]) that the expression in the left-hand member of (1) is a topological invariant, and that the value of the invariant is m, would constitute an alternative, topological proof of the extended fundamental theorem of algebra for these surfaces.

3. The extended fundamental theorem of algebra, in its analytical formulation, has been established (Reference [7] of [1]) for rational and logarithmico-rational minimal surfaces in extended Euclidean n-space, $n \geq 3$. Further, an extension of the notion of the spherical representation has been made [2] for minimal surfaces in Euclidean n-space, $n > 3$.

It would seem desirable, then, also to extend the notion of counterspherical representation to minimal surfaces in Euclidean n-space, $n > 3$, with the purpose, among others, of stating, and then topologically proving, a topological formulation of the extended fundamental theorem of algebra for rational and logarithmico-rational minimal surfaces in Euclidean n-space, $n > 3$.

4. The extended fundamental theorem of algebra, in analytical formulation, was suggested by consideration of the function $h(a;S)$ that occurs when the Nevanlinna theory of meromorphic functions of a complex variable is extended (Reference [2] of [1]) to meromorphic minimal surfaces. The positivity properties (see Section 1 of [1]) that make the extension of the Nevanlinna theory to minimal surfaces significant are lost when the same analysis is applied to other classes of surfaces.

Continuity considerations suggest, nevertheless, that perhaps the extended fundamental theorem of algebra, particularly in the topological formulation (1), is valid as a reflection of the gross topological structure of members of more inclusive classes of sufficiently regular rational surfaces, and other surfaces, in Euclidean 3-space.

5. Granted that the extension of the fundamental theorem of algebra suggested in item 4, above, is valid, then extensions of the notions of spherical and counterspherical representations of surfaces to surfaces of other classes, and to higher-dimensional varieties, in Euclidean n-space, $n > 3$, might possibly be made, partly with a view to extending the fundamental theorem of algebra still further.

6. The set of directions taken on by the normals to a minimal surface S, or equivalently the set of points on $\mathfrak{S}_0$ contained in the spherical representation of S, has been extensively studied, for example in investigations of Bernstein's theorem and of complete minimal surfaces (see, for example, [4]).

The analytic behavior of the counterspherical representation of S is vastly different from that of the spherical representation of S. Nevertheless, the range of possibilities for positive and negative coverage of a given $\mathfrak{S}_a$ by the counterspherical representation of a minimal surface of a given class might merit investigation.

7. With order and type defined for meromorphic minimal surfaces in strict analogy (Reference [2] of [1]) with their definitions for meromorphic functions of a complex variable, one might seek to construct nonplane examples of meromorphic minimal surfaces of various orders and types, and for such surfaces might investigate propositions analogous to known results for functions of a complex variable.

8. Many results in complex-variable theory, concerning the number of complex a-points that are contained in a given map f, follow from applications of the notion of normal, or compact, families. These results include, for example, Picard's great theorem and theorems concerning directions of Julia.

Analogous statements for minimal surfaces S are meaningful, and possibly true, when the number of complex a-points of f is replaced with the number of space a-points of S, plus the ratio of the sum of the areas of the spherical and counterspherical representations of S on $\mathfrak{S}_a$ to the area of $\mathfrak{S}_a$.

9. The standard proof of the second fundamental theorem in the Nevanlinna theory of meromorphic functions of a complex variable is due to L. V. Ahlfors (Reference [1] of [1]). For this proof, the plane of the map determined by the function is projected onto a spherical surface, and the members of the equation expressing the first fundamental theorem are averaged, with respect to a suitable measure function, over the surface of the sphere.

If the plane of the foregoing map is considered as being embedded in 3-dimensional Euclidean space E^3, as an instance of a meromorphic minimal surface in E^3, so that the map can be "viewed" from points not in the plane, and if this embedding space is then projected onto a 3-dimensional hyperspherical surface in 4-dimensional Euclidean space, then the equation of the first fundamental theorem, adjusted to this context, now contains a "visibility" term in addition to the familiar proximity and enumerative terms.

When the members of this last equation are averaged, again with respect to a suitable measure function, over the hyperspherical surface, the average of the enumerative term loses the crucial role that it plays in the Ahlfors proof of the second fundamental theorem, for this average is now identically zero. The role is instead taken over by the average of the visibility term.

The Ahlfors proof is simple, elementary, and extremely ingenious. The present approach is perhaps simpler, is no less elementary, and seems to require less ingenuity. For example, the derivative function, which is introduced by a clever geometric argument in the Ahlfors proof and by other devices in the original proof by Rolf Nevanlinna and in the proof by his brother Frithiof Nevanlinna, now appears automatically in the visibility function.

Perhaps this method, simplified as much as possible, should be considered as an alternative approach in the theory of meromorphic functions of a complex variable.

10. To apply the second fundamental theorem of Nevanlinna, extended to meromorphic minimal surfaces, defect and ramification functions can be defined in analogy with the complex-variable case. Because of the visibility term $H_1(r;S)$ in the inequality (25) of [1] expressing the extended theorem, the number of such functions is greater in the extended theory than in the classical theory. Upper bounds on these functions and their sums are furnished by the inequality.

An easy consequence is the following extension of the Picard theorem: If S is a nonconstant meromorphic minimal surface, then there can be at most two points a that can neither "feel" $[n(r,a;S) \equiv 0]$ nor "see" $[h(r,a;S) = 0]$ the surface; and if there are two such points, then S is a plane surface and the points are on the plane.

It would seem desirable to construct nonplane meromorphic minimal surfaces for which the bounds mentioned above are attained or approached arbitrarily closely. In general, an arbitrarily close approach can be anticipated, with the surface reducing to a plane surface in the limit.

REFERENCES

1. E. F. Beckenbach, The counterspherical representation of a minimal surface, Proceedings of the first international conference on general inequalities, Oberwolfach, 1976, pp. 277-299, Birkhäuser Verlag, Basel, Stuttgart, 1977.

2. Shiing-Shen Chern, Minimal surfaces in Euclidean space of N dimensions, Differential and combinatorial topology, A symposium in honor of Marston Morse, pp. 187-198, Princeton University Press, Princeton, N.J., 1965.

3. Richard Courant and Herbert Robbins, What is mathematics?, pp. 269-271, Oxford University Press, London, New York, Toronto, 1941.

4. Johannes C. C. Nitsche, On new results in the theory of minimal surfaces, Bull. Amer. Math. Soc. 71 (1965), 195-270.

A PROBLEM CONCERNING t-NORMS

B. Schweizer

A <u>t-norm</u> is a mapping $T: [0,1] \times [0,1] \to [0,1]$ such that: (i) $T(a,1) = a$; (ii) $T(a,b) = T(b,a)$; (iii) $T(a,b) \leq T(c,d)$, whenever $a \leq c$, $b \leq d$; (iv) $T(a,T(b,c)) = T(T(a,b),c)$.

If T_1 and T_2 are t-norms, then T_1 <u>dominates</u> T_2 if and only if

$$T_1(T_2(a,b),T_2(c,d)) \geq T_2(T_1(a,c),T_1(b,d)) \quad .$$

It is easy to show that the relation "dominates" is reflexive and antisymmetric. Prove or disprove: "dominates" is transitive. (This problem stems from R. Tardiff's study of the product of probabilistic metric spaces.)

* * * * * * *

REMARKS AND A PROBLEM ON AN INEQUALITY

Alexander M. Ostrowski

In the inequality

$$\int_0^1\int_0^1 \left|\frac{f(x) - f(y)}{x - y}\right|^\alpha dxdy \leq c_\alpha \int_0^1 |f'(x)|^\alpha dx \quad ,$$

$\alpha \geq 1$, $c_\alpha \leq c_1 = \log 4$, the determination of c_2 by Garsia, Fichera, and Sneider was discussed, and the problem of determining c_α for $\alpha \neq 1$ or 2 was pointed out.

* * * * * * *

ON CHARACTERIZING LORENTZ TRANSFORMATIONS

W. Benz

The causal automorphisms ($\leq$- automorphisms) of $\mathbb{R}^n$, $n > 2$, are known to be orthochronous Lorentz transformations up to scale factors. The result fails in case $n = 2$. Bijective mappings that map light cones onto light cones are also known to be Lorentz transformations (up to scale factors) in case $n > 2$. This result fails as well for $n = 2$. Our result (including previous ones) is that invariance of the pseudo-euclidean distance 1 characterizes Lorentz transformations for $n \geq 2$. For $n = 2$ the distance 1 can be replaced by each distance $\rho \neq 0$ (in case $n > 2$ by $\rho > 0$).

NAME INDEX

(including citations by reference number)

SUBJECT INDEX